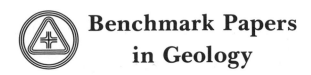

Benchmark Papers
in Geology

Series Editor: Rhodes W. Fairbridge
Columbia University

Published Volumes

RIVER MORPHOLOGY / Stanley A. Schumm
SLOPE MORPHOLOGY / Stanley A. Schumm and M. Paul Mosley
SPITS AND BARS / Maurice L. Schwartz
BARRIER ISLANDS / Maurice L. Schwartz
ENVIRONMENTAL GEOMORPHOLOGY AND LANDSCAPE CONSERVATION,
    VOLUME I: Prior to 1900; VOLUME II: Urban Regions; and VOLUME III:
    Non-Urban Areas / Donald R. Coates
TEKTITES / Virgil E. Barnes and Mildred A. Barnes
GEOCHRONOLOGY: Radiometric Dating of Rocks and Minerals / C. T. Harper
MARINE EVAPORITES: Origin, Diagenesis, and Geochemistry / Douglas W. Kirk-
    land and Robert Evans
GLACIAL ISOSTASY / John T. Andrews
PHILOSOPHY OF GEOHISTORY: 1785–1970 / Claude C. Albritton, Jr.
GEOCHEMISTRY OF GERMANIUM / Jon N. Weber
GEOCHEMISTRY AND THE ORIGIN OF LIFE / Keith A. Kvenvolden
GEOCHEMISTRY OF WATER / Yasushi Kitano
GEOCHEMISTRY OF IRON / Henry Lepp
GEOCHEMISTRY OF BORON / C. T. Walker
SEDIMENTARY ROCKS: Concepts and History / Albert V. Carozzi
METAMORPHISM AND PLATE TECTONIC REGIMES / W. G. Ernst
SUBDUCTION ZONE METAMORPHISM / W. G. Ernst
PLAYAS AND DRIED LAKES: Occurrence and Development / James T. Neal
GLACIAL DEPOSITS / Richard P. Goldthwait

Additional volumes in preparation

Benchmark Papers
in Geology / 21

——— A *BENCHMARK* ® Books Series ———

GLACIAL DEPOSITS

Edited by
RICHARD P. GOLDTHWAIT
Ohio State University

Dowden, Hutchinson & Ross, Inc.
Stroudsburg, Pennsylvania

Distributed by
HALSTED PRESS *A Division of John Wiley & Sons, Inc.*

Dedicated to my "850" class,
who did so much to judge
useful references.

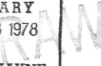
Copyright © 1975 by **Dowden, Hutchinson & Ross, Inc.**
Benchmark Papers in Geology, Volume 21
Library of Congress Catalog Card Number: 75-1396
ISBN: 0-470-31276-9

77 76 75   1 2 3 4 5
Manufactured in the United States of America.

LIBRARY OF CONGRESS CATALOGING IN PUBLICATION DATA

Goldthwait, Richard Parker, 1911-   comp.
   Glacial deposits.

   (Benchmark papers in geology ; v. 21)
   1. Glacial landforms--Addresses, essays, lectures.
I. Title.
GB581.G64   551.3'15   75-1396
ISBN 0-470-31276-9

Exclusive Distributor: **Halsted Press**
A Division of John Wiley & Sons, Inc.

Acknowledgments
and Permissions

ACKNOWLEDGMENTS

GEOLOGICAL SOCIETY OF AMERICA—*Bulletin of the Geological Society of America*
Pleistocene Geology of Western Cape Cod, Massachusetts
Till Fabric

GEOLOGY DEPARTMENT, UNIVERSITY OF MASSACHUSETTS, AMHERST—*Geology Department Contribution 10* Rhythmic Sedimentation in Glacial Lake Hitchcock, Massachusetts–Connecticut

PERMISSIONS

The following papers have been reprinted with the permission of the authors and copyright holders.

AMERICAN JOURNAL OF SCIENCE (YALE UNIVERSITY)—*American Journal of Science*
The Origin of Drumlins
Some Aspects of Drumlin Geometry

GEO ABSTRACTS LTD.—*Proceedings of the 2nd Guelph Symposium on Geomorphology, 1971*
Research Methods in Pleistocene Geomorphology, The Effect of Lithology upon Texture of Till

GEOGRAPHICAL SOCIETY OF FINLAND—*Fennia*
On the Transportation of Materials in the Esker of Laitila

GEOLOGICAL SOCIETY OF AMERICA—*Bulletin of the Geological Society of America*
Mechanisms of Clast Attrition in Basal Till

INTERNATIONAL GLACIOLOGICAL SOCIETY—*Journal of Glaciology*
The Formation and Shape of Drumlins and Their Distribution and Orientation in Drumlin Fields
Mechanism for the Formation of Inner Moraines Found near the Edge of Cold Ice Caps and Ice Sheets
Observations on a Collapsing Kame Terrace in Glacier Bay National Monument, South-Eastern Alaska

OHIO STATE UNIVERSITY PRESS—*Till, A Symposium*
Glaciotectonic Structures in Drift
Till Genesis and Fabric in Svalbard, Spitsbergen

REGENTS OF THE UNIVERSITY OF COLORADO—*Arctic and Alpine Research*
Nature and Rate of Basal Till Deposition in a Stagnating Ice Mass, Burroughs Glacier, Alaska

ROYAL SOCIETY OF CANADA—*Transactions of the Royal Society of Canada*
    An Experimental Study of Varve Deposition

SWEDISH GEOGRAPHICAL SOCIETY—*Geografiska Annaler*
    Eskers near the Casement Glacier, Alaska
    Mappings and Geochronological Investigations in Some Moraine Areas of South-Central
      Sweden
    Some Observations on Fluted Moraine Surfaces

UNIVERSITY OF CHICAGO PRESS—*The Journal of Geology*
    Development of End Moraines in East-Central Baffin Island
    Ice-Disintegration Features in Western Canada
    Ice-Thrust Ridges in Western Canada

Series Editor's Preface

The philosophy behind the "Benchmark Papers in Geology" is one of collection, sifting, and rediffusion. Scientific literature today is so vast, so dispersed, and, in the case of old papers, so inaccessible for readers not in the immediate neighborhood of major libraries that much valuable information has been ignored by default. It has become just so difficult, or so time consuming, to search out the key papers in any basic area of research that one can hardly blame a busy man for skimping on some of his "homework."

This series of volumes has been devised, therefore, to make a practical contribution to this critical problem. The geologist, perhaps even more than any other scientist, often suffers from twin difficulties—isolation from central library resources and immensely diffused sources of material. New colleges and industrial libraries simply cannot afford to purchase complete runs of all the world's earth science literature. Specialists simply cannot locate reprints or copies of all their principal reference materials. So it is that we are now making a concerted effort to gather into single volumes the critical material needed to reconstruct the background of any and every major topic of our discipline.

We are interpreting "geology" in its broadest sense: the fundamental science of the planet Earth, its materials, its history, and its dynamics. Because of training and experience in "earthy" materials, we also take in astrogeology, the corresponding aspect of the planetary sciences. Besides the classical core disciplines such as mineralogy, petrology, structure, geomorphology, paleontology, and stratigraphy, we embrace the newer fields of geophysics and geochemistry, applied also to oceanography, geochronology, and paleoecology. We recognize the work of the mining geologists, the petroleum geologists, the hydrologists, the engineering and environmental geologists. Each specialist needs his working library. We are endeavoring to make his task a little easier.

Each volume in the series contains an Introduction prepared by a specialist (the volume editor)—a "state of the art" opening or a summary of the object and content of the volume. The articles, usually some twenty to fifty reproduced either in their entirety or in significant extracts, are selected in an attempt to cover the field, from the key papers of the last century to fairly recent work. Where the original works are in foreign languages, we have endeavored to locate or commission translations. Geologists, because of their global subject, are often acutely aware of the oneness of our world. The selections cannot, therefore, be restricted to any one country, and whenever possible an attempt is made to scan the world literature.

To each article, or group of kindred articles, some sort of "highlight commentary" is usually supplied by the volume editor. This commentary should serve to bring that article into historical perspective and to emphasize its particular role in the growth of the field. References, or citations, wherever possible, will be reproduced in their entirety—for by this means the observant reader can assess the background material available to that particular author, or, if he wishes, he, too, can double check the earlier sources.

A "benchmark," in surveyor's terminology, is an established point on the ground, recorded on our maps. It is usually anything that is a vantage point, from a modest hill to a mountain peak. From the historical viewpoint, these benchmarks are the bricks of our scientific edifice.

Rhodes W. Fairbridge

Preface

This book seeks to put into one volume the most illuminating original writings on glacial deposits. We are concerned with process and origin, not with history or chronology per se. Ten volumes would not exhaust readings on the genesis of deposits; witness several recent volumes on till. To help determine the choicest contributions to glacial geology, lists were solicited from a few experienced glacial geology teachers; and most of these papers, including many contending articles, were reviewed thoroughly over two years in a seminar on glacial morphology. Thanks go to these colleagues and graduate students for their critiques. The end product, however, is my responsibility. For lack of space, articles that were simply useful compilations of hypotheses were excluded (including my own "Introduction to Till Today," 1971).

Many ideas and potential contributions lie buried within detailed studies of Pleistocene or Holocene stratigraphy in limited areas; only those dominated by a "process approach" are utilized in this volume, although all those selected stem from observations in specific areas, ranging from Poland to Alaska. No attempt was made to pick the most famous authors, although such distinguished names as De Geer and Dreimanis are included. Nongeologists who contributed are generally physicists who have been concerned with glaciology, for example, Fraser and Weertman. Emphasis is upon studies involving *living* glaciers; and, since the science of glaciology has surged ahead only recently, some lesser known authors such as Boulton, Mickelson, Drake, and McKenzie are featured. There are 30 authors in all, half of whom are Americans and only five of whom were deceased at time of publication. Scientists of six countries are represented, but there is a natural tendency to gravitate toward articles written in English and from areas north of 40°N.

This book presents only one half of the story of glacial geology; the other half, glacial erosion, is covered in a separate volume. In the same vein, varves are only briefly summarized since they are the key to still another volume—one on glacial chronology.

Richard P. Goldthwait

Contents

I. TILL AND GROUND MORAINE

II. DRUMLINS AND FLUTED MORAINE

Contents by Author

Introduction

Definition

The importance of glacial deposits is that they constitute the loose surface dirt on which many people live and seek their livelihood, notably in the northern third of the United States, in much of Canada, and in all the northern countries of Europe.

In the glaciated lands nearly all road excavations and most foundations are dug in glacial "drift." All groundwater filters through it and most small local water supplies are extracted from it. Variations in the materials, and the water in it, are of key importance in urban, highway, rail, airport, and pipeline construction. Glacial gravels and sands, sorted and handled by streams that flowed out of glaciers, constitute either the first or second chief mineral resource (by value or tonnage) in 16 of the 18 states or provinces north of the Susquehanna, Ohio, Missouri, and Columbia rivers. Drift also carries the lithologic keys, mineral signatures, and chemical traces of hidden or remote bedrock ores, as well as containing occasional placer concentrations of such heavy metals as tin, gold, and platinum. Understanding the origin of glacial deposits is the first step in economic or land-use investigation of any sort (Legget, 1973).

Glacial drift (deposits) may be as thin as a few scattered boulders, or as thick as a 200-m sequence of tills, gravels, and clays lying on bedrock. It may come directly from ice as till, or it may be carried and sorted in a glacial stream as sand and gravel, or it may settle as layers of clay and silt in a lake or in the sea. This material is all called "glacial drift" here, because it has been moved from different bedrocks far away by a free ride on a glacier. The old English term "drift" came out of the biblical explanation: transport in the giant flood that floated Noah's Ark. The specific expression *glacial drift* arose in the mid-nineteenth century from the heretic idea that long-gone glaciers carried the materials, at least partway.

How far away from the reach of the ice itself do glacial deposits extend today? The silt-clay layers and sandy deltas deposited in a lake blocked by glacier ice, or the marine till settled below shelf ice in an ocean, are both glacier-dependent. Fully half the glacial

1

deposits do have dual origin, and a double name helps identify them. Outwash for some distance down valleys is glacier-dependent, but the dependence on glaciers of deltaic deposits such as those of the Mississippi in the faraway Gulf of Mexico might be argued. Typical periglacial features, such as loess or frost-moved slope debris, are not exclusively glacial (e.g., north-central Alaska), so these will be handled in another Benchmark volume. For the details Washburn's (1973) *Periglacial Processes and Environments* is an appropriate text. It is logical that one of the best recent texts (Embleton and King, 1968) is entitled *Glacial and Periglacial Geomorphology,* for the two sorts of features are in juxtaposition and do overlap.

This is the first book to deal fully and exclusively with all glacial deposits. Glacial geology deals with all the records left by a glacier; glaciology deals with the glacier itself; and Quaternary geology is the historico-environmental record of the present ice age. The best general texts to cover all these are the evolutionary sequence by Richard F. Flint: *Glacial Geology and the Pleistocene Epoch* (1947), *Glacial and Pleistocene Geology* (1957), and *Glacial and Quaternary Geology* (1971). Yet in the last of these, only 80 pages (of 809) are devoted primarily to the *origin* of the glacial deposits. The Pleistocene Epoch of the Quaternary Period (roughly 2 million to 10,000 years ago) includes many glacial stages when most glaciers were very big and when most glacial deposits now exposed were made, so the genesis of glacial deposits is often buried in details of Quaternary history.

History of the Subject

Not until the period 1840 to 1860 did the hypothesis that huge continental glaciers had once covered northern Europe, England, and North America become a serious scientific theory. Glaciology got its crucial start in the Alps at the same time with the work of deCharpentier (1841), Hugi (1843), and Tyndall (1860). The science of glaciology made only modest headway in the late nineteenth century (Hess, 1904) and early twentieth (Ahlmann and Thorarinsson, 1937). In the first half of the present century, hundreds of regional geology studies of glacial regions were published, but they were largely descriptive (or were made in the search for aggregate resources), and origins of features were often imagined and deduced by authors who had never studied a glacier.

Suddenly after World War II glaciology gained impetus from the military need to know and understand the Arctic–Antarctic regions; theoretical and empirical ice studies blossomed forth and were accommodated by the *Journal of Glaciology.* Concomitant growth in the study of glacial deposits caught in the act of deposition has led to a surge of articles since then. Seventy-one percent of this book stems from literature published during the last three decades. The growth in Quaternary geological literature is even more spectacular.

Classification of Glacial Deposits

Nearly all authors agree that glacial deposits contain mixed and varied lithologies and/or exotic rock or mineral types. However, no two glacial geologists will agree on exactly how best to subdivide glacial deposits. Classification is artificial from its inception a century ago, and everywhere one finds "borderline" cases or arguable multiple hypotheses of origin. Usually the ice is no longer there; if present, its rate of change is slow, limiting observations of its dynamic activities and morphology. From a concensus of the literature we might agree there are two general classes of deposits:

1. *Direct glacial deposits,* notably till, a material of broad range of sizes, lacking good or extensive bedding or sorting, and commonly identified morphologically as *moraine* (Parts I to III).

2. *Indirect, aqueous deposits,* resulting in cobbles, pebbles, sand, silt, or clay of a narrow range of size in any one layer, clearly bedded and sometimes labeled *washed drift* (Parts V and VI).

The commonly used subdivision of glacial deposits given in Table 1 depends primarily on presumed or known location with respect to the active ice at the time of deposition. There is, for example, the group of *ice-contact (washed) deposits* made in or against the irregular ice of the zone of ablation (Part IV). Another group consists of the *proglacial deposits,* such as outwash or glacial lake deposits extending some kilometers outside the glacier edge at the time of deposition (Parts V and VI). Actual observations of outwash fans and lakes forming today indicate that they may extend far back under a thinning ice edge as well. So, while every subdivision has its exceptions, the classifications adopted in this book are based upon the classical names that are derived partly from the materials but mostly from shape with some inferred ice relations.

Subgroups, such as the more recently devised "disintegration deposits," bridge the gap between the moraine and the outwash, because they represent large areas of thin glacier stagnating and decaying near the former ice edge (Part V). Although these include numerous washed and banded deposits (shown in the table by asterisks), they seem to diagnose a common ice-marginal situation. Streamlined deposits are separated as a subgroup (Part II) because it is not clear or agreed that these must be made near the ice edge nor are they a necessary part of the end moraine; but they represent erosion and/or relocation of some watery till. All observers now agree that they represent shaped accumulations due to sliding and/or melting of basal ice. In other words, shape of accumulation is an important criterion in classification.

The names of many specific deposits on the right side of Table 1 have been used for more than a century. I prefer to preserve the use of old terms such as esker, kame, and drumlin, although some variations in exactly what is called a "kame," for example, has led some to seek its abandonment (Cook, 1946). It is curious how many of these terms were defined and introduced by just one man, Maxwell Close, in 1867. Since his definitions are vague and nonquantitative, it will be useful to sharpen up his language;

3

Table 1. Classification of Glacial Deposits

Dominant Materials		Conjectured Situation		Shape or Morphology
Lodgement till subglacial, basal, nonbedded, compact	*Till* till dom-inated, poorly sorted	*Ground moraine* under ice and off retreating edge	I	*Till plain* or rolling hills washboard moraine *minor moraines
		Streamlined sliding melting base	II	*Drumlin* grooved till crag-tail
Ablation till superglacial, loose lens, bedded, contorted		*End moraine* at standing or advancing ice edge	III	*Lobate/looped moraine* push/thrust boulder belt lateral/interlobate moraine *kame moraine
		Disintegration stagnant, decaying marginal area, buried ice masses	V	*Controlled/uncontrolled disintegration* dead ice knobs/rings disintegration ridge inverted lake
Glaciofluvial coarse cobble to silt, channeled or cross-bedded	*Wash* well sorted	*Ice contact* dipping, deformed, irregular beds in ice pit, channel, or tunnel	IV	*Esker* *crevasse filling chain (of kames)
				Kame *field/kame and kettle kame moraine moulin kame kame terrace/plain
Glaciolacustrine fines: fine sand to colloid clay, laminated		*Proglacial* at grade, uniform beds extending away from ice	V	*Outwash plain/fan* valley train kettled/pitted outwash *collapsed outwash
				Lacustrine/marine delta strand/raised beach glacial *varve*

*Other forms often attributed to stagnant ice disintegration.

arbitrarily, for example, a ridge of glacial sand or gravel more than 10 times as long as it is wide is identified as an esker or one of its subclasses, whereas a mound of sand/silt/gravel less than 10 times as long as wide is called a kame. In any case they are closely related in space and time. Each one may be observed in several different situations or associations.

No effort is made here to separate the sedimentary products of mountain glaciers from those of continental glaciers. Indeed the editor would claim that there is no major difference except that of confinement which results in topographically directed ice flow in mountains. Unquestionably mountain glaciers produce a very different and spectacular scenery which is dominated by erosive effects. Certain subclasses of deposits, such as stranded lateral moraines on the one hand, or interlobate moraines on the other, are each peculiar to a different kind of glacier and topography, but even that is not 100 percent true; some ice processes are common to both. More and more it is evident that ice is ice with flow properties determined by ice temperature, ice slope, and

confining pressure, and not defined by classification. Both the pattern of flow and the shape of lingering residual ice masses during deglaciation are determined by the underlying topography.

State of the Art

The more glacial geology derives answers from studies of glaciological action, the more we find that the deposition is an actual event of the last 5 to 30 percent of the time of glaciation (depending on position relative to the center and edge of glaciation). Very little that occurs in the first centuries of a long glaciation is left afterward to be seen except for some erosive effects (Goldthwait, 1973). In mountainous terrain, erosion is much more significant. To be sure, under the outer one-third of any glacier, many earlier deposits are preserved by overriding, sometimes over vast areas, and thus Quaternary studies have made a great and continuing contribution to glacial geology.

With expanding glaciological studies, and long periods for the study with repeated airphoto coverage, it may be expected that future contributions to glacial geology will clarify details of deposition that take from a decade to a century to occur. Right now, structural studies of the deformation of drift in old deposits seems to be adding much to our understanding. Perhaps the major contributions or changes in the future will come from physical studies and subglacial drilling and from underwater studies in lakes and seas contiguous to small mountain glaciers and to the huge Antarctic ice sheet. The two great unknowns are the extensive undulating floor of the continental glacier in action and the sedimentation in murky waters adjacent to the ice.

References Cited

Ahlmann, H. W., and S. Thorarinsson (1937) Vatnejokull. Scientific results of the Swedish Icelandic investigations 1936–1937: Geografiska Annaler, v. 19, p. 176–211.

Close, M. H. (1867) Notes on the general glaciation of Ireland: Roy. Geol. Soc. Ireland Jour., v. l, p. 207–242.

Cook, J. H. (1946) Kame complexes and perforation deposits: Amer. Jour. Sci., v. 244, no. 8, p. 573–583.

deCharpentier, J. (1841) *Essai sur les Glaciers et sur le terrain erratique du Bassin du Rhône:* Marc Ducloux, Lausanne, 363 p.
Embleton, C., and C. A. M. King (1968) *Glacial and Periglacial Geomorphology:* Edward Arnold, London, 608 p. (also St. Martin's Press, New York).
Flint, R. F. (1947) *Glacial Geology and the Pleistocene Epoch:* Wiley, New York, 589 p.
_____ (1957) *Glacial and Pleistocene Geology:* Wiley, New York, 553 p.
_____ (1971) *Glacial and Quaternary Geology:* Wiley, New York, 892 p.
Goldthwait, R. P. (1973) Till deposition versus glacial erosion: in *Research in Polar and Alpine Geomorphology*, 3rd Gulph Symposium, ed. B. D. Fahey and R. D. Thompson, Geo Abstracts Ltd., Norwich, England, p. 159–166.
Hess, H. (1904) *Die Gletscher:* F. Vieweg, Braunschweig, 426 p.
Hugi, F. I. (1843) *Die Gletscher und die erratischen Blöcke:* Ient and Gasmann, Colothurn, 256 p.
Legget, R. F. (1973) *Cities and Geology:* McGraw-Hill, New York, 624 p.
Patterson, W. S. B. (1969) *The Physics of Glaciers:* Pergamon Press, Oxford, 250 p.
Tyndall, J. (1860) *Glaciers of the Alps:* John Murray, London, 444 p.
Washburn, A. L. (1973) *Periglacial Processes and Environment:* St. Martin's Press, New York, 320 p.

General Texts on Glacial Deposits

Charlesworth, J. K. (1957) *The Quaternary Era with Special Reference to Its Glaciation*, v. 1, *Glaciology and Glacial Geology:* Edward Arnold, London, 591 p.
Cotton, C. A. (1942) *Climatic Accidents in Landscape-making:* Whitcomb & Tombs, Christchurch, New Zealand, 354 p. (reprinted 1943, Wiley, New York).
Lliboutry, L. (1958) *Traité de glaciologie*, Tome 1: Masson, Paris, 1040 p.
Price, R. J. (1973) *Glacial and Fluvioglacial Landforms:* Oliver & Boyd, Edinburgh, 242 p.
Thwaites, F. T. (1956) *Outline of Glacial Geology:* University of Wisconsin Press, Madison, Wis., 129 p.

I
Till and Ground Moraine

Editor's Comments on Papers 1 Through 6

This group of six selections is the longest in this book; good contributions on till could fill two volumes. From the continuing flood of studies, mostly since 1958, I selected six that first set out some basic principles.

The original Holmes article on till fabric (Paper 1) is long for this consolidated volume, so I have eliminated the sections on "Schedule of field data," "Field procedure," and "Graphic representation." Several pages of initial analysis of "composite group" are also deleted. Most of Holmes procedures have been superseded by simplified methods. The large wooden goniometer he used for axial strike and dip is more simply done now with a Brunton compass. Details of statistical analysis added since 1941 may be found in Andrews (1971) or Kauranne (1960). Many articles have come out since then concerning the origin of the fabrics (e.g., Glen et al., 1957; Harrison, 1957). Some related aspects, such as preconsolidation values in silts, were explored (Harrison, 1958). Although these do not have general use as a tool to determine former ice thickness, they do aid in identification of basal till.

Two pertinent statistical studies emerged in the late 1960s from active till laboratories: University of Western Ontario (Paper 3) and Ohio State University (Paper 4). For still others, see Gross and Moran (1971). In the first, Dreimanis and Vagners related the detailed analysis of grain size and mineralogy from known bedrock sources to discover that different minerals are glacially ground to different limiting grade sizes (see also Dreimanis and Vagners, 1969, 1971). In the second statistical paper (Paper 4), Drake studied 19 sites of lower, compact "basal" till in New Hampshire and showed how shape and rounding of some 1800 pebbles from known bedrock sources reveals the processing (abrasion and breakage) beneath a huge ice sheet. Still another "first," pioneered by Krinsley and Donahue (1968), is the use of the electron microscope to identify the processes responsible for various sand grain surface textures. Because of space limitations, this study could not be included.

The notion of two kinds of till, one emplaced by surface ablation (melt) and another by basal melt, was presented last century on hypothetical grounds (e.g., Upham, 1891). Drake finally put this concept on a statistical basis (1971), and many others contributed details (Hartshorn, 1958). Until 1970 no one had studied all the modes of till deposition taking place in the actual ice of a glacier. Almost simultaneously Boulton studied glaciers in Spitzbergen while Mickelson studied glaciers in Glacier Bay, Alaska. Boulton's most comprehensive early work (Paper 2) is included to show differences in ablation till and basal till. Some ideas were expressed in his earlier publications (1968,

1970a, 1970b). Since both young workers discovered the same rate of basal till deposition, only Mickelson's more complete report on this one subject is reproduced (Paper 5; see also Mickelson, 1971).

Finally, no treatment of the origin of till would be complete without recognition of structures in till—a subject vigorously studied. German and Danish workers recognized redeposition and contortion of earlier materials overridden by ice advance in central-northern Europe (see Slater, 1927; Virkkala, 1952). Moran's recent study (Paper 6) in Illinois is one of the first up-to-date treatments in English (see also Christiansen, 1971).

References Cited

Andrews, J. T. (1971) Methods in the analysis of till fabrics: in *Till, a symposium*, ed. R. P. Goldthwait, Ohio State University Press, Columbus, Ohio, p. 321–327.

Boulton, G. S. (1968) Flow tills and related deposits on some Vestspitsbergen glaciers: Jour. Glaciol., v. 7, p. 391–412.

———(1970a) On the origin and transport of englacial debris in Svalbard glacier: Jour. Glaciol., v. 19, no. 56, p. 213–229.

———(1970b) On the deposition of subglacial and melt-out tills at the margins of certain Svalbard glaciers: Jour. Glaciol. v. 9, no. 56, p. 231–246.

Christiansen, E. A. (1971) Till in southern Saskatchewan, Canada: in *Till, a symposium*, ed. R. P. Goldthwait, Ohio State University Press, Columbus, Ohio, p. 167–183.

Drake, L. D. (1971) Evidence for ablation and basal till in east-central New Hampshire: in *Till, a symposium*, ed. R. P. Goldthwait, Ohio State University Press, Columbus, Ohio, p. 73–91.

Dreimanis, A., and U. J. Vagners (1969) Lithologic relationship of till to bedrock: in *Quaternary geology and climate*, ed. H. E. Wright, National Academy of Science, Washington, D. C., VII INQUA Congr. Proc., v. 16, p. 93–98.

——— , and U. J. Vagners (1971) Bimodal distribution of rock and mineral fragments in basal till: in *Till, a symposium*, ed. R. P. Goldthwait, Ohio State University Press, Columbus, Ohio, p. 237–250.

Glen, J. W., J. J. Donner, and R. G. West (1957) On the mechanism by which stones in till become oriented: Amer. Jour. Sci., v. 255, no. 2, p. 194–205.

Gross, D. L., and S. R. Moran (1971) Grain size and mineralogical gradations within tills: in *Till, a symposium*, ed. R.P. Goldthwait, Ohio State University Press, Columbus, Ohio, p. 251–274.

Harrison, P. W. (1957) A clay-till fabric: its character and origin: Jour. Geol., v. 65, no. 3, p. 275–303.

——— (1958) Marginal zones of vanished glaciers reconstructed from preconsolidated pressure values of overridden silts: Jour. Geol., v. 66, p. 72–95.

Hartshorn, J. H. (1958) Flowtill in southeastern Massachusetts: Geol. Soc. America Bull., v. 69, p. 477–482.

Kauranne, L. K. (1960) A statistical study of stone orientation in glacial till: Comm. Géol. de Finlande Bull., v. 188, p. 87–97 (also in Soc. Géol. de Finlande Comptes Rendus, v. 32, p. 87–97).

Krinsley, D. H., and J. Donahue (1968) Environmental interpretation of sand grain surface textures by electron microscopy: Geol. Soc. America Bull., v. 79, p. 743–748.

Mickelson, D. M. (1971) Glacial geology of the Burroughs Glacier area: Ohio State University, Institute of Polar Studies, Rept. 40, 149 p.

Slater, G. (1927) The structure of the disturbed deposits of Moens Klint, Denmark: Proc. Roy. Soc. Edinburgh, v. 60, pt. 2, no. 12, p. 289–302.

Upham, W. (1891) Criteria of englacial and subglacial drift: Amer. Geologist, v. 8, p. 376–385.

Virkkala, K. (1952) On the bed structure of till in eastern Finland: Comm. Géol. de Finlande Bull., v. 157, p. 97–109 (also in Soc. Géol. de Finlande Comptes Rendus, v. 25, p. 97–108).

1

Reprinted from *Geol. Sox. America Bull.*, **52**, 1301–1304, 1310–1315, 1319–1321, 1324–1352, 1354 (1941)

Till Fabric

CHAUNCEY D. HOLMES

ABSTRACT

A study of the arrangement of component materials in undisturbed till, the till fabric, shows that at most localities the imbedded stones tend statistically to lie so that their long axes are parallel to the direction of glacier flow at the time of deposition. In a few localities the dominant statistical preference is for alignment at right angles to that direction. Presumably the parallel orientation was normally acquired by sliding, and the transverse orientation by rotation, and permanent deposition commonly occurred without loss of alignment. Fabric analyses indicate that stones of certain forms and degrees of roundness (enumerated in text) have a greater-than-average statistical chance for deposition either parallel to, or transverse to, the direction of transport. Such stones thus serve as guides to the direction of glacier flow and are independent of other evidence. Characteristic depositional attitudes of certain types of till stones permit inferences regarding the probable nature of the transportational environment.

INTRODUCTION

FACT OF TILL ORGANIZATION

Undisturbed till has an inherent organization. At most localities this organization manifests itself in the tendency of imbedded stones to lie so that their longest dimension or axis coincides approximately with the direction of glacier flow at the time of deposition. Such tendency is best revealed by a statistical study of the positions of at least 100 till stones from any one locality. The results can be expressed by diagrams such as the simple "rose" figure or by the contoured diagram commonly used in petrofabric studies. This makes possible the determination of ice-flow direction at places where no striae or other criteria are available. Moreover, the direction of glacier flow is known to have varied from time to time at any one place. Hence, in problems such as locating mineral deposits by tracing drift fragments to their sources (Sauramo, 1924), data from till-stone orientation may be as essential as those from striae on bedrock beneath the till.

HISTORICAL STATEMENT

Miller (1884) published probably the first critical observations on till-stone orientation. In describing "pavement boulders"[1] in the till near Edinburgh, Scotland, he stated (p. 167):

"The longer axis of the stone is often directed in the line of glaciation, and the pointed end is frequently, but not always, toward the ice."

Smaller stones may be oriented in "fluxion structures" around the larger boulders.

"It does not follow, however, that wherever we find an orientation of boulders in the till there was fluidal motion in the layer in which they lie. If the ice had a fluxion structure of its own, such boulders as were incorporated within its mass would arrange their axes conformably; and, when they lagged and came to rest and were imbedded, they might retain in many cases the arrangement that marked them when in motion" (p. 187).

---

[1] Stones with strongly striated upper surfaces, generally occurring in definite horizons in till and constituting a "pavement" on which the superjacent till rests.

Editor's Note: A row of asterisks indicates material deleted owing to lack of space. Certain figures and tables referred to in the text have also been omitted for reasons of brevity.

Bell (1888b) carried the investigation further by visiting a number of Swiss glaciers to determine whether such orientation of stones existed in transit. He concluded (p. 341)

". . . that the tendency of boulders on all glaciers is to assume a longitudial position, and that this is most observable . . . on large glaciers, where the obstructions are fewer in proportion to the mass, and produce the least disturbing effect."

Unfortunately, both Miller's and Bell's data are largely qualitative and selective.

Upham (1891) described the characteristic position of "oblong" stones in subglacial till as having their long axes parallel to contiguous striae. Flat stones were said to lie parallel with the surface of deposition. However, Upham recorded no systematic investigation.

The results achieved by these early investigators failed to attract the notice they deserved. James Geikie's classic treatise (1895, p. 15, 62) probably reflects the prevailing attitude of that time. Although mentioning that "in certain regions the large and small till stones are oriented in parallel fashion," and citing Miller's work, Geikie described till as a clay containing a "confused and pell-mell mixture of stones." For many years this concept of chaotic agglomeration remained practically unchallenged. Twenhofel (1932, p. 86) summarized more recent opinion by stating that ground moraine consists of "unstratified and unorganized material."

Papers by Richter (1932, 1933, 1936) are among the most important published works of recent date. His discoveries apparently developed as a modification of the method used by the Scottish geologists for determining the direction of glacier flow from striae on the upper surfaces of "pavement boulders" in areas where striated bedrock ledges are absent. In northern Germany, Richter found that the long axes of elongate stones tend to parallel the direction of pavement-boulder striae, even though such stones themselves bore no striae. Like Miller, he reasoned that these stones had been oriented as streamlined bodies in the glacier and had been deposited with but little or no change in orientation. A statistical grouping of long-axis orientations indicates the direction in which the glacier was moving. Later studies at Engebrae and Fondalsbrae glaciers in Norway verified his conclusions. Stones imbedded in debris zones near the end of the glacier show a statistical preference for long-axis orientation parallel with the direction of glacier flow. The majority of stones in the till beside the glacier have similar orientations. Although Richter's conclusions are essentially those of Miller and Bell and appear to be a rediscovery of the earlier knowledge, his conclusions are supported by quantitative, statistical data.

Richter noted that many of the larger till stones (small boulders) at the Norwegian localities tend to an orientation transverse to glacier flow. These were thought to have been oriented originally parallel with the ice-flow direction and to have been shoved during a temporary readvance of the glacier; or to have been rolled beneath the glacier because they projected through the lowermost shear layer (Scherpakete). But the small cobbles in the same deposit are oriented essentially parallel with the ice-flow direction. This, he believed, may have resulted from the action of later meltwater streams which were competent to reorient the smaller stones, leaving the larger ones unmoved.

In describing the Pleistocene glaciation of a part of Yellowstone National Park, Miner (1937) cited the phenomenon of surface boulders with their long axes in parallel orientation, trending diagonally across a valley, but he offered no special explanation.

Krumbein (1938, p. 273; 1939) has investigated statistically the axial orientation of till stones in the western Great Lakes region. He concluded that both the mode and the arithmetic mean of long-axis directions approximate the direction of glacier flow at the locality studied.

The present writer (Holmes, 1938) published a preliminary statement on till-stone orientation in central New York and suggested till fabric as a term denoting the space relations among the component rock and mineral fragments in undisturbed till.

PRESENT PROBLEM

The till in central New York shows the usual preferred arrangement of stones parallel with the direction of glacier flow. In addition, it contains many stones whose long-axis orientation is transverse to the known direction of flow, thus giving two statistically preferred positions of the long axes. At a few localities the transverse orientation predominates. Therefore inference as to the direction of glacier flow, based on mere statistical preference in long-axis positions at those localities, would be wholly unreliable. Richter's published data indicate that transverse orientation exists to some extent in the till of northern Germany, though he did not discuss it, and he apparently regarded it as merely accidental. The hypothesis of reorientation of stones by meltwater streams is inapplicable to the till of central New York. The till in that region has a few sandy or silty lenses but it is prevailingly clayey and lacks evidence of vigorous meltwater action. The stones in transverse position lie closely adjacent both horizontally and vertically to those aligned in other directions, and the enclosing till matrix is essentially uniform at any given locality.

The present work is a search for the factors that have caused some stones to assume the transverse position while others, deposited contemporaneously, are aligned parallel with the direction of glacier flow.

Believing that the results of this study could be applied directly in field practice, conclusions have been based on data readily ascertained and recorded in the field rather than on subsequent laboratory study of materials collected. The nature of pertinent data and the techniques used in their collection and analysis may be unfamiliar to most readers. Hence, brief explanations of these subjects are necessary.

ACKNOWLEDGMENTS

Field study was done during the summers of 1937 and 1938. In the preparation of results, the writer is greatly indebted to Professor Richard Foster Flint of Yale University. Professor Adolph Knopf and Dr. Eleanora Bliss Knopf also have given valuable suggestions and assistance.

This paper is part of a dissertation presented for the degree of Doctor of Philosophy in Yale University.

FIELD DATA AND TECHNIQUE

LOCATION AND CHARACTER OF THE AREA

Investigations were made at 31 localities distributed over the Cazenovia and Tully quadrangles and the northern part of the Pitcher quadrangle, near the margin of the Allegheny Plateau in central New York (Fig. 1). The region is hilly and strongly glaciated, with several prominent valleys leading north to the Ontario Lowland. Altitudes of station localities range from about 600 feet to 1900 feet above sea level and are located so as to include a wide variety of topographic situations. Some are within the belt of Valley Heads end-moraine (Fairchild, 1932) and in the border zone of Valley Heads drift beyond that moraine. Others are located north of the moraine zone, and several lie south of the limits reached by the Valley Heads glacier. Striae and drumloidal forms indicate that the Valley Heads glacier came into the region from the northwest. An earlier Wisconsin glacier, extending far beyond the limits of the area, came from the northeast.

*   *   *   *   *   *   *

STATION PATTERNS

General statement.—A rose diagram of the 100 or more unselected stones from any one station has its own unique features of lengths and angular relations of lines. No two are alike, and it became convenient to refer to each diagram as a fabric pattern. Thus each of the 31

13

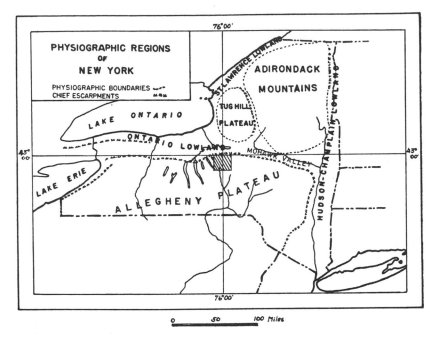

FIGURE 1.—*Map showing location and physiographic relations of area (diagonally lined) in which till-fabric study was made*

stations has its own peculiar pattern, showing one or more preferred directions of long-axis orientation. Figure 4 shows representative station patterns.

Normal "simple" and "compound" patterns.—On nearly all patterns, the longest line (or one of the longer ones where two or more are present) parallels approximately the direction of glacier flow as indicated by the nearest striae or drumloidal axis and by the topographic environment of the station. Other prominent lines may intersect the main one at angles similar to those of intersecting sets of striae commonly observed on rock ledges. At some localities a pattern derived at a given horizon in a section shows preferred orientations slightly different from those of patterns representing higher or lower zones. At one station, two such zones are separated by a sandy layer about 4 inches thick bearing comparatively few stones. At another station in moderately sandy-silty till, similar zones one or two inches thick are superposed with no perceptible boundaries. Other similar occurrences were noted. A pattern that represents only one horizon is likely to have a single prominent line indicating the direction of glacier flow and is a relatively "simple" pattern. Less prominent lines occur on either side of the main one and are progressively shorter with increasing angular distance in the

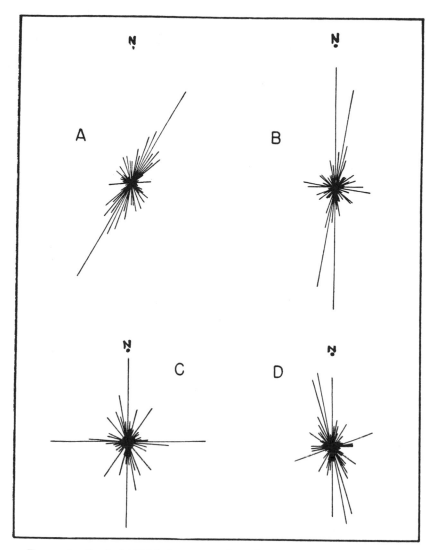

FIGURE 4.—*Typical till-fabric patterns from localities in central New York*

(A) Branch of Mud Creek valley, 1.7 miles south of Lincklaen Center, Pitcher quadrangle.
(B) Shallow through-valley, 0.75 mile northeast of Carpenter Pond, Cazenovia quadrangle.
(C) Limestone Creek valley, 1.8 miles southeast of Oran, Cazenovia quadrangle.
(D) Upland, 3.5 miles northwest of Pompey, Cazenovia quadrangle.

pattern. By contrast, many patterns represent more than one till horizon and therefore present a "compound" appearance. However, both are alike in having their longest lines coincident with the ice-flow direction, a feature normal to the majority of fabric patterns.

Transverse patterns.—A feature common to most station patterns is the presence of one or more prominent lines essentially at right angles to the inferred direction of glacier flow; and in a few patterns such transverse lines are much longer than the others. Thus arises a possible (though unlikely) alternative interpretation that in those places the glacier had flowed in a direction at right angles to that indicated by other available evidence. Moreover, a few of these stations are so situated that glacier flow could be inferred in either of the two right-angle directions indicated by the patterns. These patterns are a departure from the "normal" fabric and are conveniently designated as "transverse" patterns. The diagram in Figure 3 is known to be a transverse pattern and is discussed in detail on a later page as a "problem pattern."

Interpretations.—Superposed fabric patterns in a till section are interpreted as a record of slight shifts in direction of glacier movement while the till was accumulating beneath the glacier by a "plastering-on" process (Fairchild, 1907). Such changes in direction may or may not have been accompanied by intermittent glacier flow. A pressure-melting temperature has been shown to exist at the base of large glaciers (Hubbert, 1937). Hence constant deposition of the basal load—a "plastering-on"—seems probable except under conditions tending to depress the pressure-melting-point isogeotherm into the floor beneath the glacier.

Presumably the factors responsible for orientation of a few till stones transverse to glacier flow at many places are responsible also for the dominantly transverse orientation observed at some localities. These factors must be found in order that patterns showing dominantly parallel or transverse preference may be distinguished. The field facts warrant the conclusion that the preferred directions of long-axis orientation of unselected groups of stones are not infallible independent criteria for determining the direction of glacier flow. Further analysis is necessary.

FABRIC ANALYSIS

GENERAL METHOD

The field studies were planned to include factors believed most likely to show positive correlation with either parallel or transverse alignment of till stones with respect to the direction of glacier flow. The general plan of analysis is to select groups of stones on the basis of their ascertained characters and depositional attitudes and then to compare the orientation patterns of these selected groups. In this way the relative influence of form and degree of roundness on the behavior of stones as they were being deposited can be evaluated. Facts thus secured provide independent criteria for distinguishing between parallel and transverse orientation, and hence the direction of glacier flow. Also, inferences can be made regarding the mechanics of glacial deposition and the probable nature of the transportational environment.

The depositional attitude of a stone in undisturbed till is the position attained at the final moment of glacial transportation. The following analyses show that characters inherent in the till stones predispose them to certain depositional attitudes, but factors in the transportational environment are equally important in determining the till fabric.

COMPOSITE GROUP

General statement.—A homogeneous body of data suitable for analysis was obtained by combining the data from those stations that satisfied three requirements: (1) The station pattern must indicate, as nearly as possible, a single direction of glacier movement (like a single set of striae). (2) The direction of glacier movement at the station must be known from other evidence. (3) The distribution of long-axis dips, plotted by stereographic projection, must be reasonably symmetrical. On the assumption that the till accumulated progressively or by successive increments, a symmetric distribution of long-axis dips would indicate that the surface of deposition was essentially horizontal and therefore practically coincident with the imaginary reference plane from which dip is measured.

Of the 31 stations, 10 fulfilled these requirements sufficiently to permit combining them into a composite group, giving a total of 1180 stones. The several station patterns were combined by rotating each one until its line representing direction of glacier flow was in the conventional north-south position. Thus all lines in the composite pattern (Fig. 5A) have reference only to the known direction of glacier flow and not to compass directions. The direction from which the ice approached is indicated on this and on all subsequent diagrams (except those of Figures 6 and 23) by a dot above each diagram.

Characteristics of the composite group.—Two significant features of the composite group should be mentioned before discussing selected subgroups. First is the three-dimensional distribution of long-axis positions as shown by the contoured diagrams. Figure 5B shows that the long-axis poles tend to concentrate along two girdles. The more conspicuous girdle occupies the peripheral zone of the diagram and includes both the parallel and the transverse maxima. The other girdle, less well defined, includes the parallel maximum but lies normal to the transverse

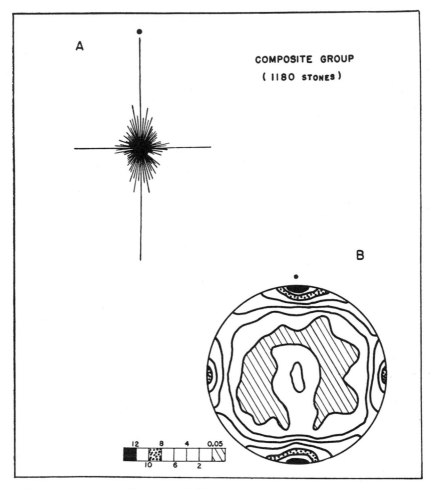

FIGURE 5.—*Rose and contoured diagrams representing combined till-fabric*
patterns from 10 localities

Each individual pattern was superposed by placing its recorded ice-flow direction in north-
south position. Thus the direction from which the ice approached is indicated by a dot above
each composite diagram.

maximum and includes a prominent concentration of steeply dipping
poles near the center of the diagram. These facts suggest that the presence of this secondary girdle may be a reliable guide to the direction
of ice flow. Inspection of the contoured diagrams for the several stations
(Fig. 6) shows that such a girdle is present in some but is indistinct in
many, possibly because of an insufficient number of stones represented
in each.

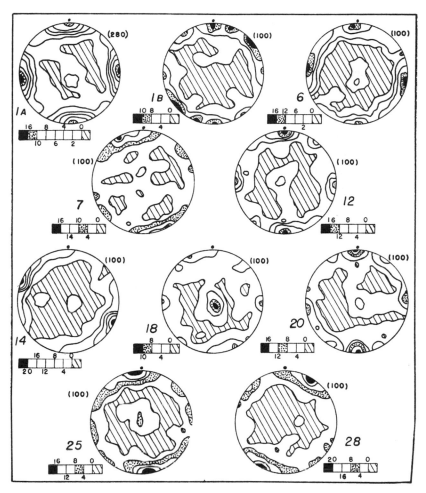

FIGURE 6.—*Contoured till-fabric diagrams of the 10 stations in composite group*

1A, 1B, 6, and so forth, are field station numbers. Number of stones represented in each diagram indicated in parentheses. True north shown by dot above each diagram. Note indistinct percent-concentration bands (girdles) across some of the diagrams, and compare with Figure 5B.

A second important feature of the composite group is the numerical distribution of axial dips through the entire range of dip. Table 2 shows this distribution for both long and intermediate axes. More than one-third of the total number of stones were imbedded with neither axis dipping as much as 10 degrees. With increasing amounts of dip, the numerical frequency declines sharply at about 20 degrees but rises again between 70 and 90 degrees.

*   *   *   *   *   *   *

INTERMEDIATE-AXIS DIP IN RELATION TO LONG-AXIS ORIENTATION

The tendency of till stones to occur with their intermediate axes dipping either less than 20 degrees or more than 70 degrees makes possible a three-fold subdivision of the composite group on the basis of amounts of intermediate-axis dip. The limits chosen are arbitrary but are designed to restrict the end members to relatively small but equal arcs. Figure 8 shows the rose diagrams of the three groups thus established.

Comparison of these diagrams with that of the composite group shows that intermediate-axis dips not exceeding 15 degrees are statistically associated with parallel orientation; that transverse orientation is likewise associated with steeply dipping intermediate axes; and, similarly,

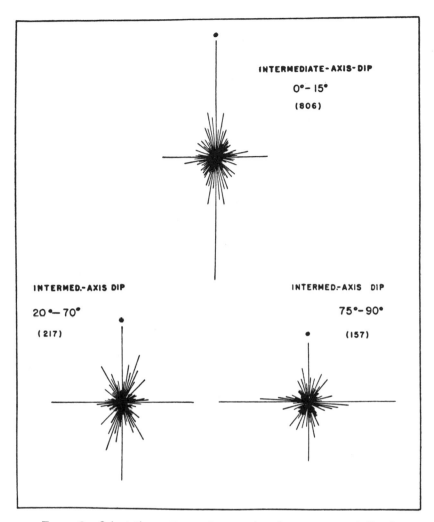

FIGURE 8.—*Orientation patterns of groups based on amounts of dip of intermediate axes*

Numbers of stones represented shown in parentheses. Comparison of the three diagrams shows that steep dip of intermediate axis is correlated with long-axis direction transverse to glacier flow; and that low dip of intermediate axis is correlated with long-axis direction parallel with that direction.

that intermediate-axis dips ranging from 20 to 70 degrees are correlated with diagonal orientation. These facts show that the relative scarcity (per unit range or interval) of long-axis dips ranging from 20 to 70 degrees (shown by Table 2) cannot be attributed to postdepositional compaction of the till. Although compaction would reduce the amount

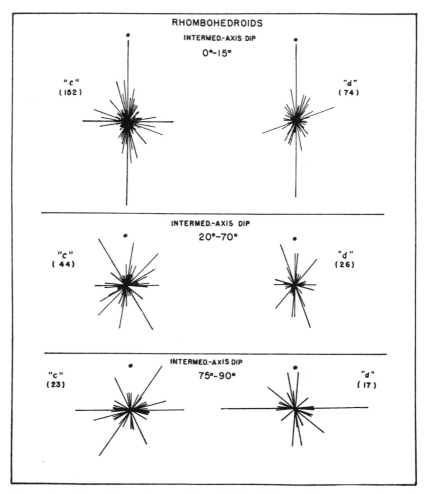

FIGURE 9.—*Orientation patterns of rhombohedroids from three groups shown in Figure 8, further classified as to roundness*

Slightly greater contrast appears among the "*d*" patterns than among the "*c*" patterns; but the contrast averages somewhat greater among these axial-ratio groups than is shown by those of Figure 8.

of tilt of imbedded stones and would be most effective within the 20 to 70-degree range, it would not at the same time exercise a selective change in long-axis direction. Hence compactness of the till with its imbedded stones is believed to be an original depositional character and seems to be logically related to gradual accumulation ("plastering-on") beneath the weight of the moving glacier; it implies a well-defined contact between the ice and its till floor.

*   *   *   *   *   *   *

22

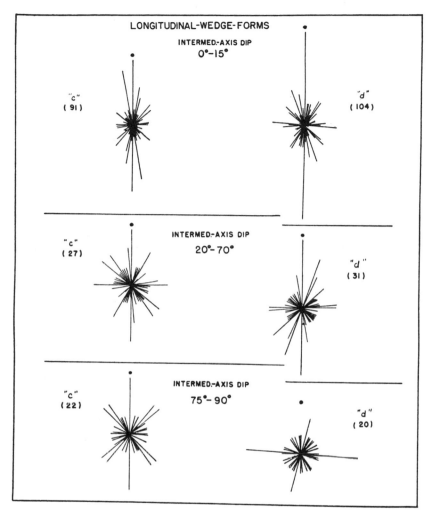

FIGURE 10.—*Orientation patterns of longitudinal-wedge-form stones from three groups shown in Figure 8*

Transversely aligned stones tend to an approximate rather than a precise alignment in that position. Patterns of lateral-wedge-forms resemble those of rhombohedroids and are not shown.

The diagrams in Figure 8 show that the attitude of the plane that includes the two longer axes of a till stone is as important as the direction of the long axis. Neither rose nor contoured diagram can show both simultaneously, but the three rose diagrams together constitute an approximation within the limits of the dip-interval selected.

The next step in analysis is to determine how the different forms and degrees of roundness of till stones influenced long-axis direction in each of the three groups just considered.

COMBINED INFLUENCE OF FORM, ROUNDNESS, AND DIP OF INTERMEDIATE AXIS ON LONG-AXIS ORIENTATION

General statement.—Figures 9 to 12 show the three groups (based on amounts of intermediate-axis dip) further subdivided on the combined bases of form and roundness, except those subgroups that are too small to warrant plotting. These diagrams present some of the greatest contrasts brought out by any basis of subdivision.

Among rhombohedroids of "*d*" roundness, not one in the 75 to 90-degree group is oriented in parallel position; while in the 0 to 15-degree group only one occupies the transverse position (Fig. 9). Therefore, on the basis of intermediate-axis dip, these two groups of sharply angular rhombohedroids are good statistical indicators of the transverse and parallel positions, respectively. Rhombohedroids of "*c*" roundness show the same tendencies but to a smaller extent. Both "*c*" and "*d*" rhombohedroids show a statistical avoidance of the parallel and transverse positions when their intermediate axes dip within the 20 to 70-degree range.

Much less contrast appears among the longitudinal-wedge-forms [3] (Fig. 10), though in the 75 to 90-degree group, roundness "*d*" presents one of the strongest transverse patterns of any wedge-form group. However, slight rounding ("*c*") established a parallel-orientation preference. Excepting this small group of sharply angular stones, wedge-forms show a consistent preference for parallel or near-parallel orientation somewhat stronger than that of other forms.

Low dip of intermediate axis is characteristic of a large majority of ovoids. (Fig. 11.) Those of roundness "*a*" and "*b*" show also an extraordinary preference for parallel orientation, while those less perfectly streamlined ("*c*") have a decided preference for transverse or near-transverse alignment. Ovoids of "*c*" roundness approach varihedroids "*c*" in general relative proportions and appear to resemble them in depositional habit.

Further appreciation of the contrasts in orientation preference already noted is gained by comparing the diagrams in Figures 9 to 12 with those of groups based only on form and roundness. Accordingly Figures

---

[3] Longitudinal-wedge-forms include all wedge-form stones except those with surfaces converging only laterally. The latter have been found to be more closely related to rhombohedroids and are not included here.

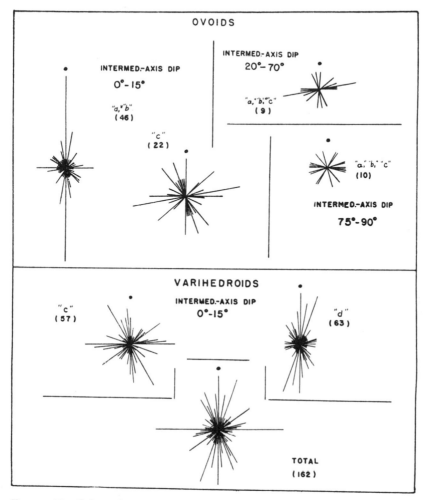

FIGURE 11.—*Orientation patterns of ovoids and varihedroids from the three groups shown in Figure 8*

In the 0-15-degree group, roundness has a greater influence on long-axis direction of ovoids than on that of other till-stone forms. Varihedroids show the least consistency in orientation of any of the till-stone forms. However, rounding transforms many varihedroids into general ovoid forms, which shows in similarity of the corresponding "c" patterns.

13 to 15 have been prepared. The long-axis orientation of discoids appears to be closely related to that of tabular stones (Fig. 15). Therefore the discoids are combined with the later stones of "c" roundness in the contoured diagram as Tabular "a," "b," and "c." . The group of least rounded ("c") ovoids is too small to be plotted stereographically by

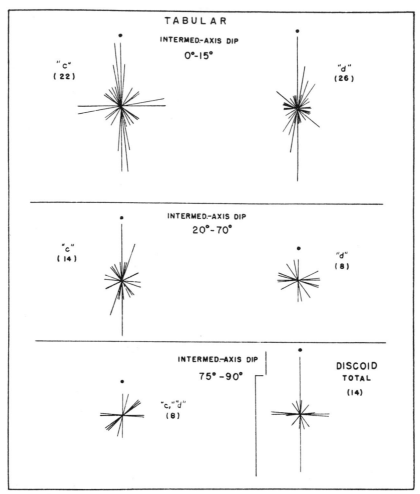

FIGURE 12.—*Orientation patterns of tabular stones from three groups shown in Figure 8*

Composite group contains only 14 discoids (mostly "*a*" and "*b*" roundness), and the discoid pattern is included here to show its similarity to that of the "*c*" tabular stones.

itself, but its general characters can be evaluated by comparing the two contoured diagrams of the ovoid group (Fig. 14).

SYMMETRY OF FORM

The diagrams show that varihedroid stones (Fig. 14) are comparatively lacking in orientation preference, apparently because of their corresponding lack of symmetry. By contrast, ovoids of "*a*" and "*b*" roundness have the strongest tendency to assume or to maintain an orientation

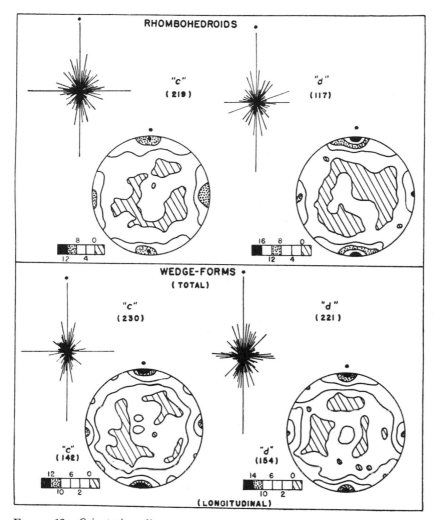

FIGURE 13.—*Orientation diagrams of rhombohedroids and wedge-forms from composite group shown in Figure 5 subdivided according to roundness*
Rose diagrams of wedge-forms include both lateral and longitudinal types. Lateral-wedge-forms are omitted from contoured diagrams.

in harmony with the direction of glacier movement. These ovoids are the most perfectly streamlined of all till-stone forms. Rhombohedroids are comparable to varihedroids in roundness, but their degree of symmetry is more nearly that of ovoids, and their degree of orientation preference is pronounced. Hence it is concluded that symmetry of form conduces to strength and uniformity of depositional orientation.

27

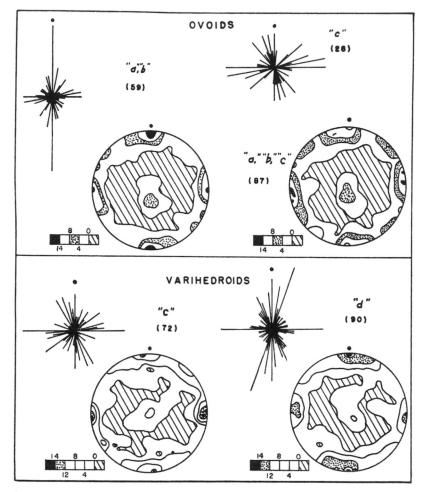

FIGURE 14.—*Orientation diagrams of ovoids and varihedroids from composite group shown in Figure 5, subdivided according to roundness*

Contoured diagram of ovoids at right includes total number of these forms in composite group. Note similarity of patterns of "c" groups.

CONTROL BY PROMINENT SURFACES OF TILL STONES

The trend of striae on till stones is significant inasmuch as such striae record the relative direction of differential movement of the stone and its ice matrix. Striae on rhombohedroids generally parallel the long axis. Because that axis parallels the sides of the stone, the relative importance of surfaces and of long-axis location in controlling the movement of a rhombohedroid in transit cannot be evaluated. But on wedge-form stones the angular relation between sides and long axis permits such

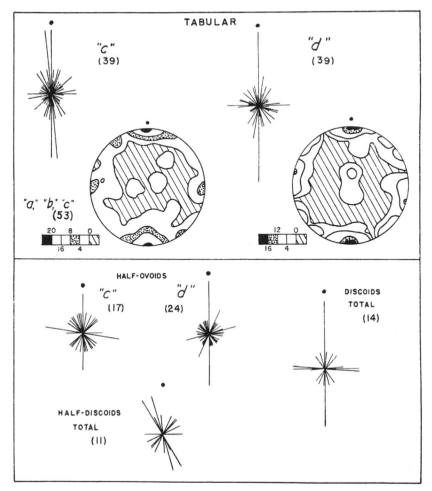

FIGURE 15.—*Orientation diagrams of tabular stones, discoids, half-ovoids, and half-discoids from composite group shown in Figure 5*

Tabular stones and half-ovoids subdivided according to roundness. Half-ovoids and half-discoids are included here for completeness. Diagram of all discoids is shown for comparison with those of tabular stones. The close similarity is apparent.

evaluation. The photograph (Pl. 1, fig. 2) shows that striae on wedge-forms have a strong tendency to parallel the edges of surfaces on which they occur and consequently are parallel to the adjacent surfaces bounded by these same edges. Scarcely any strong striae parallel the long axis of such stones. Photographs of wedge-shaped stones published by Wentworth (1936) show the same tendency. This means that, in the differential movement of stone and matrix whereby the striae originated, the stone was so oriented that adjacent sides intersecting at a given edge (or possibly two edges and three surfaces) were parallel to the direction of flow at any one time. The position of the stone shifted occasionally as control of orientation passed to other surfaces and edges. Movement in this case was by gliding rather than by rotation and evidently was due to ice pressure upon the rear surface because of greater velocity above the horizon of the base of the stone. The long axis of the stone would seldom be well aligned with the direction of movement unless the pointed end of the wedge were directed toward the ice thrust (rear) and the diverging sides adjusted to a streamlined position.

Thus wedge-form stones show that prominent surfaces exercise a strong control in orientation, but further discussion of the probable transportational behavior of wedge-forms is best deferred to a later page. Rhombohedroids seem to illustrate best the problem of orientation by gliding.

ORIENTATION BY GLIDING

Orientation controlled by prominent surfaces is assumed to be much more effective with rhombohedroids than with wedge-forms because all the four principal surfaces are (ideally) parallel to each other. As these surfaces are likewise parallel to the long axis, rhombohedroids are considered best for studying orientation control in gliding.

Figure 9 shows that rhombohedroids whose intermediate axes dip less than 20 degrees lie predominantly parallel to the direction of glacier flow. The contoured diagrams show that the long axes of most stones thus aligned are likewise essentially horizontal. This is indicated also to some extent by the chart (Table 3). These facts suggest the following interpretation: Where till is accumulating beneath the moving glacier, a stone in contact with the till floor is retarded by friction which may become great enough to overcome the forward force of ice thrust. A rhombohedroid sliding thus in parallel orientation and with the two longer axes horizontal presents the least possible cross-sectional area against ice thrust. Its third (shortest) dimension affords a minimum projection into the faster-moving ice above the floor, and a maximum basal surface is tending to adhere to the deposited till. Such an orientation is imposed by sliding (Sander, 1934) and thus may easily become the permanent depositional attitude unless the stone is shoved by another in transit before it becomes buried in the accumulating till. The great number of apparently random orientations suggests that such shoving is common, but the statistical preference of flat-lying rhombohedroids for parallel orientation is interpreted as the position inherited from their transportational orientation.

ORIENTATION BY ROTATION

The fact that the sides of rhombohedroid forms are parallel to the long axis makes these stones especially adapted to forward movement by rotation. Observed phenomena and experimental analogy suggest that they do roll forward when carried above the base of the ice.

In a recent paper, Demorest (1938) has interpreted the conditions of flow at the base of a small, recently vanished glacier in Glacier National Park. Striae on the irregular rock floor show that the ice had moved by laminar flow comparable to the movement of a viscous liquid, apparently in conformity to the laws of fluid mechanics. An experiment by Taylor (1923) illustrates the kind of movement postulated for stones carried in such a medium.

Taylor studied the motion of small oblate and prolate spheroids immersed in water glass of high viscosity. The containing vessel was cylindrical and mounted on a base capable of rotation on a vertical axis. A uniform shearing motion was imparted by a large stationary rod fastened vertically in axial position within the container. The aluminum bodies rotated in response to the shearing motion. When they had reached stable orientations, the prolate spheroids were rotating about their long axes, the latter being in a vertical position. The oblate spheroids rotated about equatorial diameters in similar orientation. Therefore, by analogy, a rhombohedroid carried above the base of the ice would roll forward on its long axis, with that axis horizontal and transverse to the direction of movement. Ovoids likewise would adjust readily to that orientation. Tabular stones would correspond to the oblate spheroids and may be expected to rotate similarly. Most striae on convex surfaces of till stones are believed to be a record of such rotation in a plastic abrasive medium.

Deposition from the rotational orientation postulated for rhombohedroids would account satisfactorily for long-axis direction transverse to that of glacier flow. This would be possible if movement ceased, as in stagnation, but would fail to explain the association of steeply dipping intermediate axes that is statistically characteristic of the rhombohedroids thus oriented. In theory, continued forward movement should be easiest and most likely when the intermediate axis has reached a dip of about 70 degrees and least likely when continued rotation has brought that axis again to a near-horizontal position. Cessation of glacier flow when the rotating stones were in edgewise position is a statistically improbable coincidence. Moreover, these stones were imbedded in the same compact till matrix with those in parallel alignment, for which all evidence indicates gradual accumulation beneath

31

the moving glacier. Inasmuch as rotation may conceivably take place at, or in slight contact with, the till floor, the following interpretation of deposition from rotational orientation is offered.

The rhombohedroid stone was rotating in transverse orientation along the till floor of the glacier. With each rotation the intermediate and short axes rose alternately past the vertical, and the intermediate axis, being the longer of the two, required a larger vertical range. Other stones carried slightly higher (and therefore moving slightly faster) entered this zone, intercepted the rising edge of the rhombohedroid, and by continued forward movement thrust the rhombohedroid edgewise into the till floor. When the intermediate axis was nearly vertical (or just past the vertical) the impinging stone was free to move past, leaving the rhombodedroid permanently imbedded.

This hypothesis is equally applicable to ovoids. Inasmuch as ovoids lack the parallel edges and prominent flat surfaces characteristic of rhombohedroids, a greater tendency to axial wabbling may be expected with consequently less accurate transverse alignment when imbedded. The orientation diagrams of ovoids confirm this expectation (Figs. 11, 14).

In contrast with rhombohedroids, wedge-forms are not well adapted to continued rolling in transverse orientation. On rotation, the pointed end would advance less rapidly than the other, and one or two complete rotations would suffice to bring the stone back into parallel alignment. Hence wedge-form stones receive more consistent abrasion by sliding than by rotation, which seems to explain the common observation that wedge-forms are the most likely to bear noticeable striae (Wentworth, 1936). The orientation diagrams indicate that sharply angular ("d") wedge-forms are equal in persistent parallel orientation to those of "c" roundness. As field observation has shown that the "d" wedge-forms are produced from larger stones by crushing, the wedge form is the cause of the presence of striae rather than the result of "controlled" glacial abrasion as interpreted by Von Engeln (1930). Demonstrable abrasion facets are rare, and the majority of those observed in central New York seem best explained as boulder-pavement phenomena.

Thrusting of stones into the till floor in the manner suggested offers an explanation for the numerous diagonal orientations shown conspicuously by many rhombohedroid and wedge-form patterns (Figs. 9, 10, 18, 19, 20). A wedge-form stone, sliding forward with the point foremost and with one of the convergent sides parallel to the direction of transport, may change from gliding to rotation. The initial rotation would bring the stone up on the other (diagonally oriented) convergent side, with the long axis in a diagonal direction. If during this movement superjacent debris prevented the rear edge from rising far, the stone would be forced down into the till beneath, with the long axis still in diagonal orientation. Likewise, a rhombohedroid in parallel orientation may be first turned until its diagonal dimension is parallel to the direction of gliding (like a double wedge). If rotation is then induced, the initial movement would take place by rolling on one of the diagonally placed sides. Superjacent debris might then intercept the rising edge and press the rhombohedroid into the till floor in permanent diagonal orientation.

The hypothesis of permanent deposition from a rotational orientation not only explains adequately the associated steep dips of intermediate axes but also suggests an explanation for the concentration of steeply dipping long-axis orientations shown by many of the contoured diagrams. Richter (1933) noted similar near-vertical orientations of long axes in northern Germany and attributed them to local vertical currents in the glacier. That hypothesis is inapplicable to the till of central New York because such stones are distributed closely among those of prevailingly horizontal positions for which all evidence indicates gradual accumulation of the till beneath the moving glacier. But the possibility of rotation on an axis other than the longest, with deposition by thrust into the till floor, seems worthy of analysis.

A stone moving forward in rotational orientation (transverse) may assume a parallel orientation on coming in contact with the till floor if forward movement continues. Should the stone then be raised slightly above the floor, as by overrunning a gently sloping obstacle, rotation about the intermediate axis could be induced, and the stone might then be thrust into the till by pressure from superjacent debris which would intercept the rising rear end. When the stone had reached a near-vertical position, the impinging stone would be free to slide past.

To test this hypothesis of rotation on an axis other than the longest, Figure 16 was prepared. Figure 16A represents the preferred orientation of the intermediate axes of stones whose long axes were dipping 20 degrees or more. A long-axis dip exceeding 15 degrees (at any of the 10 stations selected for the composite group) probably means that the stone came to its final position from some movement other than the normal one imposed by sliding. Figure 16A shows a reasonably strong percent-concentration in horizontal transverse position, indicating that many of the stones could have been rotating on their intermediate axes, with the latter essentially horizontal, when their transportational move-

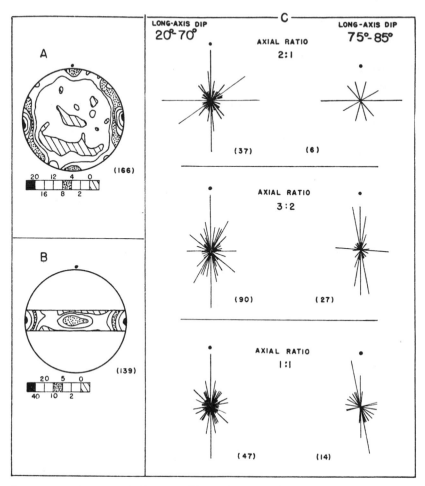

FIGURE 16.—*Orientation diagrams illustrating possible deposition from rotation on one of the two shorter axes*

(A) Diagram showing orientation of intermediate axes of stones whose long-axis dips were 20 degrees or more. The majority of these intermediate axes were in horizontal transverse orientation, which is the normal position of rotation.

(B) Orientation diagram of intermediate axes of stones whose long-axis directions fell within a range of 10 degrees from the direction of glacier flow. Data from four representative stations are shown.

(C) Subgroups based on amounts of long-axis dip and on relative axial lengths. These diagrams should be compared with those in Figure 17, in which these subgroups are included.

ment ceased. Figure 16B is a similar diagram combining four representative stations and showing the percent-concentration of intermediate-axis poles of all stones whose long-axis directions fell within a range of 10 degrees from the direction of ice movement. The diagram indicates that the great majority of these intermediate axes were essentially horizontal and could have served as axes of rotation. Most of the remaining intermediate axes were essentially vertical.

The near-vertical axes are especially significant in connection with Figures 8 to 12 because the shortest axes of those stones were occupying the position normally preferred by the intermediate axes (that is, in horizontal transverse position). Hence the long axes also must have been essentially horizontal. Of the 139 stones represented in Figure 16B, 21 have intermediate-axis dips of 70 degrees or more and comprise the near-vertical group. Of these, seven have a difference of less than one centimeter in the lengths of the two shorter axes, and only five have a difference of more than 2 centimeters in this respect. Therefore many instances of steep dip of intermediate axis are to be explained by the small difference in the lengths of the two shorter axes whereby the shortest may occupy temporarily the position usually assumed by the intermediate axis. Such an explanation probably accounts for most of the parallel-oriented stones represented in the strongly transverse patterns of Figures 8 to 12.

The conclusion indicated by Figure 16A, B is that till stones may rotate to some extent on their intermediate axes, and possibly also on their short axes if the latter are of nearly the same length as the intermediate axes.

Figure 16C and Figure 17 afford additional evidence in support of this conclusion by showing how relative elongation of form may influence depositional orientation. Striae on elongate stones (especially rhombohedroids) are generally parallel to the long axis. However, nearly equidimensional rhombohedroids characteristically acquire intersecting sets of striae arranged parallel to the length (side) and width (end) of the stone. These facts suggest that, where transportational conditions favor sliding, an elongate stone adjusts readily to parallel orientation, while a subequidimensional stone is subequally stable when sliding in either parallel or transverse orientation, presumably controlled by the nearly equal side- and end-surfaces. Likewise, should the mode of forward movement change from sliding to rotation, elongate stones would shift to transverse orientation more readily than would those of subequidimensional proportions.

Figure 17 shows the composite group subdivided on the basis of relative axial proportions (axial ratio) and of size. Stones whose long axes are two or more times the length of the intermediate axes are classed as elongate and for convenience are referred to by the symbol *2:1*. Those

35

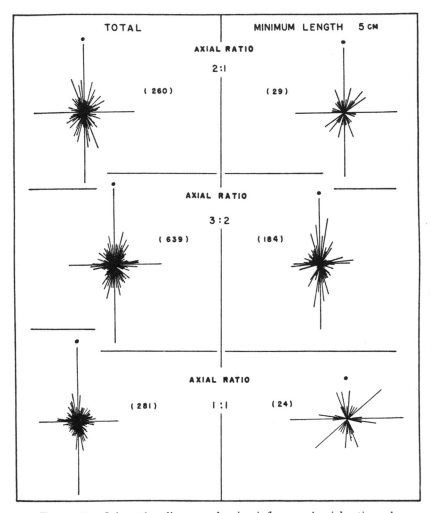

FIGURE 17.—*Orientation diagrams showing influence of axial ratio and relative size*

The three diagrams at the left show composite group subdivided according to relative axial lengths. Stones having long axes 5 cm. or more in length are shown in corresponding diagrams at the right. Note in what ways the larger stones contribute to the total subgroups, especially those of *2:1* and *1:1* axial ratio.

whose axial lengths differ by less than one centimeter (or as near as 7:6) comprise the subequidimensional group, represented by the symbol *1:1*. All other stones are of intermediate axial ratio and are designated by the symbol *3:2*. Except the large *1:1* stones, which have strong transverse preference, the diagrams reveal a slight, though gradual, increase

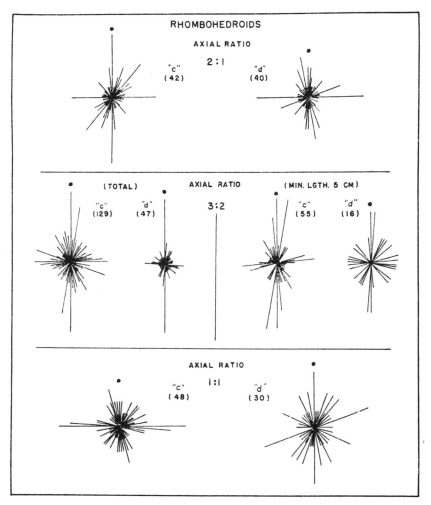

FIGURE 18.—*Orientation diagrams of rhombohedroids from the subgroups (total)*
shown in Figure 17

Large stones of *3:2* axial ratio are plotted separately. Influence of roundness on orientation
preference is especially marked in stones of *2:1* and *1:1* axial ratio.

in preference for parallel orientation with decreasing difference in axial
lengths. From these groups, the stones (213 total) with long-axis dip
exceeding 15 degrees are represented in Figure 16C, grouped according
to axial ratio and amounts of long-axis dip. The tendency to increased
preference for parallel orientation with decreasing difference in axial
lengths is appreciably stronger than that shown by the corresponding
groups in Figure 17.

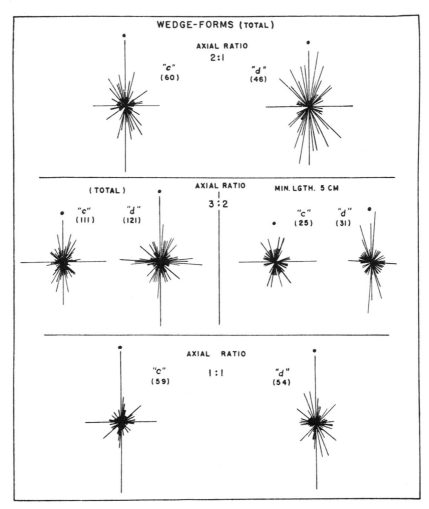

FIGURE 19.—*Orientation diagrams of wedge-form stones from the subgroups (total) shown in Figure 17*

Large stones of *3:2* axial ratio are plotted separately. Among these subgroups, rounding appears to have contrasting effects on orientation preference.

The fact that the *1:1* stones in Figure 16C show a notably strong preference for parallel position, while the amounts of long-axis dip suggest deposition from rotation on the intermediate axis, indicates the probability that the rotational movement began after parallel orientation had been acquired by sliding. Assuming the same transportational history for the other groups in Figure 16C the greatest proportionate loss of parallel orientation before deposition occurred appears in the *2:1* group, while the *3:2* group shows an intermediate degree of change. The relative influence of form in this connection is suggested by reference to the contoured diagrams in Figures 13 to 15. Diagrams of wedge-form stones show a comparative absence of a girdle normal to the horizontal transverse position, showing that these stones do not lodge readily from rotation on the intermediate axis. This supports the conclusion that any rotation of wedge-forms is more likely to bring them again into parallel orientation. By contrast, the comparatively strong girdle in the ovoid diagrams seems to reflect the relative ease with which a parallel-oriented ovoid may begin rotation on its intermediate axis. If partially imbedded in vertical position, its rounded form would be less subject to dislodging than would the more angular stones under similar conditions.

Summarizing, the following conclusions seem warranted: (1) Stones in glacial transportation may gain and lose parallel orientation at least once, and probably many times, before their final deposition. (2) A parallel-oriented stone in sliding movement may be induced to begin rotation. Normally the initial rotation begins about the intermediate axis, but the long axis soon swings into transverse position and becomes the axis of rotation. (3) The more elongate a stone is, the more readily the long axis assumes transverse orientation.

AXIAL RATIO, FORM, AND ROUNDNESS IN RELATION TO LONG-AXIS DIRECTION

Further analysis of the three groups shown in Figure 17 is significant in showing how axial ratio and roundness may influence preferred depositional orientation of the several till-stone forms. These preferences are illustrated by Figures 18 to 22; though for the small form-groups only the axial-ratio diagrams are given.

Sharply angular ("*d*") rhombohedroids of *1:1* and *3:2* axial ratio (Fig. 18) show a much stronger preference for parallel orientation than is shown by the corresponding groups of "*c*" roundness; but this difference in preference is reversed in stones whose axial ratio is *2:1*. Hence the orientation stability afforded by prominent flat surfaces (discussed previously) evidently does not outweigh the liability of sharp edges and corners of an elongate stone to impinge on obstacles instead of sliding past them. This interpretation supports the belief that elongate stones are the more responsive to conditions tending to change their transportational orientation. On a *1:1* stone, a given amount of rounding destroys a greater percentage of flat "bearing surface" than it would on an elongate stone of equal cross-sectional area, thereby permitting easier rotation and consequently greater chance of deposition from the rotational orientation. The same amount of rounding on an elongate stone would create only a sled-runner effect, enabling the stone to slide past obstructions.

Another possible factor in the orientation stability of angular stones may be suggested hypothetically by assuming a rhombohedroid oriented by sliding. Its forward velocity is less than that of the ice passing over it. Forward movement of the ice necessitates plastic adjustment at both front and rear, with accompanying pressure-melting and regelation. Adjustment at the front end may take place within a conical zone immediately ahead of the stone. Flow of ice across the top and off the distal end establishes a pressure gradient upward on that end, with the lowest pressure at the upper leading edge of the stone. Hence that edge is the principal locus of regelation. The more sharply angular the edge is, the more concentrated is the zone of regelation. Ice may freeze to that part of the stone, and the general effect of lowered pressure along the upper edge should make the ice in that zone a little firmer than that beneath it. Thus the firm ice may serve as a horizontally asymmetric prow tending to elevate the front end of the stone and to prevent rotation. With slight rounding of the edge, regelation would be no longer sharply concentrated, and the stabilizing effect would be lost.

To apply this principle to the sharply angular *1:1* wedge-forms, it seems necessary to assume that the pointed end faced the ice thrust, with regelation along the broad end of the wedge. Ice thrust against the divergent wedge surfaces might result in more plastic adjustment and less forward movement of the stone, but the long axis would be in close alignment with the direction of transport. The rose diagram (Fig. 20) shows that such alignment is realized to an exceptional degree. If this hypothesis of regelation along a sharp edge is even slightly operative in the manner suggested, it may be the determining factor in holding many angular stones in parallel alignment while finer debris gathers about them, whereas less angular stones would be rolled forward and assume a transverse orientation.

The strongly transverse patterns of the *1:1* ovoids illustrate the tendency of subequidimensional stones devoid of flat surfaces to be deposited from a rotational orientation (Fig. 21). In contrast, ovoids of *2:1* axial ratio show equal persistence in parallel orientation, presumably due to the absence of projecting edges and corners which might otherwise catch on obstructions or on faster-moving debris and thereby cause the stone to turn. Hence these ovoids evidently have a greater resistance to loss of parallel alignment than is characteristic of the *2:1* group as a whole. Ovoids of *3:2* axial ratio show depositional tendencies intermediate between those of the other axial ratio groups.

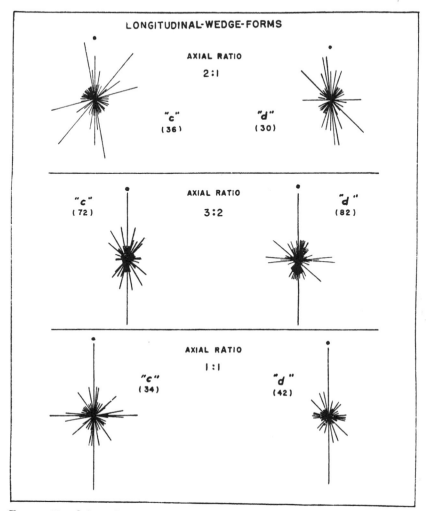

FIGURE 20.—*Orientation diagrams of wedge-form stones, omitting those with surfaces converging only laterally*

Comparison with Figure 19 (in which these stones are included) shows marked differences in the *2:1* and *3:2* groups and only slight contrast in the *1:1* groups.

The depositional behavior of tabular stones deserves brief mention. Although as a group they have a fairly strong tendency to deposition in parallel alignment, they are less likely to be imbedded in horizontal position than other forms are. (*See* Table 3.) Stones with neither long nor intermediate axis dipping more than 15 degrees include 64 per cent of all ovoids; 57 per cent of all varihedroids; 53 per cent of all rhombohedroids and wedge-forms; and only 42 per cent of all tabular stones

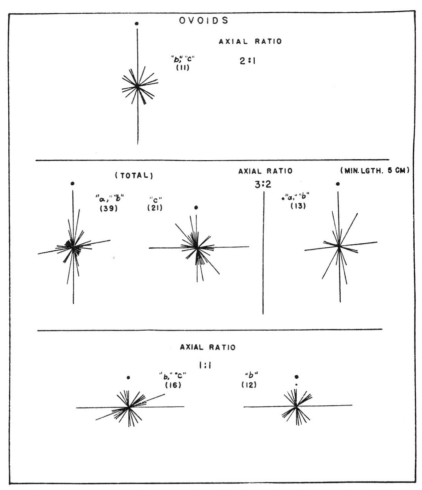

FIGURE 21.—*Orientation diagrams of ovoids from the subgroups (total) shown in Figure 17*

Note marked contrast between the *2:1* and *1:1* groups and the influence of roundness in the *3:2* groups.

(including discoids). These percentages have qualitative significance, for the order of their arrangement is that of the relative thickness of representative stones, that is, length of the shortest axis compared with the lengths of the other two.

Interpreting from Taylor's experiment (described previously), a tabular stone carried above the floor of the glacier would tend to rotate on its long axis. The difference in vertical range required by the two shorter axes at each rotation is greater for tabular stones than for any

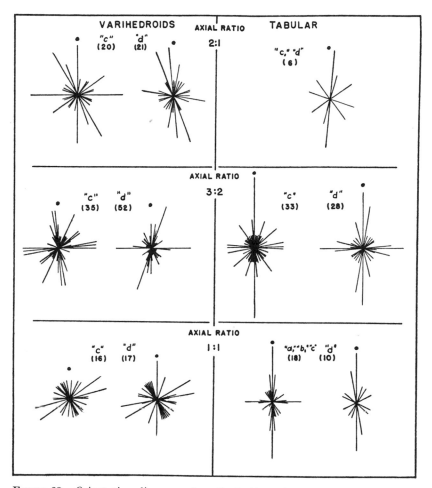

FIGURE 22.—*Orientation diagrams of varihedroids and tabular stones from sub-groups (total) shown in Figure 17*

Note characteristic absence of strong alignment in both parallel and transverse positions in most of the varihedroid patterns.

other form, and therefore the longer one (intermediate) would be more likely to encounter superjacent debris in rising past the vertical than would the corresponding axis of any other form. In the debris-laden base of the glacier, such interference may greatly impede rotation. However, the tendency to steeper axial dips suggests deposition by thrust into the till floor, and the preference for long-axis direction parallel to the direction of ice flow then implies at least partial rotation on the intermediate axis. These considerations support the inference that conditions preventing free rotation tend to compel orientation by sliding and constitute optimum conditions for thrust into the till floor if rotation begins.

CRITERIA FOR RECOGNITION OF ORIENTATION POSITIONS

The foregoing analyses show that till stones differ in statistical preference of depositional position according to their varying inherent characters, modified by the varying conditions of glacial transportation. The variety of groups studied permits the designation of stones whose statistical chance of deposition either parallel with the direction of glacier flow or transverse thereto is somewhat greater than average. Many groups seem to offer possibilities, but the following are considered the best guide forms.

Parallel position: (R. indicates roundness.)
1. Subequidimensional stones R. *"d,"* with long axes less than 5 centimeters, except varihedroids.
2. All subequidimensional rhombohedroids R. *"d."*
3. All subequidimensional longitudinal wedge-forms R. *"d,"* with intermediate-axis dip not steeper than 70 degrees.
4. The following forms with intermediate-axis dips not steeper than 15 degrees:
   a. Longitudinal wedge-forms R. *"c"* and *"d,"* except those of *3:2* axial ratio R. *"c."*
   b. Rhombohedroids R. *"d."*
   c. Ovoids R. *"a"* and *"b."* except subequidimensional.
   d. Tabular R. *"c"* and *"d."*

Transverse position:
1. Subequidimensional rhombohedroids R. *"c."*
2. All subequidimensional stones with long axes at least 5 centimeters in length, except rhombohedroids R. *"d:"* and wedge-forms.
3. Ovoids R. *"c."*
4. Rhombohedroids R. *"c"* and *"d"* with intermediate-axis dips of 75 degrees or more, except subequidimensional R. *"d."*

This list of guide forms contains some duplication. For instance, a subequidimensional longitudinal wedge-form stone, roundness *"d,"* may be included in three groups if its length is less than 5 centimeters and if its intermediate axis dips no more than 15 degrees. The statistical value of such stones is probably greater than that of stones qualifying for only one group because they combine a greater number of properties that give orientation stability.

The limitations of these guide forms must be clearly recognized. Their orientation preference is only statistical and relative. Many of them are not abundant in an unselected group, and by the law of chance the first few encountered may possibly show any orientation except their statistically preferred one. Table 5 shows that a frequency of 20 per cent in parallel position, or of 15 per cent in transverse position, is attained in only a few groups.

Pattern habit of the several forms must also be considered. For example, wedge-form stones of *"c"* roundness are likely to be oriented in near-parallel rather than precisely parallel position. Tabular stones of *"c"* roundness show a similar tendency. The value of ovoids of roundness *"c"* as transverse indicators lies largely in their tendency to near-transverse orientation.

The *1:1* rhombohedroids illustrate a critical problem in strong orientation preferences because rounding from *"d"* to *"c"* reverses their preference. The reversal probably requires a certain range in the rounding process, and therefore stones in this transition stage are probably worthless as guides. A variation of the problem appears in groups selected on the basis of intermediate-axis dips (Figs. 9-12). The intervening dip-interval is 50 degrees (20 to 70), and the groups seem well separated. But they really grade into each other as the two shorter axes of the stones approach each other in length. The control exercised by axial-length difference disappears if the two axes are equal.

As mentioned previously in connection with contoured diagrams, the presence of a girdle developed in a vertical or near-vertical plane is an important guide to the parallel position, though at most stations it is not sufficiently well defined by 100 stones to be of diagnostic value. However, a rough correlation seems to exist between a strongly developed transverse orientation pattern and the presence of a girdle normal to the transverse position. These are logical consequences of deposition following rotational movement.

The criteria for distinguishing the parallel and transverse positions have been tested by applying them to actual cases. Results of the tests are given on the following pages. Figure 24 shows the regional relationships of the till-fabric data.

TESTS OF CRITERIA FOR DISTINGUISHING ORIENTATION POSITIONS

General statement.—A few stations in the central New York area yielded orientation diagrams that were anomalous or ambiguous. Hence these problem stations are ideal for testing the practical value of the criteria selected for determining the direction of glacier flow.

In the following analyses, no special weight is given to those indicators or guide forms that fall in more than one guide group. Each is recorded but once.

Station 15.—Station 15 is situated 0.6 mile northwest of Carpenter Pond. in the west central part of the Cazenovia quadrangle. The exposure is a road cut in one of the many till mounds on the west slope of a shallow through-valley. The rose pattern from this station is regular and symmetrical, and evidence from striae and topography is sufficient to show the relation of this pattern to the direction of ice

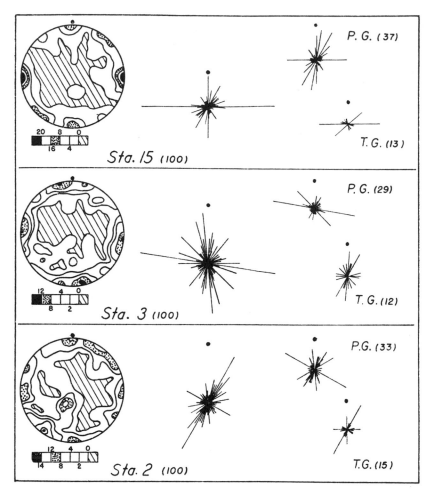

FIGURE 23.—*Analyses of anomalous patterns from three localities*

Parallel and transverse guide forms (indicated by P.G. and T.G., respectively) are shown separately. Their contribution to the complete station pattern indicates clearly the direction of glacier flow. True north is shown by a dot above each diagram.

flow, but the pattern is so different from most others that it was rejected in making up the composite group.

Figure 23 shows the various diagrams pertaining to this station. The contoured diagram shows an unusually strong east-west maximum, with almost no suggestion of a girdle across the diagram connecting the two parts of this maximum. However, an imperfect girdle is developed in a vertical plane normal to the east-west position. On the basis of these facts, the pattern is one of strong transverse orientation.

The guide forms are plotted separately. Each of the five stones oriented N. 10° E. is a guide to the parallel position, and two of them are included in more than one guide group. The average of intermediate-axis dips at this station is very low, and the five parallel guides in transverse position appear there because of their low intermediate-axis dips. Even so, the statistical effect of these forms on the entire pattern is to show clearly that the stones oriented east-west are in transverse position. Therefore the analysis indicates that the orientation of stones at this station presents a strong transverse pattern. All field evidence confirms this interpretation.

Station 3.—Station 3 is located on the south bank of Limestone Creek, one mile northeast of New Woodstock (central part of Cazenovia quadrangle). The valley has a general east-west trend. Striae about half a mile to the southwest indicate ice movement due south; and about one mile northwest, striae are aligned S. 30° E. No evidence was found within the valley to show whether the ice flowed across it or along its axis.

The rose diagram of this station (Fig. 23) is characteristic except for its slight asymmetry. Though its longest line nearly parallels the valley trend, nothing indicates whether that line represents the parallel or the transverse position. The contoured diagram shows a slight tendency to general southward dip but offers no clear solution to the problem of ice-flow direction.

The parallel indicators show that the depositing ice flowed principally up-valley (eastward) but probably shifted its direction occasionally to a more southeasterly course. Seven of the eleven stones oriented N. 85° W. are good guides to the parallel position. Likewise, four of the six stones aligned N. 45° W. are parallel guides. The transverse indicators, though few in number, are in general agreement with evidence from the parallel guides.

The lithology of the till in this part of the valley changes abruptly about half a mile farther upstream (east), marking the limit of drift brought into the valley by ice from the northwest.

Station 2,—Station 2 is located on the south bank of Middle Branch Tioughnioga Creek, one mile southeast of Sheds (southeastern part of Cazenovia quadrangle). The valley slopes to the west, and the station is less than half a mile within the limits of drift brought from the northwest. This genetic relation of the till was not fully known at the time of field study, and the singularly symmetrical pattern (Fig. 23) was then interpreted as the work of ice flowing from the northeast, which had preceded the later flow from the northwest. But after the till relation was known, the diagonal position of the pattern with respect to the valley axis left considerable uncertainty as to the direction of glacier flow.

The contoured diagram suggests the presence of girdles other than the peripheral one, but not clearly enough for safe interpretation.

The indicators of the parallel and transverse positions show clearly that the pattern is one of prevailing transverse orientation. Five of the six stones aligned N. 60° W. are good parallel indicators, and the sixth is fairly good. Therefore the depositing ice was moving toward the southeast as it reached that part of the valley.

Conclusion.—Pattern analyses of these problem stations show that the guide forms give correct results in all cases where these results can be checked by other evidence. This gives assurance that the criteria are dependable where other evidence is lacking. Parallel-position guides are more numerous in an unselected group of 100 stones and give more decisive results than transverse-position guides. Decisiveness of results generally corresponds to the degree of simplicity and symmetry of the

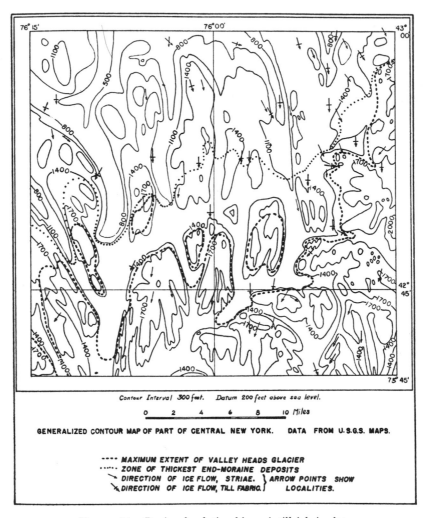

FIGURE 24.—*Regional relationships of till-fabric data*

Cazenovia quadrangle (northeast), Tully quadrangle (northwest), and parts of Pitcher quadrangle (southeast) and Cortland quadrangle (southwest). The Valley Heads glacier moved southeast, and the earlier glacier toward the southwest. Strong local topographic control of ice-flow direction is apparent.

pattern to be analyzed. This correspondence seems primarily related to the constancy in direction of glacier flow during the time of deposition.

SUMMARY OF CONCLUSIONS

The conclusions drawn from this study of till fabric are believed to be valid for till in any region where the conditions of glaciation were comparable to those of central New York. Comparative studies from other regions seem highly desirable. The principal conclusions for the central New York area may be summarized as follows:

(1) The ground-moraine till accumulated gradually beneath the moving glacier, but with occasional local and temporary intervals of erosion that gave rise to boulder-pavement phenomena. Fabric patterns from successive layers of till record occasional shifts in direction of glacier flow similar to those shown by intersecting sets of striae on rock ledges.

(2) Glacier ice in which the stones were carried moved as a plastic solid flowing in obedience to the laws of fluid mechanics. Stones carried above the floor were subject to rotation about an axis lying in a plane of uniform shear and normal to the direction of glacier flow. Ordinarily the longest axis was the axis of rotation, though under some conditions, probably following temporary orientation by sliding, rotation about the intermediate or the shortest axis probably occurred. Stones rotating on or near the floor were subject to shoving into the deposited till by thrust from higher and faster moving debris. Many of the stones thus imbedded remained permanently in that position.

(3) Stones in contact with the glacier floor (or possibly along a well-defined shear plane in the ice) moved by sliding. Normally both long and intermediate axes were parallel to the floor plane, with the long axis aligned in the direction of movement. However, prominent flat surfaces, especially on wedge-form stones, commonly exercised more control in orientation by sliding than did the trend of the long axis. The controlling surfaces were aligned parallel to the direction of movement.

(4) Inherent characters such as form, roundness, size, and relative axial lengths, predispose a stone to certain kinds of movements and thus influence the likelihood of deposition in certain diagnostic attitudes or positions. Some types of stones have a relatively high statistical probability in depositional attitude, with the long axis aligned either parallel to or transverse to the direction of glacier flow. Thus it is possible to determine the direction in which the glacier moved while the deposit was accumulating. This method of ascertaining the direction of ice flow is independent of other evidence and can be used directly in the field.

(5) Minor differences in till-stone characters, such as slight rounding, have a profound effect on the depositional orientation preference of some types of stones (for example, Fig. 18). This in turn suggests that the number of such influences has not been fully measured and that a more detailed classification of till stones will probably reveal further signifi-

cant relations. Delicate adjustment to slight degrees of change is characteristic of glacial behavior, and the enormous forces of the continental glacier seem to have responded even to very small differences in those factors commonly grouped under the term structure.

WORKS TO WHICH REFERENCE IS MADE

Bell, Dugald (1888a) *On some boulders near Arden, Lochlomond,* Geol. Soc. Glasgow, Tr., vol. 8, pt. 2, p. 254-261.

——— (1888b) *Additional note to Mr. Bell's papers (p. 237-261),* Geol. Soc. Glasgow, Tr., vol. 8, pt. 2, p. 341.

Demorest, Max (1938) *Ice flowage as revealed by glacial striae,* Jour. Geol., vol. 46, p. 700-725.

Fairchild, H. L. (1907) *Drumlins of central western New York,* N. Y. State Mus., Bull. 111, 443 pages.

——— (1932) *New York moraines,* Geol. Soc. Am., Bull., vol. 43, p. 627-662.

Geikie, James (1895) *The great ice age,* 2d ed., London, D. Appleton Co., 850 pages.

Holmes, C. D. (1938) *Till fabric* (abstract), Geol. Soc. Am., Bull., vol. 49, p. 1886-1887.

Hubbert, K. M. (1937) *Theory of scale models as applied to the study of geologic structures,* Geol. Soc. Am., Bull., vol. 48, p. 1459-1519.

Knopf, E. B. and Ingerson, Earl (1938) *Structural petrology,* Geol. Soc. Am., Mem. 6, 270 pages.

Krumbein, W. C. (1939) *Preferred orientation of pebbles in sedimentary deposits,* Jour. Geol., vol. 47, p. 673-706.

——— and Pettijohn, F. J. (1938) *Manual of sedimentary petrography,* New York, D. Appleton-Century Co., 549 pages.

Miller, Hugh (1884) *On boulder-glaciation,* Royal Physical Soc. Edinburgh, Pr., vol. 8, p. 156-189.

Miner, Neil (1937) *Evidence of multiple glaciation in the northern part of Yellowstone National Park,* Jour. Geol., vol. 45, p. 636-647.

Richter, Konrad (1932) *Die Bewegungsrichtung des Inlandeis reconstruiert aus den Kritzen und Längsachsen der Geschiebe,* Zeitschrift Geschiebeforschung, Bd. 8, p. 62-66.

——— (1933) *Gefüge und Zusammensetzung des norddeutschen Jungmoränengebietes,* Abh. geol.-paleont. Inst. Greifswald, Bd. 11, p. 1-63.

——— (1936) *Gefügestudien im Engebrae, Fondalsbrae und ihren Vorlandscdimenten,* Zeitschrift Gletscherkunde, Bd. 24, p. 22-30.

Sander, Bruno (1934) *Fortschritte der Gefügekunde der Gestein-Anwendung, Ergebnisse, Kritik,* Fortschritte der Mineralogie, Kristallographie, und Petrographie, vol. 18, p. 111-170.

Sauramo, Matti (1924) *Tracing of glacial boulders and its application in prospecting,* Comm. géol. Finlande, Bull. 67, 37 pages.

Taylor, G. I. (1923) *The motion of ellipsoidal particles in a viscous fluid,* Royal Soc. London, Pr., vol. 103 A, p. 58-61.

Tester, A. C. (1931) *The measurement of the shape of rock particles,* Jour. Sedim. Petrol., vol. 1, p. 3-11.

Twenhofel, W. H. (1932) *Treatise on sedimentation,* Baltimore, Williams and Wilkins, 926 pages.

Upham, Warren (1891) *Criteria of englacial and subglacial drift,* Am. Geol., vol. 8, p. 376-385.

Von Engeln, O. D. (1930) *Type form of faceted and striated glacial pebbles,* Am. Jour. Sci., 5th ser., vol. 19, p. 9-16.

Wadell, H. (1932) *Volume, shape, and roundness of rock particles,* Jour. Geol., vol. 40, p. 443-451.

Wentworth, C. K. (1923) *A method for measuring and plotting the shapes of pebbles,* U. S. Geol. Survey, Bull. 730 C, p. 91-102.

—————— (1936) *An analysis of the shapes of glacial cobbles,* Jour. Sedim. Petrol., vol. 6, p. 85-96.

204 Swallow Hall, University of Missouri, Columbia, Mo.
Manuscript received by the Geological Society, February 27, 1941.
Presented before the Geological Society, December 29, 1938.

Plate 1.—Stones from Till in Central New York

Figure 1. Typical till stones

Illustrating forms and degrees of roundness. About ⅙ natural size. All have remnants of original concave surfaces. Note low range of "c" roundness and nearly complete transformation of numbers 4 and 8 to ovoid form.

1	} Tabular	"c"	10	} Wedge-form	"c"
2		"c"	11		"d"
3		"d"	12	Discoid	"b"
4		"b"	13	} Ovoid	"a"
5	} Rhombohedroid	"c"	14		"b"
6		"c"	15		"c"
7		"d"	16	} Varihedroid	"c"
8	} Wedge-form	"b"	17		"d"
9		"c"			

Figure 2. Wedge-form stones

Showing striae parallel with prominent edges rather than with long axes of stones.

51

FIGURE 1

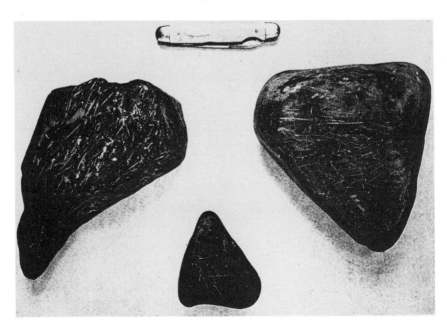

FIGURE 2

STONES FROM TILL IN CENTRAL NEW YORK

2

Till Genesis and Fabric in Svalbard, Spitsbergen

Geoffrey S. Boulton

ABSTRACT

Svalbard glaciers are characterized by their transport of considerable volumes of englacial debris, which are released from the ice to form three main types of till. *Flowtill* is released as a fluid mass from the englacial-debris load when this is exposed by down-wasting of the glacier surface; *melt-out till* is deposited by slow melting out of the top surface of masses of dead ice covered by a stable overburden; *subglacial lodgement till* is material released from the basal ice either by pressure melting against bedrock obstructions, or by melting of debris-rich ice masses which have become stagnant beneath the moving glacier sole.

Englacial stone orientations are described, but these are rarely preserved, except, in part, in melt-out tills and in certain subglacial tills. Flowtills often except, in part, in melt-out tills ad in certain subglacial tills. Flowtills often show an upper, more fluid element in which fabrics both transverse and parallel to flow develop, while the lower part shows a parallel fabric with up-slope, *a*-axis imbrication. Melt-out tills have fabrics in which *a/b* planes tend to lie in the plane of deposition, but which reflect azimuthally the former englacial fabrics. Subglacial tills show variable fabrics dependent upon the shape of the underlying bedrock surface, although where this is planar, parallel fabrics are dominant. All three main types of till can show systematic regional fabric patterns, although fabrics at one site can rarely be taken to represent the direction of glacier movement.

INTRODUCTION

Till fabric analysis is one of the most commonly used techniques in glacial geology, although very little is known about the actual process by which tills acquire these fabrics. This difficulty is shared by the field of glaciology as a whole, for there are indeed very few observations of the ways in which tills are deposited by modern glaciers. Over the years, however, a considerable body of hypothesis has been built up regarding the mode of origin of tills and of the sequences in which they occur, especially in the lowland areas of Europe and North America. Much of this hypothesis is entirely speculative and little based on studies of those modern glaciers which are now producing similar sequences. Thus has been built up the hypothetical model that compact, fine-grained tills are, almost by definition, subglacial in origin, while observations of thin moraines on the surfaces of valley glaciers (Sharp, 1949) have added an upper element, the washed, ablation till of supraglacial origin (Flint, 1957).

The number of observations on tills of known origin are very few, and there has been a certain reluctance to use even these few results. Perhaps this reluctance is based on the fact that relatively easily accessible alpine valley glaciers in both Europe and North America provide rather poor depositional models for Pleistocene ice caps on lowland plains.

The ice caps and valley glaciers of Svalbard have long been known to deposit tills of very considerable thickness over wide areas, and to produce multi-till sequences similar to those of Pleistocene age in lowland areas of temperate latitudes (Garwood and Gregory, 1898; Lamplugh, 1911; Gripp, 1929). In view of this similarity, a project was undertaken in 1964 to identify, as far as possible, the modes of origin of these tills and the sequences in which they lie, and to investigate the relation between till genesis and character. It is possible to be very specific about the origin of tills within the Svalbard sequences, simply because they are often so intimately related to the ice from which they are derived.

Svalbard glaciers tend to transport considerable volumes of debris in an englacial position, and at their margins, where there is often a strong upward component of flow, this debris is released at the glacier surface as it wastes down, forming a cover of supraglacial till (Boulton, 1968). As the thickness of this till builds up, ablation of the underlying ice is increasingly inhibited, resulting, if the glacier is retreating, in very large areas of supraglacial till, which tend to acquire a hummocky surface as a result of differential ablation. Fluvial or lacustrine activity in this marginal zone also produces thick supraglacial accumulations of stratified sediments between ice-cored hummocks and ridges, which, when sedimentation ceases, are covered by thick deposits of till continuously moving off the ice-cored ridges (*flowtill*) (Fig. 1).

Masses of buried stagnant ice, with its high debris-content, are also an important source of another type of till. This is till which melts out slowly from beneath overburden, either at the top surface or bottom surface of buried stagnant ice, and which retains much of its original englacial struc-

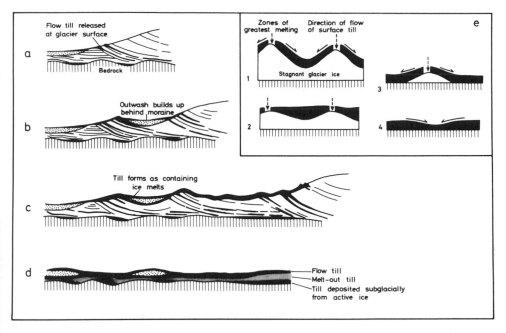

Figure 1. A schematic diagram showing the typical development of till sequences in Svalbard (a-c). d is a prediction of the resulting sequence if all buried ice were to melt, while e shows the changing pattern of differential ablation and flow which produces flow till sheets of low relief and allows buried ice to melt. Absence of further flow beyond stage e (2) would lead to a hummocky till surface.

ture (*melt-out till*). A third type of till is that which is deposited beneath active ice (*lodgement till*). A series of schematic diagrams (Fig. 1a-1c) contrasts the ways in which these three different types of till develop, and Figure 1d illustrates the probable result if all the buried ice were to melt; the left-hand side of Figure 1d shows the type of sequence which would develop where supraglacial fluvial and lacustrine deposits were extensive, and the right-hand side shows the type of sequence which would develop where they were not.

This classification of tills is presented in Table 1, arranged so as to show their relationship with the existing terminology. This classification has been chosen because the three kinds of tills—flowtill, melt-out till, and lodgement till—have fundamentally different stratigraphic implications, and because they are relatively easy to distinguish.

TABLE 1

Classsification of Tills

Classification Adopted Here	Origin	Existing Classification
Flowtill	} Supraglacial	Ablation Till
Melt-out Till		
Lodgement Till	Subglacial	Lodgement Till

The intention in what follows is to describe the stone orientation fabrics of tills whose origin is known, in order that this data can be applied to tills of unknown origin. In some cases the process of deposition is so easily studied that fabric variation is readily predicted; in other cases, where the mode of deposition cannot be so closely specified, considerable sampling problems are met with.

Collection and Treatment of Data

It has not been possible at any one site to standardize the number of measurements made, the methods of sampling, the accuracy of these measurements, or the types of stones sampled. For instance, some subglacial till pockets contain very few stones. Stone shapes and dimensions vary with varying bedrock. In some cases it was possible to collect oriented till samples and to measure stone orientations in the laboratory, but this was impossible with englacial stones; and in all cases the number of measurements, and probably accuracy, varied with temperature.

Much of the three-dimensional directional data is presented by Schmitt equal-area projection of the lower hemisphere. This is then contoured by a method suggested by Kamb (1959), which has the dual advantages of presenting the data in a visual form and giving an indication of the statistical significance of apparent orientation peaks. In this method the area of the counter used in the conventional contouring procedure is chosen so that data-point densities can be contoured at the desired levels of significance. I have adopted contour intervals of 2σ and a significance level of 3σ. Significant maxima have then been analyzed individually using Fisher's (1953) function for probability density on a sphere (Steinmetz, 1962), which gives orientation of the resultant vector (A = angle of dip, D = direction of dip) its magnitude (R) in relation to the total number of observations (N), a radius of circle of confidence (Θ), and a precision parameter (K). This overcomes some, although not all, of the difficulties of attempting to analyze spherical distributions, which are often non-normal, by methods which are only suitable for the analysis of spherical normal distributions (Green, 1962).

Several conventions are necessary. The direction normal to the glacier front, parallel to the general axis of advance of the glacier, is termed the "parallel" direction, while the "transverse" direction is at right-angles to this. In the local deformational system, which may be unrelated to the parallel and transverse directions, the direction of major tectonic transport is denoted A, with B and C as intermediate and minor axes, respectively. These terms are used simply because till-stone orientations have often been assumed to be due entirely to direct glacier-generated stresses related to the parallel and transverse directions, and thus it is useful to compare the local stress situation responsible for a till fabric with these two directions. Th long, intermediate, and short axes of till clasts are referred to as the a, b, and c axes, respectively; prolate particles are those in which $b \simeq c$ and $a/b \geqq 2$; blades are those in which $a: b: c \geqq 3: 2: 1$; and plates are those in which $a \simeq b$ and $a/c \geqq 2$.

ENGLACIAL TRANSPORT

By far the largest part of the materials which make up the three main types of till are transported englacially to their point of deposition. The original clast-orientation fabrics of englacial debris are partly preserved in melt-out tills, and may, in special cases, influence the final fabrics of supra-glacial flowtill and subglacial lodgement till. Englacial-debris fabrics are thus of some importance.

Figure 2. Basal ice in the compressive zone on Makarovbreen. Thrust planes cut the ice foliation planes and boulders are rotated so that a-axes lie parallel to the direction of movement along the thrust planes. Ice movement is from left to right and boulders are generally of ellipsoidal shape.

The greatest amount of englacially transported debris is derived from the glacier bed and can be divided into two main types: masses of unlithified subglacial sediment which have been incorporated *en bloc* into the glacier, in which the included ice is only interstitial, and, by far the more important, particulate suspensions in glacier ice. These particles, which vary in size throughout the range of sizes normally found in till, may be merely aggregations of smaller particles, or they may be individual grains. The concentrations of these suspensions vary from below 10 percent to above 50 percent by volume in the parallel-to-foliation bands in which they tend to occur.

Observations of the orientations of englacial clasts at ten sites in areas of planar, sub-horizontal, or dipping foliation show that the *a/b* planes of blade- and plate-shaped particles or the *a*-axes of prolate particles tend to lie within the foliation plane (Fig. 3). The orientation of *a*-axes within this plane appears to depend upon the local stress situation. In the snout area of the glacier Makarovbreen in Ny Friesland, there is a transition from a zone of extending flow to a zone of compressive flow (Boulton, 1970a). In the zone of extending flow, the ice foliation is sub-horizontal, parallel to the glacier substratum, and *a/b* planes of stones in the basal debris-rich layers lie parallel to this foliation, while long axes lie parallel to the direction of flow. In the terminal compressive zone, the foliation has a steeper, up-glacier dip, and although the *a/b* planes of blade- and plate-shaped stones still lie in the foliation planes, the *a*-axes of prolate and blade-shaped stones now tend to lie transverse to glacier movement (Fig. 3). These observations

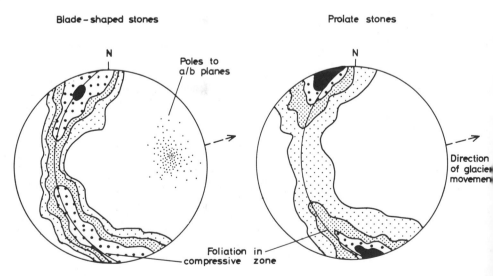

Figure 3. Transverse *a*-axis orientations in a zone of compressive flow, Makarovbreen. Note the stronger orientation of prolate stones, Contours at 2σ intervals. For blade-shaped stones; A = 4°, D = 353°, N = 100, R = 67.8, θ = 9°40′, K = 3.1. For prolate stones; A = 3°, D = 342°, N = 100, R = 76.6 θ = 7°46′, K = 4.2.

suggest the possibility that stone orientations are strongly affected by the change from a tensile- to a compressive-stress system, and that the direction of maximum extension of the triaxial strain ellipsoid has had a controlling influence on *a*-axis orientation.

Further evidence for this thesis was provided by observations at the northern margin of Aavatsmarkbreen in Oscar II Land where the measured geometry of the strain ellipsoid was compared with the orientation of englacial-debris fragments and other structures. This glacier flows rapidly into the sea in a westerly direction, but at its northwestern boundary, it flows up-slope (Fig. 4), as a result of which the glacier foliation planes have a steep up-glacier dip (of as much as 60°) and are truncated by even more steeply dipping thrust planes. Structures within the ice indicate considerable parallel compression and transverse extension, an assessment in accord with the measured orientation of the strain ellipsoid (Nye, 1959). Up-thrust slices of debris-rich basal ice are responsible for the outcrops of numerous debris bands on the glacier foreslope, a number of which were

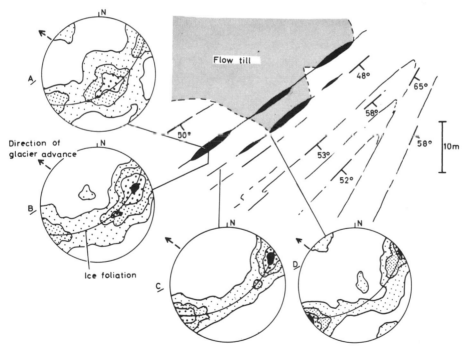

Figure 4. Structural map of part of the margin of Aavatsmarkbreen, Oscar II Land, showing foliation planes in debris-rich ice. The principle strain-rate component at the glacier surface was oriented at 42.5° east of north. A and B show the orientations of the *a*-axes of blades and prolate particles in the middle and at the margin of a lense of debris which contains only interstitial ice. C and D show long axis orientations of blades and prolate particles of debris suspended ice. Contours at 2σ intervals. (A); A = 58°, D = 123°, N = 70, R = 33.3, θ = 18°12′, K = 1.88. (B); A = 15°, D = 68°, N = 70, R = 37.8, θ = 16°14′, K = 2.14. (C); A = 7°, D = 59°, N = 70, R = 40.8, θ = 14°45′, K = 2.36. (D); A = 0°, D = 51°, N = 70, R = 39.7, θ = 15°12′, K = 2.28.

investigated. Two bands containing englacial debris as particulate suspensions yielded clast-fabric analyses showing transverse peaks of *a*-axis orientation of varying strengths, and *a/b* planes of blade- and plate-shaped stones lying in the ice-foliation planes, which dip at angles varying from 48° to 65° (Fig. 4c-4d). In addition, a number of debris bands consisting of lenses of till which contain merely interstitial ice, crop out on the glacier surface. In places several of these lenses lie along the line of a single foliation plane. Orientation measurements made on particles lying along the plane of contact between glacier ice and debris and within the body of the debris band gave rather different results. In the first case, *a*-axes of prolate particles were oriented with a mean vector within 20° of the long-axis of the strain ellipsoid, with a slight down-dip component (Fig. 4b), while in the second case, there was a diffused down-dip orientation maximum and a weak secondary peak at right angles to this direction (Fig. 4a).

The discontinuous lenticular nature of some debris banks in which the ice content is merely interstitial in nature has been noted before (Boulton, 1968). Dort (1967) described englacial bands from Antarctica consisting of sand which had originally been deposited by wind on the upper part of the glacier; where these were exposed near the glacier snout, the bands formed lenticular rods with their axes parallel to glacier movement, and it was suggested that they had originally formed parts of more extensive sheets which had broken up by transverse elongation. The debris bands referred to on Aavatsmarkbreen, comparable to *boudins* in metamorphic rocks, could have reacted to stress in a relatively brittle fashion, being pulled apart by lateral extension of the more plastic ice matrix. In this case, the dominantly transverse fabric at the debris-ice interface would result from relative movement of ice over the upper and lower surfaces of the bands, while, as long as the materials forming the bands were relatively rigid, there would be little or no re-orientation of particles during transport within this zone. The parallel fabrics could either be derived from the time before the debris was picked up *en bloc* from the glacier bed, or from some earlier englacial stress system.

These observations suggest that lateral extension in a zone of longitudinal compression can have a considerable effect on englacial-stone fabrics. The effect of this extension, however, will depend upon the magnitude of the simple shear strain in a vertical plane within the ice. Glen, Donner, and West (1957) based their theoretical analysis of the orientations of stones within glacier ice upon the assumption that parallel-to-foliation shear would be dominant. If this were true, and the glacier were to flow down a smooth channel of uniform cross-section, then it would seem most plausible that the stones would take up orientations such as those predicted by Glen, Donner, and West (1957). If, however, lateral extension occurs within the glacier, stone orientations would be influenced by both major components of flow. Another possibility is that, especially in the snout area of the glacier, shear could be restricted to narrow zones. On Makarovbreen, relict thrust planes are completely obliterated in the basal two to three meters of the glacier, suggesting the presence of intense shear at this level, whereas at higher levels such structures survive considerable transport without being deformed. This suggests that shearing parallel to the foliation planes is relatively unimportant and therefore that clast-rotation is also unimportant at these higher levels. The distinction between zones of shear and zones of triaxial compression is well-marked on Makarovbreen. Between high-angle thrust planes in the ice, a/b planes of blades and plates lies along the ice-foliation planes. In contrast, those stones which are intersected by thrust planes (which also cut and displace the ice foliation) have been rotated so that their a/b planes lie within the thrust plane, at a high angle to the ice foliation, while the a-axes of these stones tend to be parallel to the direction of movement along the thrust plane. A similar phenomenon was reported from Dunerbreen in Ny Friesland (Boulton, 1968), where stones lying in a zone of shear had a preferred a-axis orientation parallel to movement along the shear planes, whereas away from the shear planes, the a-axes of blade-shaped stones merely lay in the foliation plane but had no significant preferred orientation within this plane. Thus it seems likely that, along shear planes, stone orientations are controlled by the movement of one ice surface across another, which tends to produce a-axis orientation parallel to movement.

The effect of shape on a-axis orientation is seen in Figure 3, where a-axes of prolate particles are much more strongly grouped about the axis of maximum strain than are the a-axes of blades. At three localities, where there was neither marked parallel extension (no crevasses at surface) nor compression (horizontal foliation present, with no obvious thrust planes), there was no significant preferred orientation of a-axes of blades, and only a very weak preferred orientation of a-axes of prolate particles.

Other observations of englacial-clast fabrics show variations similar to those reported here. Donner and West (1957) reported measurements on steeply dipping bands of englacial till from the margin of Vestfonna and the ice cap on Nordaustlandet, Svalbard. The fabric maxima varied greatly; some were parallel to the glacier advance, some were parallel with a secondary transverse peak, and some were transverse with a secondary parallel peak. Unfortunately, structural relationships were difficult to determine, and there was no indication whether the englacial "till" contained merely interstitial ice or whether it consisted of debris suspensions in ice. Harrison (1957) also reported fabrics determined on a block of englacial till from the ice-cap margin at Thule, Greenland. Particles within the till showed both parallel and transverse preferred orientation, and the poles to the a/b planes of blades and plates were normal to ice-foliation planes. Measurements by Richter (1936) and by Galloway (1956a) also recorded parallel englacial fabrics.

FABRICS ACQUIRED DURING SUPRAGLACIAL DEPOSITION

Flow Tills

Supraglacial tills develop on those glaciers in which there is a strong upward component of movement in the terminal area, and in which there is a thick englacial-debris sequence. Ablation of the glacier surface in the terminal zones exposes the englacial debris at the surface, where, on melting out, it produces a highly fluid till. Movement of this till down the glacier surface can give rise to a supraglacial cover over wide areas. When the thickness of this till cover increases to more than about three centimeters, the rate of ablation of the underlying ice decreases. The thickness of this supraglacial till may be increased from two sources: by further flows from up-slope positions, and by accretion of till from the underlying melting ice. Thus a supraglacial till may be composed of two elements: an upper allochthonous element, which has been subjected to sub-aerial processes, and a lower autochthonous or parautochthonous element, which has not been exposed at the surface. If the thickness of till accumulating on the glacier surface increases to more than two to three meters, the depth of summer melting is equalled and no further accumulation of autochthonous till takes place at the ice/till interface, although allochthonous till flows may still cause accumulation of till at the surface. Thus very large areas of sediment-covered glacier ice may survive glacier retreat.

There are three basic flow processes which affect supraglacial tills such as those described above: mobile, liquid flow; semi-plastic flow; and down-slope creep, the nature of the flow depending on the nature of the slopes and the water content of the till. These three processes are considered separately below.

Mobile, Liquid Flow

The first type of flow affects those tills which have just been released sub-aerially directly onto glacier ice, or where an excessive amount of water occurs on the surface of an existing till. The water content of these tills is very high, above the liquid limit, and very mobile downslope flows are produced. These flows are rarely more than 20 cm thick, and boulders and stones tend to settle through them, producing a distinction between a lower, stony, relatively slowly moving element, and an upper, stone-free, rapidly moving element, with velocities in this upper element exceeding one meter per hour. The frequent release of water at the surface of these flows produces normal water sorting, and thus they may acquire a stratification. The flows tend to be elongate, with a lobate front. In the body of the flows, stones tend to move with their a/b planes parallel to the underlying surface and a-axes parallel to the direction of transport, whereas, in the nose of the flow, a-axis orientations develop that are transverse to the flow direction.

Semi-plastic Flow

The second type of movement occurs on more stable slopes, where changes in surface loading or pore-pressure cause the shear strength of the till to be exceeded, and failure takes place along an arcuate slip-plane, down

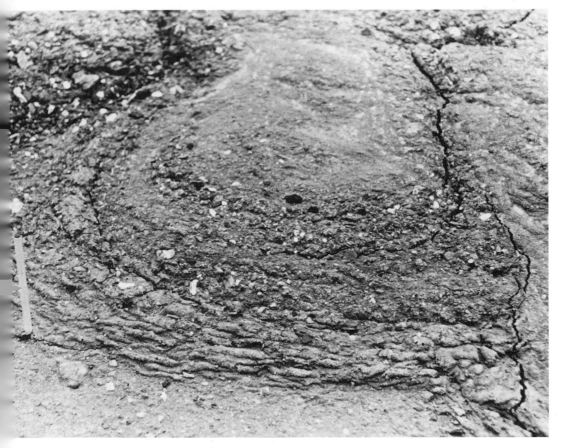

Figure 5. The surface of a small lobate till flow, showing compression at its front with transverse-to-flow stone orientations. In the body of the flow, parallel-to-flow orientations dominate. When sectioned it has a structure similar to that shown in Figure 7B.

which the till moves as a semi-plastic mass. The slip-plane is usually the junction between unfrozen and frozen till, or unfrozen till and glacier ice. The till moves downslope as a lobate flow, and the slip plane becomes an arcuate scar (Boulton, 1968, Fig. 11), the back wall of which is oversteepened and unstable, and therefore continues to feed the flow and to enlarge itself, often to considerable size. These failures are important in early summer, when the ground is thawing and there are rapid increases of pore-pressure.

Observations on the surfaces of flows show that stone orientations vary systematically in different parts of the flow. A very small-scale flow which illustrates this is shown in Figure 5. The variation in orientations of blade-shaped stones in an ideal flow is shown in Figure 7a in a schematic diagram (made up from many observations), which is very similar to the observations of Lundqvist (1949). In the body of the flow, rapid longitudinal extension and vertical settling produce a preferred orientation of a-axes in the direction of flow and a/b planes which either are parallel to the flow surface or dip toward or away from the direction of flow, depending on the shape of the underlying bed. In the nose of the flow, longitudinal compression tends to produce a-axes oriented transverse to flow and a/b planes that dip up-flow, whereas, at the lateral margins, a/b planes dip toward the axis of the flow with a-axes oriented parallel to the flow direction. If the water content is low, dips are higher and vice versa, except for flows in constricted channels which tend to show high-angle a/b planes throughout.

Internally, most of these flows show little sign of stratification as a result of washing by excess water, although streaking out of inhomogeneities does sometimes produce some pseudo-stratification. However, on the surfaces of some flows, especially soon after the initial movement (failure), excess water does occur, resulting in some liquid flow that tends to produce stratification (Fig. 6). In such a flow, the stratification, though it may be limited, is very valuable in that it will reflect any subsequent deformation of the flow. Figure 7b is one flow of a series in vertical sequence which has acquired some stratified horizons as a result of washing, and whose structure is thus revealed. The flow structure is essentially that of a flat-lying isoclinal fold, with a well-marked basal shearplane. Stone-orientation fabrics were determined at four points within the fold. In order to sample only a small area, oriented samples of 300-500 cc were collected, and the small stones (between one and five centimeters) were measured in the laboratory. Measurements of a-axis orientation produced A-direction peaks of varying strength throughout the flow (near the top surface, in the middle, and at the bottom surface of the flow), whereas in the fold nose, a strong B-direction peak was produced. Considerable variation occurred, within the flow in the orientations of blades and plates, which were imbricated up-

64

Figure 6. A flow till which carries evidence of surface washing. Some of the laminae are isoclinally folded.

flow at the base, formed a transverse stereographic girdle in the middle, had a down-flow imbrication at the top, and in the nose had a high-angle,

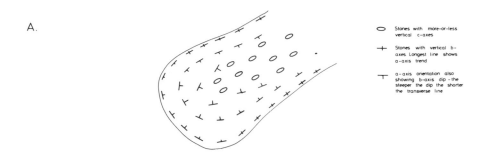

Figure 7A. Idealized lobate till flow showing the pattern of orientations which would be taken up by blade-shaped particles.

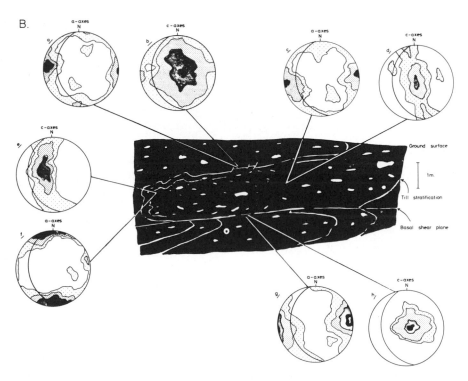

Figure 7B. The pattern of a-axis (prolates and blades) and c-axis (blades and plates) fabrics from a section in an old lobate till flow in which washed horizons reveal the structure. Contours at 2σ intervals. (a); Major peak: $A = 7°$, $D = 227°$, $N = 55$, $R = 36.91$, $\theta = 20°56'$, $K = 2.98$. Minor peak: $A = 4°$, $D = 189$, $\theta = 26°14'$, $K = 2.6$. (b); $A = 75°$, $D = 112°$, $N = 70$, $R = 30.8$, $\theta = 19°42'$, $K = 2.4$. (c); $A = 2°$, $D = 268°$ $N = 80$, $R = 47.2$, $\theta = 13°18'$, $K = 2.4$. (d); $A = 86°$, $D = 67°$ $N = 70$, $R = 36.9$, $\theta = 16°31'$, $K = 2.1$. (e); $A = 66°$, $D = 271°$, $N = 70$, $R = 38.2$, $\theta = 15°54'$, $K = 2.2$. (f); Major peak: $A = 4°$, $D = 181°$, $N = 64$, $R = 47.5$, $\theta = 10°30'$, $K = 3.8$. Minor peak: $A = 3°$, $D = 77°$, $N = 13$, $R = 9.2$, $\theta = 18°42'$, $K = 3.2$. (g); $A = 7°$, $D = 82°$, $N = 80$, $R = 56.3$, $\theta = 10°21'$, $K = 33.3$. (h); $A = 77°$, $D = 102°$, $N = 70$, $R = 50.5$, $\theta = 10°51'$, $K = 3.5$.

66

up-slope dip, though slightly oblique to the direction of movement. A very similar fabric distribution has been reported in flat-lying isoclinal folds in Norfolk tills (Banham, 1966).

Downslope Creep

On slopes where the supraglacial-till cover is stable, tills are composed of two elements: an upper allochthonous element, which has flowed into place by a combination of both liquid and plastic flow, and a lower autochthonous or parautochthonous element derived from the underlying ice, which has not been exposed to sub-aerial processes. There are many exposures in which this distinction cannot be recognized, but in others the two elements are clear: (1) an upper allochthonous element, which shows signs of washing (loss of silt-clay as in the conventional ablation till, or stratification) over (2) a lower element which tends to be compact and unstratified (Fig. 8).

Figure 8. The lower, parauthochtonous part of a flow till, which has not been subjected to sub-aerial processes. Flow from right to left, note imbrication.

The majority of such compact tills, although they appear to be quite stable, are in fact undergoing slow downslope creep under stresses less than the shear strength of the material, and with relatively low water contents. Where the underlying ice surface is melting, creep velocities are relatively high, on the order of one to five centimeters per month, whereas, when melting does not penetrate to the glacier surface, velocities are much less, undetectable except over several years' observations, although shear planes at the freezing surface and deformation of segregation-ice layers indicate that creep is taking place.

The parautochthonous lower element of a flow till has a characteristic

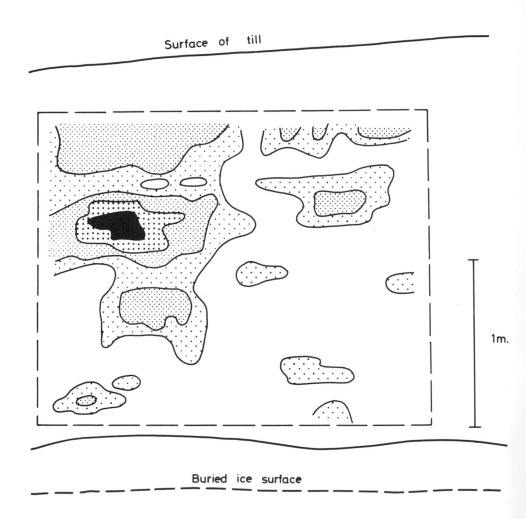

Figure 9. The frequency of occurrence of transverse-to-flow a-axes in a supraglacial flow till in which a-axes are predominantly parallel to flow. Stones selected on the basis of a 10 cm. grid, and contoured at 20, 30, 40, and 50 cm intervals.

fabric in which the *a*-axes of prolate stones tend to lie parallel to the direction of slope and to be imbricated in an up-slope direction. The *a*/*b* planes of blade-shaped stones tend to lie horizontally and, if the *a*/*b* ratio is high, the *a*-axis also tends to be oriented parallel to the direction of slope, but if the *a*/*b* ratio is low, *a*-axis orientation is not so well controlled. Thus fabric diagrams made from two samples taken from a flow till might show significantly different *a*-axis fabric patterns, with respect to the direction of flow, if one contained a higher percentage of blade-shaped stones than of the other shape. Indeed, flow till fabric diagrams made up of the *a*-axes of blades often show a pronounced partial horizontal girdle, whereas a fabric diagram made from measurements on prolate particles shows a single parallel peak.

One of the controls on the fabric of the parautochthonous element of a flow till is the original englacial fabric. This is largely preserved in a till which has just melted out, but not yet moved, though such fabric changes relatively rapidly during downslope creep. An exposure at the margin of Valhallfonna, an ice cap in Ny Friesland, shows glacier ice beneath flow till and intersecting the junction of the two, a distinctive englacial band of red-brown debris dipping up-glacier at 52°, which is easily distinguished

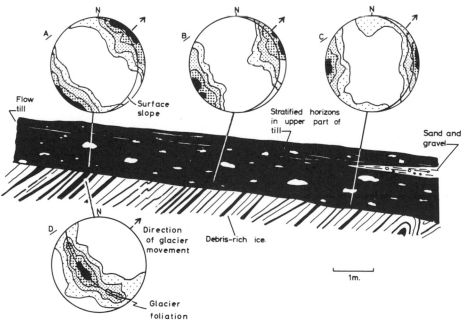

Figure 10. A flow till derived from debris in the underlying ice. The stereograms show the change in clast fabrics resulting from melting of the containing ice, and those which develop subsequently as a result of flow. Contours at 2σ intervals. (A); $A = 3°$, $D = 44°$, $N = 80$, $R = 50.2$, $\Theta = 12°15'$, $K = 2.65$. (B); $A = 1°$, $D = 241$, $N = 80$, $R = 56.1$, $\Theta = 10°59'$, $K = 3.3$, (C); $A = 2°$, $D = 268°$, $N = 80$, $R = 60.9$, $\Theta = 8°54'$, $K = 41.1$. (D); $A = 51°$, $D = 237°$, $N = 80$, $R = 62.7$, $\Theta = 8°22'$, $K = 4.6$.

from the other debris with its gray-black color. This red-brown debris can also be traced in the overlying till for a distance of 10 meters downslope from the point at which the debris band intersects the till-ice junction (Fig. 10), presumably indicating 10 meters of downslope creep since the debris band started to contribute to the base of the till. Stereograms show the fabric in the red-brown debris-band (Fig. 10d), in the till immediately above this band (Fig. 10a), and after three meters and six meters of downslope flow (Figs. 10b, 10c). Initially there is a clear azimuthal similarity in preferred orientation between stones in the till and in the englacial band, but subsequent downslope flow changes this orientation (Θ) by 40°, and the further the till flows, the greater the strength (R) of orientation becomes.

Spatial Variation in Flowtill Fabrics

Variation in flowtill fabric tends to be greater in the vertical direction than in the horizontal. A flowtill occurring above ice or above bedded sediments may show a lower parautochthonous element with a strong parallel fabric acquired during slow creep, and an upper allochthonous element showing a much more variable fabric, acquired initially as the result of accumulation of a series of relatively rapidly flowing tongues, and perhaps modified subsequently by creep. Alternatively, the flowtill may consist entirely of the upper element, or may even interfinger with stratified sediments, as a result of interplay between fluvial and lacustrine deposition and mudflow deposition (Boulton 1968, p. 408).

An example of vertical variation was seen in an exposure of 2.5 m of compact, fine-grained till resting on dead ice at the margin of Aavatsmark-breen in Oscar II Land. From its position, this was obviously a flowtill, although if the ice were to melt and the till were to be deposited on bedrock, it would show none of those structures, such as washed horizons or stratified lenses, which occasionally occur in flowtills. The till contained a large number of small prolate stones in a fine-grained matrix. By means of a net-sampling system applied over an area of two and a half meters by two meters, the a-axis orientations of 286 prolate($a/b \geq 2$) particles with a-axes of between two and five centimeters were measured and their positions plotted on a plan of the site. The resultant fabric showed two clear peaks of a-axis orientation, one oriented normal to the glacier front, and one parallel to it. Each stone was then correlated with either the parallel or transverse peak and the densities of stones in each mode were then contoured using a 2σ contour interval (Fig. 9). The result showed that significant transverse peaks tended to lie in certain well-defined domains, rather than being randomly distributed, and that there were few significant transverse peaks in the fabric diagram for the lower part of the till. This diagram does *not* show the orientation strengths, nor does it imply that a fabric analysis of a small part of the lower till would not show a significant transverse element. It simply shows unstratified variation for this minimal sampling density. The most probable explanation of this variation is that the fabrics in the lowest meter of the till have developed as a result of creep, while those in the uppermost meter and a half represent the effects of the accumulation of a series of either thin flows with different transport directions or lobate flows whose fronts are represented by the transverse peaks.

In some sections, however, there is no clear differentiation into obvious parallel and transverse peaks; an orientation at one horizon may be consistent laterally, but in a vertical direction there may be considerable azimuthal variation. This may be brought about by a succession of mudflows, which in some cases may be separated by thin fluvial or lacustrine beds. Flowtill fabrics also show considerable areal variations, where an important factor is the change in both direction and inclination of the slope down which the tills move, because of slow melting of buried ice. Tills initially form on the glacier foreslope, but as the active glacier margin retreats, that slope is changed by ablation and tends to develop a slope toward the glacier on its proximal side (Fig. 1a-1b). In its subsequent development and enlargement, such ice-cored moraine may be "controlled" with a ridge system parallel to the glacier front, or "uncontrolled," a random arrangement of ridges and hillocks. But no matter which, the configuration of the till surface is constantly changing in response to patterns of differential ablation and till flow (Fig. 1e). During this process of adjustment, the upper part of the till deforms most easily and thus changes its fabric most readily, while the lower part may well retain a fabric derived from its initial flow down the glacier foreslope. Indeed, if the prediction from Jeffery's (1922) model is correct, that the a/b planes of particles in laminar flow are horizontal, then it should be possible to identify a flow which formed on a slope of different inclination from its present slope.

The term *controlled* used above in reference to a moraine system is used in the sense defined by Gravenor and Kupsch (1959), as a feature which reflects in plan the pattern of some glacial structure from which it is derived. Most "controlled" ice-cored moraine ridges tend to lie parallel to the glacier margin probably because of the fact that they are derived from series of debris bands in the underlying ice which themselves lie parallel to the glacier margin (for further discussion, see Boulton, 1968). In such a moraine, much of the flow in the supraglacial till will be either directly toward or directly away from the glacier. Thus, as flowtill fabrics tend to show peaks either parallel or transverse to flow, so in an area of controlled ice-cored moraine, flowtill peaks will tend to lie parallel or transverse to glacier movement. Areal variations in the fabrics of tills formed supraglacially in a zone of "uncontrolled" ridges would in contrast tend to be random.

71

Other observations on solifluction and mudflow fabrics show results comparable to those reported above. Glen, Donner, and West (1957) determined long-axis fabric orientations at five sites at the margin of Vestfonna, Nordaustlandet, in what they termed "ground moraine," although this was in fact supraglacial flow till and fabrics were produced by downslope flow. At all sites there was a dominant parallel peak, two of which showed secondary transverse peaks. Rudberg (1958), measuring long-axis orientations on solifluction slopes, determined a downslope peak of *a*-axis orientation with up-slope imbrication, a result similar to that of Caine (1968), who measured orientations of boulders in a periglacial blockfield (although he also found a secondary transverse peak).

Melt-out Tills

The entirely autochthonous element which occurs at the base of some stable flow till sheets is different from the overlying till in that it retains part of the original englacial fabric. Most melt-out tills which initially lie between a thin flowtill cover and glacier ice are likely to be deformed later by creep and thus to become flowtills (Fig. 10), but much of the debris contained within dead-ice masses overlain by thick overburden could well be released as melt-out till on final disappearance of the ice. Whether or not this process would take place would depend upon whether the overburden was sufficiently thick to inhibit flow and sufficiently permeable to allow water to escape from the melting ice, so that effective consolidation of the melt-out till could take place. If water was not able to escape, high pore pressures within the till and differential loading under an overburden of variable thickness (see Fig. 1) could produce flow or diapirism within the till. The effect of the diapirism in tills is well seen in the tills of north-east Norfolk, England (West and Banham, 1968), where it is recognized by the involution of the till stratification (in an unstratified till, recognition would be very difficult).

Melt-out tills may form either at the base of stagnant ice, or at its surface, by slow melting *in situ*. In Svalbard at present, only surface melting occurs. However, although the depth of seasonal thawing is only two to three meters on the average, melting of buried ice surfaces can occur at considerable depth beneath supraglacial streams or lakes, and thus melt-out tills can occur beneath thick stratified fluvial and lacustrine sequences. Their stone-orientation fabrics are essentially similar, in the plane of deposition, to the fabrics of the englacial debris from which they are derived, although they show somewhat weaker maxima, presumably because of disturbance and interactions between stones during deposition (Boulton, 1970b, Fig. 2; and Fig. 10 above). However, the angles of dip of long axes tend to be reduced during deposition by amounts which depend upon the englacial concentration of the debris from which they are derived. If this concentration is high, the original three-dimensional orientation will tend to be preserved; on the other hand, if this concentration of debris is low, vertical distances will be considerably reduced and only the projections of these

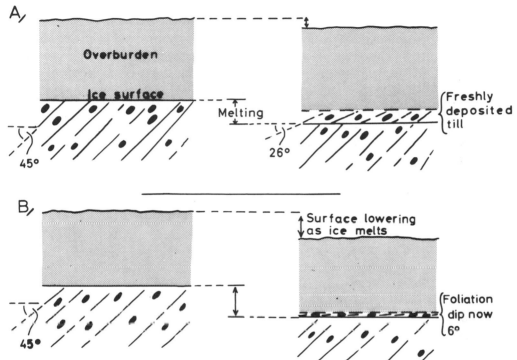

Figure 11. A. schematic diagram showing melting beneath an overburden of a mass of debris-rich ice with a debris content of 50 percent by volume and a foliation dip of 45°. The resultant melt-out till shows an a/b plane dip of 26°.
B. The production of melt-out till from ice with a debris content of 10 percent by volume. a/b planes now dip at 6°.

orientations onto the plane of deposition will be preserved (Fig. 10 and Fig. 11). Most melt-out tills develop from ice in which the foliation has a high up-glacier dip; however, because most debris bands have a relatively low ice content, this up-glacier foliation will be reflected only rarely in melt-out tills. This contrasts with the conclusion drawn by Harrison (1957). Similarly, in such a zone, a high proportion of stones show transverse orientations, and thus melt-out tills could be expected to show flat-lying a-axes and a/b planes, and commonly a transverse component of a-axis orientation.

FABRICS ACQUIRED DURING SUBGLACIAL DEPOSITION

Subglacial Lodgement Till

The thin marginal zones of many Svalbard glaciers which terminate on land are probably frozen to their substratum (Palosuo and Schytt, 1960), but some more rapidly flowing glaciers, which terminate in or near the sea, actively move over their beds in the terminal zone, and often give opportunity for inspection of the moving glacier sole. The lowest elements of the englacial load are plastically held within the ice and protrude through this sole; and, coming into contact with underlying bedrock, they are probably responsible for much of the striation and abrasion on bedrock surfaces (Carol, 1947). Where the glacier sole loses contact with the bedrock surface, above steep-sided cavities, the glacier sole itself can be seen to be heavily striated, and the a-axes of particles embedded within it lie parallel to these striations. Observations of the glacier sole were made in tunnels, in ice caves, and at the bases of ice cliffs at 17 localities. Subglacial till was present at less than half of these; at the others, the glacier rested upon smoothed and striated bedrock or deformed unlithified fluvial and lacustrine sediments. Most subglacial tills here have a drumlinoid or fluted form. Only at a limited number of localities was there any direct evidence for the processes of deposition. Two basic processes of till deposition from debris-rich ice were recognized from these exposures, although some tills may have originated by other processes for which no evidence was found. These processes were: the slow melting out of till from debris-rich ice which had become stagnant beneath the active glacier, and deposition from active ice as a result of pressure melting of the glacier sole as it passed over bedrock obstructions. The fabrics of these tills appeared to develop both from the process of deposition and from subsequent subglacial deformation. Although subglacial observations were restricted to snout areas, many glaciers have retreated considerable distances throughout recent centuries so that certain of what are now marginal subglacial deposits may have formed under a much thicker ice cover.

Lodgement Tills Released from Ice by Pressure Melting

Two localities showed definite evidence of lodgement tills actively being released as a result of pressure melting against an obstruction, and this process may well account for other subglacial till accumulations. At one locality, 25 m beneath the margin of Aavatsmarkbreen in Oscar II Land, till was actively accumulating on both sides of a bedrock knob (Fig. 12). On the up-glacier side, the debris-rich glacier sole was moving over both till and bedrock, and melting was active at this sole. The result was to

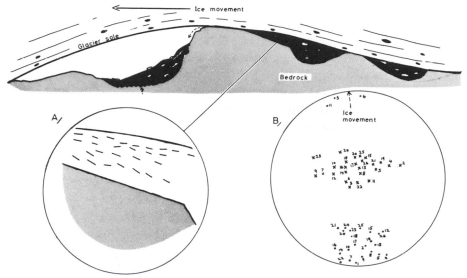

Figure 12. Section showing subglacial lodgement till built up around the flanks of a rôche moutonée. Compiled from observations made in several intersecting subglacial tunnels. A shows the a-axis dips of the magnetic susceptibility ellipsoids for 27 specimens collected from the subglacial till. The specimens are numbered from left to right, and downwards through four rows. B is a stereographic plot of the a-axes (dots) and c-axes (crosses) of the susceptibility ellipsoids for the specimens, numbered as described above.

release from the ice particles which had hitherto been partially or entirely englacial. Some of this material, a thin layer beneath 0.2 and 0.5 cm thick, seemed to be actively moved along by shearing beneath the glacier sole at about 0.7 cm/day (determined by markers). On the down-glacier side of the bedrock knob, the glacier lost contact with bedrock and thin till was oozing out like toothpaste from the ice/bedrock contact, to slump and slide down the lee side of the knob, at the foot of which was a considerable accumulation of such slumped material. This is slump deposit, so only a very indistinct bedding is acquired, presumably because much of the material falls as a semi-cohesive mass, although some sorting does occur as a result of water movement over the depositional slope. Pebbles on this slope tend to lie with their flattened surfaces on the slope and their long axes parallel to the direction of slope.

A better consolidated till had accumulated on the up-glacier flank of the obstruction; as there is no obvious subglacial source for this till, and as pressure melting and till accumulation are actively occurring on its surface, there is a strong possibility that the whole accumulation has built up by this same process. Irregularities in the underlying bedrock surface have been filled in, thus producing a smooth subglacial profile partly in rock and partly in till. Cavities can obviously be filled in, in much the same way as outlined above for the slump deposit on the down-glacier flank of the bedrock obstruction, but subglacial till has also been built up immediately above smoothed and striated bedrock surfaces, suggesting that only part of the till deposited by pressure melting is carried along at the glacier/substratum interface, subsequently to be forced into cavities, and that a small residue is actively being plastered permanently onto the substratum. The plastered-on surface of this till in contact with ice is striated; clasts lying on it show a-axis orientations parallel to ice movement along the plane, while a/b planes show a partial transverse stereographic girdle with a mode normal to the till surface. (See theoretical treatment: Nobles and Weertman, 1971; this volume.)

In order to investigate the small-scale variations in the fabric in this till, small samples were collected at the intervals shown in Figure 12. For each of these samples, three measurements of the anisotrophy of magnetic susceptibility were determined, which, when computed, gave a susceptibility ellipsoid whose three axes reflect the physical orientations of magnetic grains within the samples. These orientations were then checked against the microfabrics determined by cutting mutually perpendicular thin sections from four of the till samples. A good agreement resulted, leading to the assumption that magnetic fabrics are equivalent to microfabrics (for details of the method, see Hamilton and Rees, 1965, and Rees, 1965). As can be seen from Figure 12, the magnetic fabric shows a tendency for long-axis orientation to acquire an up-glacier imbrication with depth, and for normals to a/b planes to give a single peak rather than to describe a transverse girdle, as was found near the till/ice contact. The reason for this, I would suggest, is that initial plastering-on produces a parallelism of a-axes to the depositional surface and a transverse girdle for c-axes, whereas the shearing effect transmitted to the till lying below this surface tends to stabilize c-axes and to make a-axes dip up-glacier. The addition of till derived from the glacier by successive increments means that the components of the lowest parts of the till have been subjected to shearing stresses for the greatest time and have thus probably been most effectively reoriented by stress. Several other determinations of the susceptibility ellipsoid, made at another locality on the same glacier where such till seemed to be forming, showed similar features, but in some cases there was a tendency for down-glacier imbrication of a-axes. In addition, incomplete studies of a-axes in the till for the same locality show strong transverse orientations, because the till is being compressed against the up-glacier flank of a bedrock obstruction inducing transverse elongation within the till. Indeed, at this locality, till fabrics appear to reflect the bedrock configuration, rather than the direction of ice movement.

Melting observed beneath the accessible parts of Svalbard glaciers appears largely to be pressure melting produced by obstructions, a process which may thus be responsible for the common occurrence of till-plastered rôches moutonées or rock-cored drumlins. If, however, basal melting were common at the margin of a glacier with a large englacial debris load, then the very processes referred to above could be responsible for the deposition of a widespread lodgement-till sheet on a planar surface.

Lodgement Tills Released from Debris-Rich Stagnant Ice beneath an Active Glacier

This origin can be ascribed with certainty to tills at three localities, and may apply to others. The occurrence and some fabrics of these tills are described in a recent publication (Boulton, 1970b).

Basal debris-rich ice, less plastic than cleaner ice at higher levels, can become essentially stagnant when its forward movement is blocked by bed-rock obstructions over which the glacier is forced to rise. There is usually a well-marked plane of *décollement* separating the stagnant and active ice, and the former may well undergo further folding and deformation as a result of shear stress transmitted across this plane. Thus the fabrics normally characteristic of basal, highly sheared ice may well be changed if this same

Figure 13. Subglacial lodgement till deposited as a result of melting out of a mass of debris-rich stagnant ice beneath the active sole of Nordenskioldbreen in Bunsow Land. Note the sub-horizontal sheer planes. X-X marks the glacier sole.

ice becomes an essentially subglacial stagnant mass. Such a subglacial till was described from the glacier Nordenskiöldbreen (Boulton, 1970b), in which the till fabric showed elements inherited from an englacial-fabric maximum transverse to flow and some elements induced by glacier over-riding. These latter elements were found in those parts of the till which had reacted to glacier overriding by the formation of a series of sub-horizontal shear planes, which had, where they intersected till stones, re-oriented many of them until their *a*-axes lay parallel to movement along the shear planes, which itself was parallel to glacier movement (Fig. 13). Thus, where the till contained few shear planes, it showed a strong transverse fabric peak with a minor parallel peak; but, where the shear-plane density was very high, the transverse peak had been almost entirely replaced by a parallel peak. Presumably, post-depositional re-orientations of the type described in the previous section are also possible in these tills.

Fluted Lodgement Till

Fluted lodgement till was investigated at two localities, and bedrock was investigated at another. Most ground-moraine flutings lie either on the lee side or on the up-glacier side of large boulders, in contrast to those observed by Hoppe and Schytt (1953), or Galloway (1956), in which till accumula-tions tended to occur only on the lee side of boulders. The flutings described by Hoppe and Schytt (1953), and also by Price (1969), tend to be low, relatively continuous ridges rarely more than 50 cm high. The ridges ob-served in Svalbard tend to form less extensive though taller ridges, up to two meters high, although this latter contrast might result from the gener-ally large size of the boulders which form their core (Fig. 14). As in most tills exposed to the atmosphere, the surface is washed clear of much fine material, and many surficial boulders tend to show orientations derived from downslope movement. Measurements on the surfaces of flutings at the points at which they emerge from beneath the glacier indicate a weak though definite *a*-axis orientation of particles greater than 10 cm and less than 30 cm in diameter parallel to the direction of the fluting. However, pits dug within the fluted ridges showed more complex orientations. A pit on the lee side of a large boulder yielded a fabric in which the *a*-axes tended to dip up-glacier at 45°; a pit on one lateral flank of the same ridge showed *a*-axes dipping away from the ridge crest at 25°, while a pit on the other flank revealed *a*-axes with a 70° dip oblique to glacier movement. Yet an-other pit, at the down-glacier extremity of the same ridge, did not yield an obvious preferred orientation. Some of these data are rather difficult to interpret. The parallel orientation at the surface is obviously produced by the drag of ice along the surface of the ridge, but the fabrics below are a problem. They could have formed by development of the mechanism sug-

78

gested by Galloway (1956b), as a result of subglacial flow of till under differential overburden pressures into the cavities which formed both in front of and behind large boulders. In this case, the complex fabrics would merely reflect the complex pattern of the plastic deformation within the till as it was forced into these cavities, and, indeed, two different fabrics could both be interpreted as reflecting such flow, with a strong component upward toward the rim of the cavity. Alternatively, the ridges might have formed in a manner similar to that suggested for the tills in the two preceding sections, though the deposition of the till would be around boulder cores rather than rôches moutonnées.

These two processes are fundamentally different. In the first, the fluting is post-depositional and affects till already deposited from the ice by some other process and presumably in some other subglacial zone. This till may even have been deposited by a previous ice advance. In the second case, the flutings form as the direct result of a depositional process. Both proc-

Figure 14. Fluted lodgement till beyond the northern margin of Söre Buchananisen on Prins Karls Forland.

79

esses, of course, could play a part, but such a combination would be entirely coincidental.

The plausibility of the first hypothesis is elegantly demonstrated by an exposure at the margin of Aavatsmarkbreen. Bedrock at part of the margin of the glacier is represented by a series of gently dipping Tertiary clays containing thin siltstone bands. At one locality these have been deformed into a series of flutings, approximately two meters high, trending parallel to glacier movement. The internal bedding is a complex anticlinal structure, with an axial plane trending parallel to the ridge direction, and with many minor folds on its limbs displaying complex, competent/incompetent relationships between siltstone and clay. These structures have clearly formed subglacially. These are not eroded, transported masses, because their internal bedding parallels the surface; thus they are *in situ* and not erratics, representing deformed bedrock rather than materials deposited by the glacier. On this basis I would suggest that differential pressures, developed subglacially because of cavitation, had forced the clays and silts to flow into cavities which were probably present on the lee side of large boulders. Essentially similar mechanisms of flow of unfrozen till from areas of high to low confining pressures have been postulated for the formation of till ridges parallel to the ice front (Andrews and Smithson, 1966; Price, 1969).

In summary, five subglacial processes have been recognized which produce preferred orientations; drag of the glacier sole over a till surface, confined deformation by flow of till in response to the shear stress induced by overriding ice, brittle dislocations induced by the same stresses, unconfined downslope movement in subglacial cavities, and flow of till under a subglacial pressure gradient into subglacial cavities.

SUMMARY OF MAIN FABRIC TRENDS

All of the three main types of till described above show systematic areal fabric trends, which are related, directly or indirectly, to the direction of glacier movement, except perhaps for flow tills which have developed on an uncontrolled topography. Variations in flow till fabrics are basically related to the shape of the underlying ice surface, variations in melt-out till fabrics are related to the glacial stress system, and variations in subglacial lodgement till fabrics are related to the mode of deposition and the form of the surface onto which the till is deposited. All three types are likely to show considerable between-site variation, irrespective of any change in the direction of glacial movement. At individual sites, considerable vertical fabric variation can occur in all three types; in contrast, lateral variation is unlikely to be so great, except in the case of the lodgement till on an irregular substratum. The lower parts of flowtills tend to show systematic up-slope imbrication of a/b planes, whereas the a/b planes of melt-out till stones tend to lie in the horizontal plane.

The Use of Till Fabrics

Fabrics have commonly been used to give two rather different kinds of information: direction of movement of the depositing glacier, and the mode of origin of the till. In the former, that part of the till fabric which reflects the large-scale flow must be abstracted from that part of the fabric which may have been strongly influenced by local factors, such as mode of deposition, bedrock topography, local stresses within the ice, etc. In the latter case, in contrast, the purpose is to attempt both to build up a clear picture of the fabric in order to assess the mode of deformation of the till, and then to fuse this with other local details in order to infer a mode of origin, either on the basis of what is known about actual glacial deposition or (if this provides no guide) on a speculative basis.

In the first case, regional, between-site sampling is necessary in order to infer a systematic trend. If such a trend exists, one can be reasonably sure of identifying either the parallel or transverse direction. Should a till have both parallel and transverse maxima, however, extra care must be taken, for if these are all assumed to be parallel, then an essentially random pattern could be interpreted as a systematic one. The very nature of till fabrics and of the processes responsible for these fabrics imply that a single fabric taken at one locality has only a slight probability of reflecting exactly the local direction of ice movement; thus many detailed patterns of ancient ice movement inferred from the premise that a fabric lies in the parallel direction are basically unsound. In addition, it is of course axiomatic that all the fabrics should be taken from what is known to be the same till. Regional variance in till fabrics from a known till sheet is such that attempts to distinguish tills of different age by regional sampling are very suspect (c.f. West and Donner, 1956).

Unfortunately, some tills are only accessible at a small number of sites, in which case the inherent variability of till fabrics makes identification of the parallel direction extremely difficult. In some cases, such as that of fluted lodgement till, this is possible; but even if within-site fabric variation is small, identification of the parallel direction from one site may be a questionable procedure. Single sample fabrics from one site may be even less sound, as is obvious from the data from Svalbard, and as has been well demonstrated by Young (1969).

Within-site sampling also presents problems, as illustrated by several examples from Svalbard. A poorly designed sampling scheme for the flow till shown in Figure 8 could have revealed either fabric peaks parallel or transverse to flow, or a fabric with peaks both transverse and parallel to flow. The same is true for the subglacial till shown in Figure 13, in which there is a transverse *a*-axis orientation between joint planes and a parallel maximum along joint planes. Magnetic-susceptibility methods have a great potential in this context, in that they make possible the design of an adequate sampling scheme in a clay till, irrespective of the positions, sizes, and shapes of stones. Such methods also both reduce the size of the sample area so that it covers only one structural domain, and are less time-consuming than microscopic fabric methods, although they do depend upon the presence of elongate magnetic minerals within the till. The recognition of different structural fabric domains within a till and the separate analysis of these (Baird, 1962) could play a very useful role in interpretation of glacial motion.

The second main use of till fabrics, that of determination of till genesis, is obviously complex when one considers the number of variables involved. In addition, clast-orientation fabrics themselves provide only crude reflections of these variables, simply because of the relatively small number of characters included in fabric analyses, long-axis orientation, orientation strength, and influence of stone shape at any one locality. I would suggest that the best way of using fabrics for this purpose is in conjunction with other methods of till analysis, which, taken together, would probably give a much better indication of the general mode of origin. If, on the basis of these other criteria, it is possible to infer a subglacial origin, then till fabrics could be used to specify origin still further (Andrews and Smithson, 1966; Drake, 1971, this volume).

An attempt to infer genesis by till fabric alone seems to me to be extremely difficult, as so many different processes can produce similar results. Thus, although work such as that of Holmes (1941) and of Harrison (1957) is extremely valuable, their approach, using till fabric alone, has limited application. A possible approach using fabrics alone would be to undertake a thorough structural analysis of the stone orientations so as to be able to reconstruct the detailed strain pattern within the till, although this is not needed in stratified tills where the nature of strain is easily recognized. This would involve, perhaps, 200 values from each sample rather than the 50 to 100 which are normally adequate. One of the difficulties here lies in the problem of recognizing complex, three-dimensional orientation patterns, for the usual statistical analyses of till-fabric data assume a spherical normal distribution, and even the assessment of bimodal peaks by these methods involves some approximations. Possible methods which could be applied are those recently described by Kelley (1968), in which complex spatial patterns can be used as orientation models.

ACKNOWLEDGMENTS

Thanks for logistic support are due to W. B. Harland, and to R. H. Wallis and P. Maton, director and leaders of the Cambridge Spitsbergen Expedition. W. B. Horsefield, D. Smith, M. R. Rhodes, and C. Nash gave field assistance, and N. Hamilton provided valuable advice on the use of magnetic anisotropy methods.

REFERENCES

Andrews, J. T., and Smithson, B. B., 1966, Till fabrics of the cross-valley moraines of North Central Baffin Island, Northwest Territories, Canada: Geol. Soc. America Bull., v. 77, no. 3, p. 271-90.

Baird, A. K., 1962, Superposed deformation in the central Sierra Nevada foothills east of the Mother Lode: Univ. California Pub. in Geol., v. 42, p. 1-70.

Banham, P. H., 1966, The Significance of till pebble lineations and their relations to folds in two Pleistocene tills at Mundesley, Norfolk: Proc. Geologists Assoc., v. 77, p. 469-74.

Boulton, G. S., 1968, Flow tills and related deposits on some Vestspitsbergen glaciers: Jour. Glaciology, v. 7, no. 51, p. 391-412.

————, 1970a, On the origin and transport of englacial debris in Svalbard glaciers: Jour. Glaciology, v. 9, no. 56, p. 213-29.

————, 1970b, The deposition of subglacial and melt-out tills at the margins of certain Svalbard glaciers: Jour. Glaciology, v. 9, no. 56, p. 231-45.

Carol, H., 1947, The formation of rôches moutonées: Jour. Glaciology, v. 1, no. 2, p. 57-62.

Donner, J. J., and West, R.C., 1957, The Quaternary geology of Brageneset, Nord austlandet, Spitsbergen: Norsk Polarinstitutt Skrifter, no. 109, p. 9-16.

Dort, W. J., Jr., 1967, Internal structure of Sandy Glacier, South Victoria land, Antarctica: Jour. Glaciology, v. 6, no. 46, p. 529-40.

Fisher, R. A., 1953, Dispersion on a sphere: Royal Soc. London Proc., ser. A, v. 217, p. 295-305.

Flint, R. F., 1957, Glacial and Pleistocene geology: New York, Wiley, 589 p.

Galloway, R. W., 1956a, The structure of moraines in Lyngsdalen, North Norway: Jour. Glaciology, v. 2, no. 20, p. 730-33.

————, 1956, Rides de façonnement glaciaire sur une moraine de fond dans le Lyngsdal, Norvege septentrionale; Rev. Géomorph. Dynamique, no. 11-12, p. 174-77.

Garwood, E. J., and Gregory, J. W., 1898, Contributions to the glacial geology of Spitsbergen: Quart. Jour. Geol. Soc. London, v. 54, p. 197-227.

Glen, J. W., Donner, J. J., and West, R. G., 1957, On the mechanism by which stones in till become oriented: Amer. Jour. Sci., v. 255, no. 3, p. 194-205.

Gravenor, C. P., and Kupsch, W. O., 1959, Ice-disintegration features in western Canada: Jour. Glaciology, v. 67, no. 1, p. 48-64.

Green, R., 1962, Available methods for the analysis of vectorial data: Jour. Sed. Petrology, v. 34, no. 2, p. 440-41.

Gripp, K., 1929, Glaciologische und Geologische Ergebnisse der Hamburgischen Spitzbergen-Expedition 1927. Naturwissenschaftlicher Verein in Hamburg: Abhandlungen aus dem Gebiet der Naturwissenschaften, Bd. 22, Ht. 2-4, p. 146-249.

Hamilton, N., and Rees, A. I., 1965, The anisotrophy of magnetic susceptibility of rocks by the torque method: Jour. Geophy. Research, v. 67, p. 1565-72.

Harrison, W., 1957, A clay-till fabric: its character and origin: Jour. Geology, v. 65, no. 3, p. 275-308.

Holmes, C. D., 1941, Till fabric: Geol. Soc. America Bull., v. 52, no. 9, p. 1299-1354.

Hoppe, G., and Schytt, V., 1953, Some observations on fluted moraine surfaces: Geog. Annaler, ser. A, v. 35, no. 2, p. 105-15.

Jeffery, G. B., 1922, The motion of ellipsoid particles immersed in a viscous fluid: Royal Soc. London Proc., ser. A, v. 102, no. 715, p. 161-79.

Kamb, W. B., 1959, Ice petrofabric observations from Blue Glacier, Washington, in relation to theory and experiment: Jour. Geophys. Research, v. 64, no. 11, p. 1891-1904.

Kelley, J. C., 1968, Least squares analysis of tectonite fabric data: Geol. Soc. America Bull., v. 79, p. 223-40.

Lamplugh, G. W., 1911, On the Shelly moraine of the Sefström Glacier and other Spitsbergen phenomena illustrative of British glacial conditions: Proc. Yorkshire Geol. Soc., v. 17, p. 216-41.

Lundqvist, G., 1949, The orientation of block material in certain species of flow earth: Geog. Annaler, H. 1-2, p. 335-47.

Nye, J. F., 1959, A method of determining the strain rate tensor at the surface of a glacier: Jour. Glaciology, v. 3, no. 25, p. 409-19.

Palosuo, E., and Schytt, V., 1969, Till Nordostlandet med den Svenska Glaciologiska Expeditionem in Terrasta: no. 1, p. 1-19.

Price, R. J., 1969, Moraines, sandur, kames and eskers near Breidamerkurjökull Iceland: Trans. Inst. Brit. Geogr., no. 46, p. 17-43.

Rees, A. I., 1965, The use of anisotropy of magnetic susceptibility in the estimation of sedimentary fabric: Sedimentology, v. 4, p. 257-83.

Richter, K., 1936, Gefugestudien im Engebrae, Fondalsbrae und ihren Vorlandsedimenten: Zeitschr. für Gletschrk., v. 24, p. 22-30.

Sharp, R. P., 1949, Studies of superglacial debris in valley glaciers: Amer. Jour. Sci. v. 247, no. 5, p. 289-315.

Steinmetz, R., 1962, Analysis of vectorial data: Jour. Sed. Petrology, v. 32, no. 4 p. 801-12.

West, R. G., and Banham, P. H., 1968, Short field meeting on the north Norfoll coast: Proc. Geol. Assoc., v. 79, pt. 4, p. 493-512.

West, R. G., and Donner, J. J., 1956, The glaciations of East Anglia and the eas Midlands. A differentiation based on stone orientation measurements of the tills: Quart. Jour. Geol. Soc. London, v. 112, p. 69-91.

Young, J. A. T., 1969, Variations in till microfabric over very short distances: Geol Soc. America Bull., v. 80, no. 11, p. 2343-52.

Errata

The caption for Figure 2, page 57, should read as follows: "Zone of extention in the basal ice of Makarovbreen. Ice foliation lies parallel to the glacier bed, a/b planes of stones lie in the foliation plane, and long axes lie parallel to ice flow."

Page 78, line 15, should read: ". . . and *fluted* bedrock. . . ."

3

Reprinted from *Research Methods in Pleistocene Geomorphology*, E. Yatsu and A. Falconer, eds. (Proc. 2nd Guelph Symp. Geomorphol., 1971), 1972, pp. 66–82

THE EFFECT OF LITHOLOGY UPON TEXTURE OF TILL

A. Dreimanis and U.J. Vagners

INTRODUCTION

Texture of till depends upon several factors, most important being the following four:

(1) lithologic composition of both matrix and clasts;

(2) comminution and possible sorting of rock and mineral fragments during the transport of drift by glacial ice and its meltwaters prior to desposition of till; the comminution depends greatly upon the mode of glacial transport - superglacial, englacial, or basal;

(3) process of deposition, for instance by lodgment, basal melting, ablation, etc. - see Dreimanis (1969, Tab. 1) or Dreimanis and Vagners (1971 Fig. 2) for classification of tills in relation to transport and deposition of drift material;

(4) postdepositional changes.

This discussion will consider two factors only: lithology and comminution during glacial transport by an ice sheet - where the transport is mainly basal or englacial. The till samples investigated were taken from southern and central Ontario (Fig. 1) thus making it possible to consider a variety of lithologies, as bedrock of this area consists of Precambrian igneous and metamorphic rocks in the north and Paleozoic nonmetamorphic sedimentary rocks in the south. Regional glacial movement has been mainly from the north, or north-east, with additional lobal movements radiating out of the Great Lakes' depressions in southern Ontario.

Field examination suggests that most samples were from basal till. In order to avoid excessive amounts of local material incorporated in the tills studied, the sampling was done at least 0.6 m above the base of till. (Still two of the samples turned out to be local tills: No. 16 and one in the traverse III.) Nonoxidized material was preferred, and in no case leached or even partly leached till, or till containing visible secondary minerals was taken. The granulometric analyses were done by sieving and hydrometer method, and the Wentworth scale was applied for establishing boundaries between various particle size grades (Wentworth, 1922). In the first 25 large samples (1 cb.m average) the particle sizes ranged from 25.6 cm to less than 2 μ in others (Vagners, 1969) from 32 mm to less than 1 μ.

86 The lithologic composition of each Wentworth grade was determined

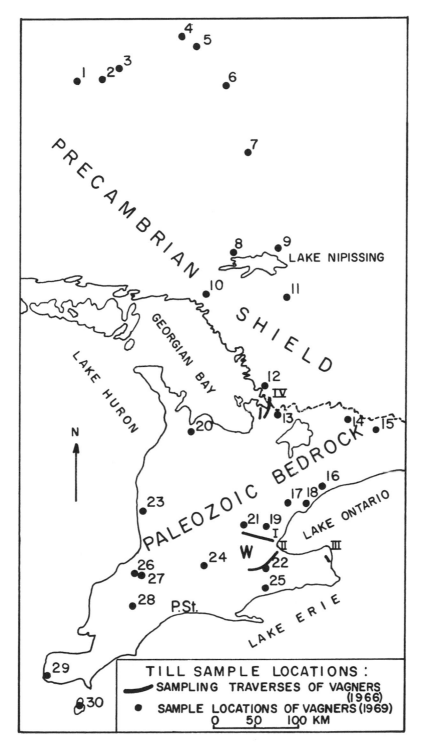

Fig. 1 Locations of till samples discussed in this paper: dots-individual samples, heavy lines-samples along traverses, P.St.-area of Port Stanley Till, W-area of Wentworth till.

separately. Detailed descriptions of sampling sites, laboratory procedures, and the results of analyses of individual samples are in Vagners (1966 and 1969). Typical results of analyses, selected from the above 55 sites of investigation will be used for illustrating this discussion, by using mainly frequency polygons, histograms and cumulative curves; the Wentworth grade scale is shown in milimeters and in its logarithmic transformation, the phi units (Krumbein, 1934).

Most published analyses of tills deal with the texture of till matrix only, with 2 mm or 4 mm as its upper boundary. The clasts of boulder, cobble and pebble sizes are usually excluded, mainly for practical reasons: determination of all the particle sizes present in till would require extremely large samples. Because of the exclusion of clasts (pebbles, cobbles, boulders), the usual terms "texture of till" or "granulometric composition of till" should be supplemented by the word "matrix" added after "till". It should be realized also that cummulative granulometric curves of the same till will look quite different, depending upon the coarsest particle size chosen (Fig. 2).

Bimodal distribution of lithic components

Texture of the entire till, including not only matrix, but also clasts, is governed mainly by the rules of bimodal distribution of its lithic components, minerals and rocks (Dreimanis and Vagners, 1969 and 1971). In the simplest case, if a till consists of fragments of a monomineralic rock only, for instance dolostone, and its mineral dolomite, the rock fragments are represented by at least one mode in the particle size grades, coarser than 0.1 mm, while its mineral grains cluster in another mode in the coarse silt size (Fig. 3). The more the rock (dolostone) is comminuted, the relatively larger becomes the mineral mode: in Fig. 3 it is obvious that the mineral mode increases on the account of the rock mode(s) with increasing distance of glacial transport.

As till usually consists of several lithic components, its granulometric composition is very often multimodal (Figures 4 and 5). Even though these illustrations represent only a part of tills with 25.6 cm and 3.2 cm as the upper boundaries of the particle sizes investigated, the multimodal nature of till and its dependence upon the modes of different lithic components is obvious.

If averages of several cumulative curves of lithologically different tills are plotted either on the Rosin and Rammler's "law of crushing" paper (for instance the composite of 38 samples of

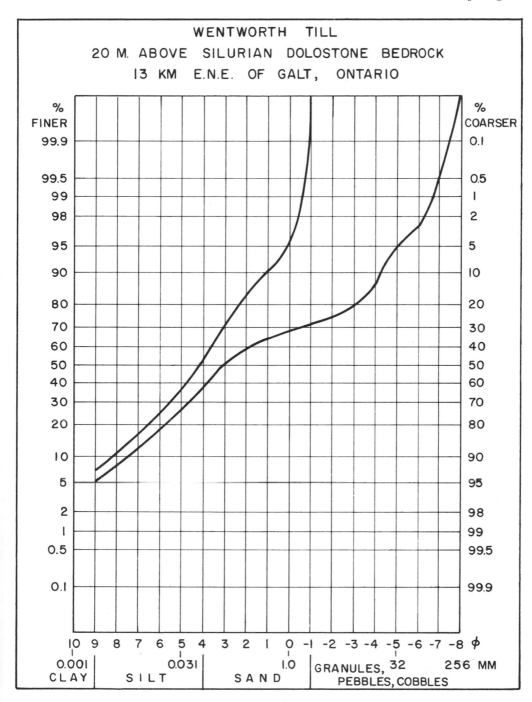

Fig. 2 Cumulative granulometric curves of a Wentworth till sample depending upon the upper boundary chosen.

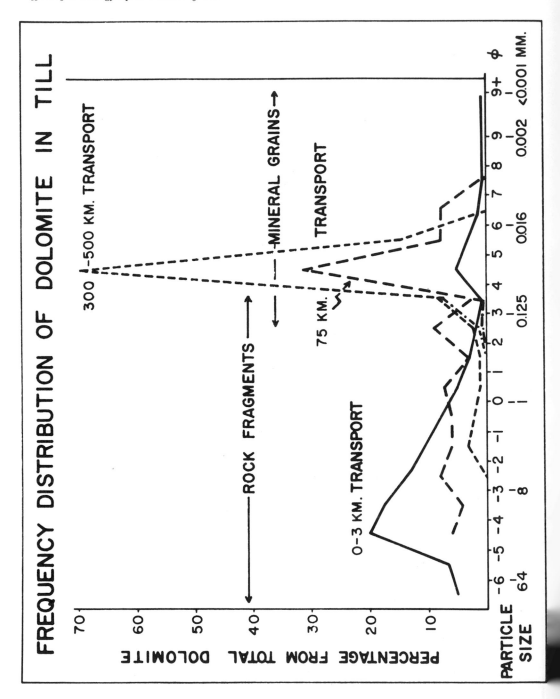

Fig. 3 Frequency distribution of dolostone-dolomite in three selected till samples from Hamilton Niagara area (Traverses II and III, see Fig. 1). After Dreimanis and Vagners (1971 Fig. 4).

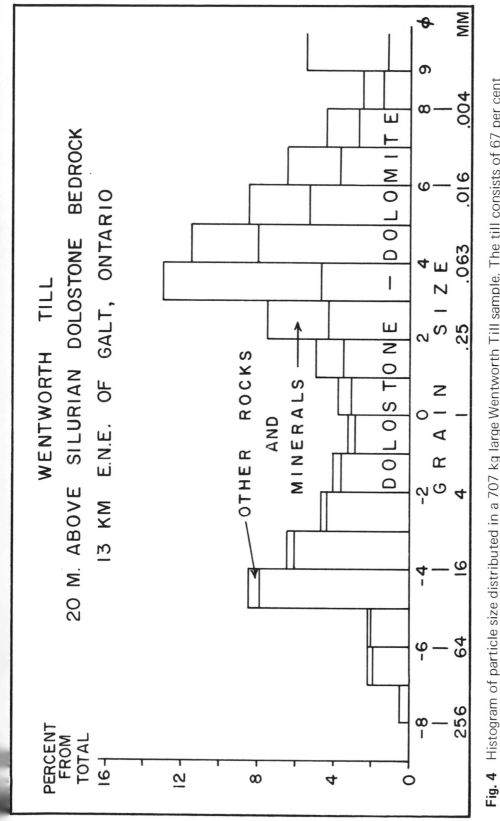

Fig. 4 Histogram of particle size distributed in a 707 kg large Wentworth Till sample. The till consists of 67 per cent dolostone-dolomite, 29 per cent igneous, metamorphic and fine grained clastic sedimentary rocks and their minerals, and 4 per cent limestone-calcite.

91

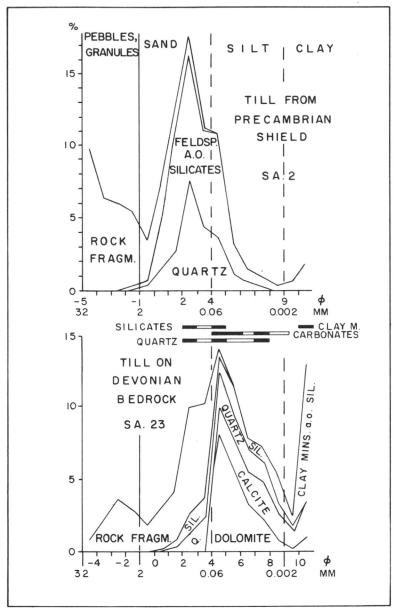

Fig. 5 Frequency distribution of rock fragments and minerals in two till samples: upper boundary – 32 mm. Sample no. 2 was collected about 50 km W. S. W. of Timmins over volcanic bedrock downglacier from an extensive area of coarse grained igneous and metamorphic bedrock; the 2.5 kg large sample consists of 42% rock fragments, 30% feldspars, 21% quartz, 3% heavy minerals, 4% fine grained silicates. Sample No. 23 was collected 5 km S.E. of Goderich in Devonian limestone bedrock area; the 1.6 kg large sample consists of 25% rock fragments 22% of each of dolomite and clay minerals, 14% calcite, 13% quartz, 4% feldspars and heavy minerals. The horizontal bars between both diagrams indicate terminal grades of most common minerals (compared with Fig. 6).

Goldthwait's lower till in Elson, 1961) or on the similar log probability paper (Adams' Cornwall till averages and grading limits in Lo and Roy, 1969 and Dreimanis' ranges of two major groups of tills in Ontario in Flint, 1971), nearly straight-line cumulative curves may be produced. They may also result, if a till consists of a variety of rocks and minerals mixed together in such proportions that their modes overlap over a wide range. Such well proportioned natural mixtures are rare, and therefore the suggestion of Elson (1961) that most tills may be recognized by straight-line curves on Rosin and Rammler's "law of crushing" paper, applies seldom to individual till analyses.

Texture of till matrix

Terminal grades of minerals. Texture of till matrix only - finer than 2 mm - depends mainly upon the terminal grades (Fig. 6) of its constituent minerals (Dreimanis and Vagners, 1969, and 1971). A terminal grade is the final product of glacial comminution, and its particle size range for each mineral depends upon the original sizes of the mineral grains while still in rocks, and upon the resistance of each mineral against comminution during the glacial transport. Comminution is accomplished by crushing, abrasion, and various other mechanical means. As the effect of abrasion decreases with decreasing particle sizes, crushing or splitting by other causes along cleavage planes and other zones of weakness inside the mineral grains and between them influences greatly the ease of comminution towards the terminal grade. Fig. 3 shows that the mineral dolomite begins to concentrate in its terminal grade already from the very beginning of its incorporation in glacial drift, 0-3 km from its source, and the particle size range of the terminal grade of dolomite remains the same over a distance of transport of several hundred kilometers. According to Vagners (1969), even such hard, but relatively brittle minerals as garnets and feldspars are comminuted to their terminal grades after glacial transport of 80-180 km, but they do not become comminuted further beyond the terminal size range even after a transport of several hundreds of kilometers.

The coarsest-grained terminal grades - in the sand and coarse silt sizes - are formed by most common igneous and metamorphic minerals: feldspars, quartz and heavy minerals (amphilbole, pyroxenes, garnets, etc.), as seen in Fig. 6. Main minerals of sedimentary rocks, except for quartz from sandstone, a rock not abundant in the area of investigation, have formed their terminal modes in finer particle sizes (Fig. 6): calcite, dolomite, and quartz from silt-stones and shales in the silt size; clay minerals were found

93

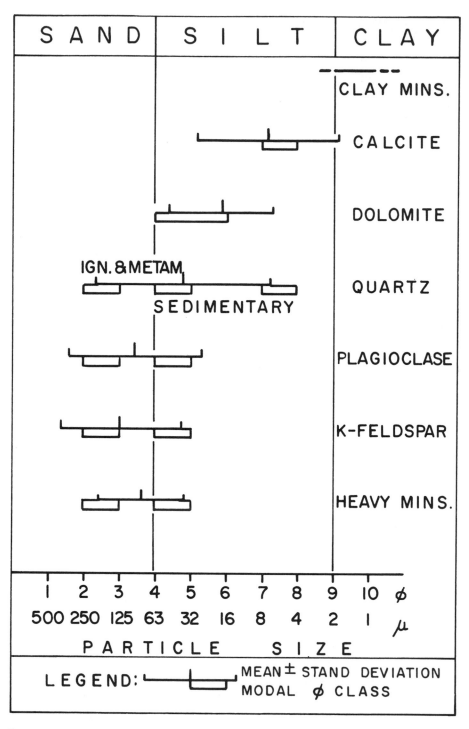

Fig. 6 Terminal grades — means, standard deviations and modal phi classes — of selected minerals in basal till. After Vagners (1969).

mainly in the clay size, but the range of their terminal size was not investigated. Thomas, 1969, found, that the sediments of Lake Erie which have derived largely from the tills of the adjoining land, contain their detrital clay minerals predominantly in the less than 2μ size. Several minerals are present in two or three modal classes, for instance quartz - in three. Judging from the results of Vagners' (1969) till analyses, the coarsest and probably also the intermediate modal classes (2-3 ϕ and 4-5 ϕ) of quartz in Ontario have derived mainly from igneous and metamorphic rocks, while the finer grained quartz grains (modal classes of 7-8 ϕ and 4-5 ϕ) come from the fine-grained clastic sedimentary rocks, particularly silt-stones (see typical examples in Fig. 5). It appears, that the terminal grades of those grains of quartz which were already relatively fine, e.g. of silt size, in the original rock, are similar to their original sizes. In other words, fine quartz grains must be resistant to comminution. However, quartz grains coarser than 0.25 mm are so common in the plutonic igneous and some metamorphic rocks over the Canadian Shield, that they should be well represented in tills collected in that area, if the original particle size rather than comminution would have governed all particle sizes of quartz in tills. In the 12 till samples collected over the Precambrian terrain (Vagners, 1969), only an average of 16% of quartz (by weight) is present in all the coarse particle sizes (over 0.25 mm), while 24% is already in the 0.25-0.125 mm grade which is the coarsest of the three modal size classes of quartz.

Till matrix. The upper boundary of frequency polygons of quartz, feldspars, and heavy minerals (included in "other silicates" in Fig. 5) is close to the usual upper boundary used for granulometric till analyses on this continent: 2 mm. Therefore Dreimanis (1969) suggested that 2 mm may be considered as a natural boundary between till matrix and clasts, because most tills contain a fair to a large amount of above minerals. However, if minerals of limestone, dolostone, shale and siltstone which are the most common sedimentary rocks are considered, then the sand-silt boundary (0.06 mm) appears to be a more natural dividing line between the matrix and clasts, as it separates the terminal grades of minerals from rock fragments. For tills rich in sandstones, the clast-matrix boundary may vary, depending upon the particle sizes of the quartz derived from sandstone.

Rock fragments. Most discussions of lithologic composition of till matrix, including the recent ones by both authors (Dreimanis and Vagners, 1969 and 1971, Dreimanis, 1969), deal with minerals only, as they are main constituents of the matrix. However, if the upper boundary of till matrix is placed at 2 mm, which is commonly done when analysing tills, then also rock fragments should be considered, particularly in the coarse particle size grades. The coarsest Wentworth grade of till matrix, 2-1 mm, always contains more rocks (67-98%) than minerals. As seen in Table 1 which summarizes Vagners (1969) data, minerals already become dominant in the grades finer than 1 mm in tills consisting of igneous and metamorphic materials: these particle size fractions consist mainly of the terminal grades of various silicate minerals (compare with Fig. 6). However, in till matrix collected over the sedimentary Paleozoic bedrock, rock fragments are more abundant, and minerals become dominant only beginning with 0.25 mm, in most samples. Again, this is a function of the upper boundary of the terminal grades - in this case the terminal grades of the sedimentary minerals, even though some of the tills Nos. 13-30 contain also fair amount of igneous and metamorphic minerals.

Highest admixture of rock fragments is found in matrix of those tills which contain large amounts of medium-hard fine-textured rocks, such as shale, siltstone, limestone, slate, for instance in local tills consisting mainly of these rocks. A typical example is sample No. 16 which was taken from the area of Collingwood Shale bedrock and consists nearly entirely of fragments of this shale (Vagners, 1969). Till matrix of No. 16 contains 58% sand, 28% silt and merely 14% clay.

Predicting the texture of till matrix. When considering all the various components of till matrix in Ontario including rock fragments finer than 2 mm, a variety of cumulative granulometric curves may develop (Fig. 7). If the rock fragments are not very abundant, and the predominant minerals of till matrix are known, its texture may be predicted. For instance, most igneous and metamorphic minerals will produce a silty sand-till matrix, that is with sand as the dominant and silt as the second most common constituent (Fig. 8: I+M).

Till derived mainly from limestone and dolostone will have a predominantly silty matrix, though it will contain also an appreciable amount of sand size particles of limestone and dolostone, and some clay size carbonates. In Figure 8, the average cumulative curve (L+D) for tills Nos. 21-25 from limestone and dolostone bedrock area, is coarser than expected for a calcite and dolomite rich till, because in addition to the 37% calcite and dolomite it contains also about 30% minerals from igneous and metamorphic rocks, brought southward from the Canadian Shield by englacial transport.

TABLE 1

PERCENTAGE OF TILL SAMPLES CONTAINING MORE ROCKS THAN MINERALS
IN INDIVIDUAL PARTICLE SIZE GRADES OF TILL MATRIX

(Data from Vagners, 1969)

Particle size grade	Tills collected on Precambrian Shield (samples No. 1-12)	Tills collected over Paleozoic sedimentary bedrock (samples No. 13-30)
2-1 mm	100%	100%
1-0.5 mm	30%	94%
0.5-0.25 mm	10%	55%
0.125-0.25 mm	0	38%
less than 0.125 mm	0	6% (?) *

*) Only one sample, collected 1-2 m over Collingwood shale 5 km north of Pickering, Ontario, contains more rock fragments (shale) than minerals in these fine grades. This is a local till, consisting mainly of shale (100% shale in 4-8 mm grade); the boundary between minute shale fragments and individual clay minerals was not determined.

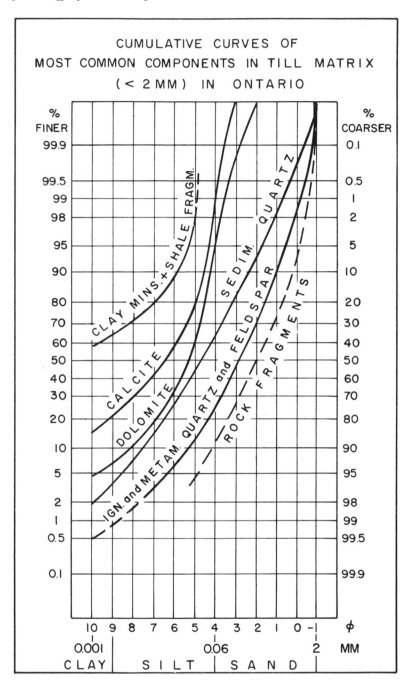

Fig. 7 Cumulative granulometric curves of most common components of till matrix (<2mm) in Ontario. The curves are averages of selected tills containing at least 6% of the corresponding component in the matrix: rock fragments from Nos. 1-30; igneous and metamorphic quartz — Nos. 1-12; sedimentary quartz — Nos. 16-19 and 26-30; dolomite — Nos. 20-25, 27, and 29; calcite — Nos. 13-15, 18, 20, 21, 23, 25-27, 29, 30; clay minerals — Nos. 16-19.

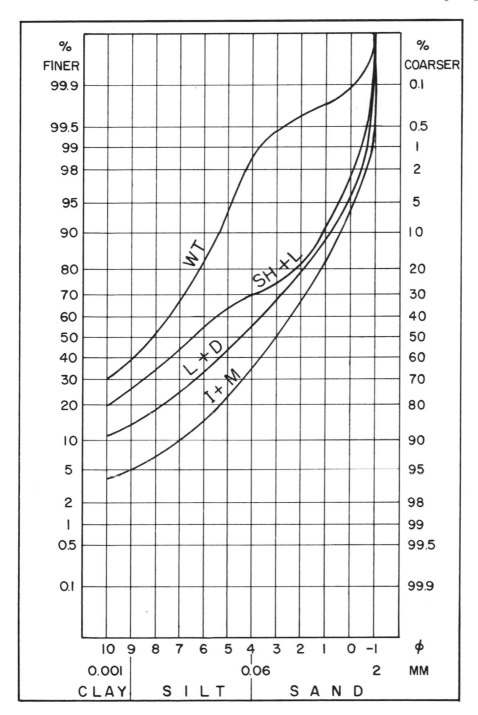

Fig. 8 Cumulative granulometric curves of the matrix of tills derived mainly from igneous and metamorphic rocks — I + M (Nos. 1-12), limestone and dolostone — L + D (Nos. 21-25), shale and limestone — SH + L (Nos. 26-28). The uppermost curve WT is an average of five waterlaid tills from the Port Stanley Drift in sourthwestern Ontario.

If shale is the only constituent, it may produce a silty clay till, if most shale fragments are soft and therefore crushed rapidly to their constituent minerals. In Ontario such shales are rare, as even the relatively soft shale formations e.g. Hamilton Shale, contain calcareous interbeds. Thus in Figure 8 the (SH+L) curve is an average for three tills (Nos. 26-28) collected from a Devonian shale area, but two of them (Nos. 26 and 27) were collected over interbedded shale and limestone of Hamilton Formation. Main constituent of the matrix of these three tills is rock fragments (an average of about 40%), mostly shale and limestone. Clay minerals, micas and tiny shale fragments of fine silt size are about 28%, carbonate minerals and silt-size quartz of sedimentary origin about 15% each, and minerals of igneous and metamorphic origin 3%. Abundance of silt-size sedimentary minerals combined with predominantly sand-size rock fragments has produced a sandy silt-till matrix, overshadowing the clay-and fine silt-size clay minerals.

Though knowledge of the mineralogical composition gives an indication of the type of texture of till matrix, the rock fragments which are particularly abundant in tills derived mainly from sedimentary fine-textured rocks, will generally cause the till to be coarser textured than would have been predicted by the mineral composition alone.

Incorporated stratified drift. Specific concentrations of minerals unrelated to the rules governing the glacial activities, may be found in tills which contain large amounts of incorporated sorted drift material, for instance the Port Stanley Till of south-western Ontario is either clayey or silty depending upon incorporation of lacustrine clays or silts, while the Wentworth Till (Karrow, 1963) is very sandy, particularly where outwash or kame sands and gravels have been incorporated. Waterlaid tills, formed most probably underneath an ice shelf, are usually very silty in the Port Stanley Drift of southwestern Ontario (Fig. 8: WT), containing large proportions of water-deposited silt-size particles.

CONCLUSIONS

Texture of till depends upon two major lithologic components: rock fragments and their constituent minerals. Most rock fragments occur in the coarse particle size classes: tills derived from igneous and metamorphic rocks contain them in particle sizes coarser than 1 mm, but tills derived mainly from sedimentary rocks - in the particle sizes coarser than 0.25 mm. Most textural analyses of tills in North America are done on till matrix, finer than 2 mm. Matrix of tills derived from igneous and metamorphic rocks depends mainly upon the terminal grades of feldspars and quartz, and therefore is coarse textured, with sand and coarse silt as main particle sizes. Matrix of tills derived mainly from sedimentary rocks is texturally more varied. Texture of these tills depends not only upon terminal grades of sedimentary minerals which are mainly of silt and clay sizes, but also upon the abundantly present fragments of medium hard sedimentary rocks, mainly of sand size. Besides, in southern Ontario, some tills contain igneous and metamorphic minerals up to one-third of the weight of their matrix. Texture of those tills which contain large proportion of incorporated stratified drift, depends mainly upon the dominant particle sizes of the incorporated material.

ACKNOWLEDGMENTS

The authors are grateful to the National Research Council of Canada for supporting their research by grant A4215, and to Mrs. R. Ringsman for drawing the illustrations.

REFERENCES

Dreimanis, A., 1969, Selection of genetically significant parameters for investigation of tills: Zeszyty Naukowe U.A.M., Geografia 8, Poznan, p. 15-29.

Dreimanis, A., and Vagners, U.J., 1969, Lithologic relationship of till to bedrock: in Wright, H.E. Jr. (ed.) Quaternary geology and climate, Nat. Acad. Sci., Publ. 1701, p. 93-98.

Dreimanis, A., and Vagners, U.J., 1971, Bimodal distribution of rock and mineral fragments in basal till: in Goldthwaite, R.P. (ed.): Till, A Symposium, Columbus, pp. 237-250.

Elson, J.A., 1961, The geology of tills: Proceed 14th Can. Soil Mich. Confer., Nat. Res. Counc. Can., Assoc. Com. Soil and Snow Mech., Techn. Mem. No. 69, p. 5-36.

Flint, R.F., 1971, Glacial and Quaternary geology, John Wiley and Sons, New York, 892 p.

Karrow, P.F., 1963, Pleistocene geology of the Hamilton-Galt area: Ont. Dept. Mines Geol. Rept. No. 16, 68 p.

Krumbein, W.C., 1934, Size frequency distributions of sediments: J. Sed. Petrology, v. 4, p. 65-77.

Lo, K.Y., and Roy, M., 1969, Rock breakage as related to some engineering and geologic processes, with discussion by Dreimanis, A.: Proceed. 22nd Can. Soil Mech. Confer., Queen's University, Dept. Civil Engin. Res. Rept. No. 67, p. 61-93.

Thomas, R.L. 1969. A note on the relationship of grain size, clay content, quartz and organic carbon in some Lake Erie and Lake Ontario sediments. Jour. Sed. Petrology, v. 39, p. 803-809.

Vagners, U.J., 1966, Lithologic relationship of till to carbonate bedrock in southern Ontario: unpublished M.Sc. thesis, Univ. Western Ontario, London, Ontario, 154 p.

Vagners, U.J., 1969, Mineral distribution in tills, south-central Ontario: unpublished Ph.D. thesis, University of Western Ontario, London, Ontario, 277 p.

Wentworth, C.K., 1922, A scale of grade and class terms for clastic sediments: J. Geol., v. 30, p. 377-392.

4

Reprinted from *Geol. Soc. America Bull.*, **83**, 2159–2165 (1972)

LON D. DRAKE *Department of Geology, The University of Iowa, Iowa City, Iowa 52240*

Mechanisms of Clast Attrition in Basal Till

ABSTRACT

Sources of 1,852 basal till pebbles from sites in east-central New Hampshire were identified. Measurements of pebble size, shape, roundness, breakage, lithology, weathering, and distance from source were determined.

Roundness, breakage, and size distributions of the pebbles, when compared to their distance from source, indicate that a state of dynamic equilibrium is reached between the processes of abrasion and crushing. The average pebble survives long enough to abraid to a roundness of .5 on Krumbein's scale, then is crushed again. The early attainment of this equilibrium, within a mile of the source, suggests that the clasts go through at least several cycles of crushing and abrasion per mile of transport. After many miles of transport, certain pebble shapes gradually evolve as stable forms. Spheres are the most stable, discs and rods less so (in that order), and blades are least stable in the subglacial environment. Comparisons to experimental data for the same rock types suggest that abrasion and crushing are about equally effective mechanisms in the production of subglacial till. The pebbles appear to be deposited in multiples of 6 to 7 mi from their source.

INTRODUCTION

The mechanisms of transportation and deposition of clastic material at the base of an actively flowing continental ice sheet have defied observation and monitoring. This study of clasts in basal till was made to determine some properties of rock fragments along their route and to deduce the mechanisms that lead to their condition. Clasts of pebble size are sufficiently large to recognize their source and small enough to obtain many specimens. The area of study in east-central New Hampshire was selected because: (1) The products of only one glaciation are dominant. (2) The basal till is clearly distinguished from ablation till (Drake, 1971). (3) The bedrock is made

up of many small igneous and metamorphic units which serve as recognizable sources for the till pebbles (Fig. 1). (4) Bedrock striations are abundant and indicate a fairly uniform flow direction of ice across the area (Fig. 2). One hundred pebbles were collected from each of 19 basal till sites (Fig. 2). The length of the A, B, and C axes of the pebbles was measured, and the roundness, breakage, weathering and lithology were determined. The distance upglacier to the nearest bedrock source area for each type of rock was measured. A total of 1,852 pebbles was used; the remaining 48 were too weathered for measurement or could not be reasonably identified as to source. All the samples were selected from fresh gray basal till, below present-day frost line (5 ft deep).

Detailed bedrock mapping is completed for much of the area (Smith and others 1939; Quinn, 1937, 1941, 1953; Moke, 1946; and Kingsley, 1931). The bedrock geology of the area is generalized on Figure 1 and obtained in part from Billings (1956). A set of bedrock samples was collected for comparison to the till clasts.

CHANGE IN ROUNDNESS FROM SOURCE

Table 1 shows the change in roundness of pebbles at increasing distance from their source as determined by visual comparison chart of Krumbein (1941, p. 64–72).

The resulting distribution is almost symmetrical about a roundness of 0.5. The horizontal component of the distribution represents the extent to which the pebbles are abraided and crushed to less than pebble size. Less than 0.1 percent of any lithology remains beyond 21 mi of its source. Goldthwait (1968, Fig. 7B) found that the distance from the southeast edge of Red Hill to the most distant edge of his 0.1 to 1.0 percent contour in the Red Hill boulder train is 22.5 mi. Apparently pebble and boulder distributions decrease at similar distances in east-central New Hampshire.

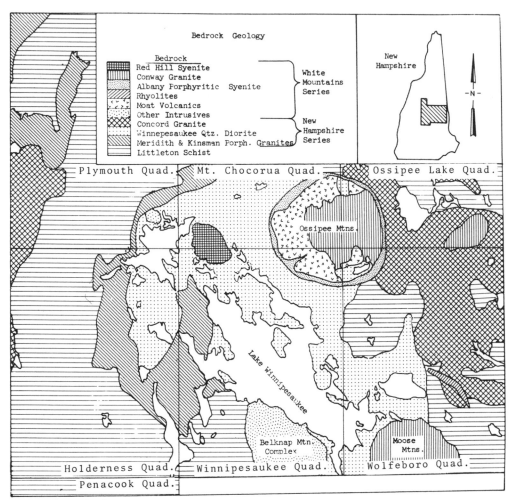

Figure 1. Bedrock geology (*after* Smith and others, 1939; Quinn, 1937, 1941, 1953; Moke, 1946; Kingsley, 1931; Billings, 1956).

The percentages on Table 1 show that more than two-thirds of the pebbles have a roundness of 0.5 within a mile of their source. The distribution of pebbles about the mean roundness of .5 remains roughly the same through each mile from the source (Compare columns on Table 1). This indicates two alternatives: (1) The pebbles go through many cycles of change each mile so that the similarity in form seen in each successive mile from the source represents their dynamic equilibrium condition. (2) No changes take place with regard to pebble roundness.

The first alternative is preferred by the author because 70 percent of the pebbles are fresh hard rock and probably possessed a roundness of 0.1 when plucked from their bed-

rock source, as does freshly broken rock. If this assumption is correct, then the average roundness of 0.5 must be maintained by repeated crushing and abrasion each mile of transport, and 0.5 may be considered the dynamic equilibrium state for the rounding of pebbles in basal till.

The second alternative is less likely because it requires that the pebbles have an initial average roundness of 0.5 at their source and then their abundance is decreased merely by dilution. With freshly broken rock, about 96 to 99 percent of the fragments have a roundness of 0.1 on Krumbein's scale. An average initial rounding of 0.5 could possibly be the result of incorporation of large quantities of pre-

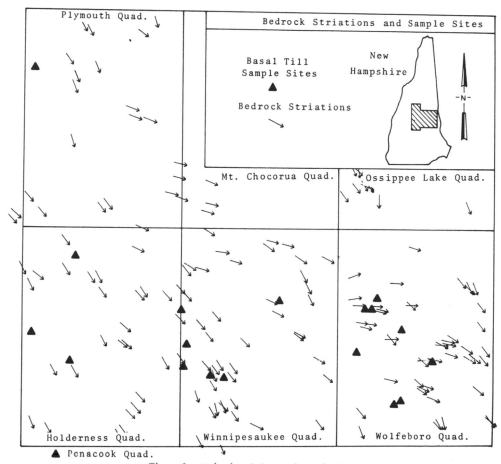

Figure 2. Bedrock striations and sample sites.

weathered material into the basal load of the glacier. Unweathered cores of firm rock could be stripped of their weathering rims, leaving a well-rounded pebble. In this study, 30 percent of the pebbles were weathered or possessed a partial or complete weathering rim. Most of the bedrock now exposed in the area is un-weathered. For these reasons it is unlikely that the majority of the pebbles possessed an initial roundness of 0.5, although one cannot eliminate the possibility of a well-rounded preglacial source.

The distribution on Table 1 is slightly skewed toward lesser roundness during the first few miles. Large pebbles, cobbles, and boulders, after being removed from their source and crushed, should produce some pebble-sized fragments which are then subject to further rounding and crushing. Farther from their

source (6 to 9 mi), the values are skewed toward higher roundness, suggesting that the supply of larger fragments is being depleted.

BREAKAGE OF BASAL TILL PEBBLES

Each pebble falls into one of three selected categories of breakage: (1) pebbles with relatively fresh surface breaks, less than or equal to 0.3 on Krumbein's scale (1941, Pl. I); (2) pebbles with the last break worn to a roundness 0.4 or greater; or (3) pebbles with no indication of earlier breakage.

Freshly broken pebbles disappear most rapidly and those showing no break persist the farthest (Table 2). This seems best explained by assuming that many of the pebbles go through three categories, then are crushed to start the cycle over again. Those pebbles in

TABLE 1. PEBBLE ROUNDNESS VERSUS DISTANCE FROM SOURCE (IN PERCENT OF TOTAL)

Roundness	0	1	2	3	4	5	6	7	8	9	11	12	15	19
Number of pebbles	1134	134	211	39	13	6	122	65	22	9	11	22	6	4
Well-rounded														
.9		.1												
.8		.4												
.7	1.4		.2				.2	.2						
.6	8.9	.9	2.4	.4	.5	.1	2.4	1.1	.4	.3	.2		.2	
.5	27.7	3.4	6.1	.7	.2	.2	3.1	1.7	.7	.2	.2	.8		.2
.4	13.2	1.7	2.4	.8			.9	.4	.1		.2	.4	.1	
.3	4.9	.7	.3	.2										
.2	2.9	.3						.1						
.1	1.7	.2												
Angular														

Miles from source

TABLE 2. PEBBLE BREAKAGE VERSUS DISTANCE FROM SOURCE (IN PERCENT OF TOTAL)

		0	1	2	3	4	5	6	7	8	9	11	13	15	21
3	No break	29.3	3.3	6.5	1.2	.6	.3	2.0	1.1	.3	.6	.5	.3	.4	.1
2	Worn break	14.5	2.8	2.7	.9			4.1	2.5	.6			.5		
1	Fresh break	16.9	1.8	2.1	.2			1.3	.3				.6		

Miles from source

category 3 then persist the longest because after they are crushed, the fragments are of less than pebble size and so will not reappear in the fresh-break category 1 for the next mile. This interpretation also explains why categories 1 and 2 disappear from the plot at the same distance (between 14 and 15 mi). When the pebbles in category 3 were crushed in the vicinity of 15 mi from their source, the fragments were less than pebble size, whereas those pebbles already in categories 1 and 2 continued to abrade to category 3. This suggests that the rate of abrasion is sufficiently rapid so that in the average time available between crushings, the average pebble can reattain a roundness of 0.5.

The above interpretation requires that pebble size decrease away from the source. The conventional measure of clast size is obtained by sieving, which essentially measures the B axes of groups of fragments, but does not allow for the isolation of individual clasts. Therefore, the B axis of each pebble was measured as an index of its size. Table 3 lists percentages of pebbles plotted as the distance (in miles) of a pebble from its source versus the length of the B axis (size) of the pebble. Note that the pebble size (as measured by the B axis) decreases regularly away from the source. The relative distributions within roundness and breakage categories (Tables 1 and 2) within the

first mile from the source, are already proportional to those distributions at greater distances. This suggests that many pebbles go through a minimum of several cycles of crushing and abrasion per mile.

PREFERENTIAL PEBBLE SHAPE PRODUCTION

The notion that glacier ice can produce certain shapes of pebbles can be traced to Von Engeln (1932), who suggested that flatiron-shaped pebbles and cobbles were exceptionally durable. Wentworth (1936), placed Von Engeln's hypothesis in doubt and it is rarely considered today. Holmes (1960) suggested that the number of ovoids increased markedly in the direction of ice flow, and wedgeforms and rhomboids decrease. Three difficulties in utilizing Holmes' data are present. (1) His pebble-shape groups (ovoid, wedgeform, rhomboid) are not likely to have the same meaning to different people. (2) In his study area, the ice-flow direction is normal to the bedding of broad bands of sedimentary rocks—hence, the noted change in pebble shape and development of bedding and jointing planes in the sedimentary rocks. (3) The bedding also imparts a strong anisotropy to most of the pebbles and automatically favors flattened shapes.

In my study these difficulties are minimized

106

TABLE 3. PEBBLE SIZE VERSUS DISTANCE FROM SOURCE (PERCENT OF TOTAL)

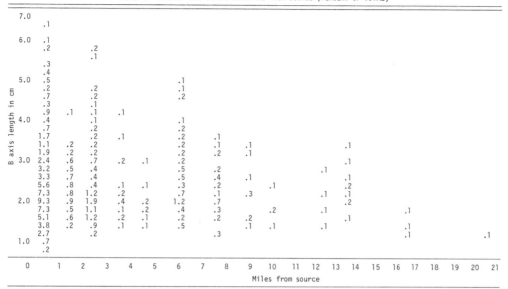

because the Zingg (1935) shapes have become a quantitative standard for representing pebble shapes, the parent sources are highly uniform in grain size and mineral content, and most of the bedrock types are either isotropic or only weakly anisotropic (Drake, 1970, p. 1356–1361).

The shapes of fragments resulting from artificially crushing the bedrock were compared with the shapes of pebbles present in the basal till. Assuming that the shapes of fragments freshly quarried by basal ice are equivalent to those resulting from artificially fracturing the rock, then this comparison should indicate which shapes would form during transportation.

To determine what the artificially crushed shapes of the lithologies present in the till would be, the percentage of pebbles of each lithology present in the basal tills was calculated. Utilizing the crushed shape standards established for the area by Drake (1970), the percentage of the 4 Zingg shapes produced by each lithology was compiled and then proportioned by the percentage of that lithology present. Over 92 percent of the pebbles were from standardized lithologies. Another 5 percent of the pebbles were other varieties of the standardized lithologies (for example, of the 6 varieties of Littleton Schist recorded, only 2 were standardized by Drake, 1970). The shapes produced by the few percent of remaining lithologies were estimated from the rock type they most closely resembled, in terms of grain-size and the degree of anisotropy. Listed below is the total percentage of each shape that would be present if the proportions of rock types in the till were artificially crushed. The actual shapes that were produced in the basal till from the same proportions of rock types are then compared.

	Percent Spheres	Percent Blades	Percent Rods	Percent Discs
Crushed Bedrock	18.2	26.0	17.2	36.4
Basal Till Pebbles	27.4	14.4	14.4	39.7

If the crushed bedrock is considered as the original shape, then the basal till spheres increased by one-half; blades decreased by one-half; rods decreased by one-sixth; and discs increased one-tenth.

A second method of evaluating shape production is to consider the survival rates of different shapes as they are transported farther from their source. Table 4 lists the percentage of pebbles plotted as Zingg shape against the distance in miles of the pebbles from their source. The ordinate is the 4 Zingg shape classes, and the values within are percentages of pebbles present in each coordinate category. From Table 4 it can be seen that blades disappear (> .01 percent) within 9 mi, rods and discs extend to 14 mi, and spheres persist to 21

Table 4. Pebble Shape Versus Distance From Source (%)

	0	1	2	3	4	5	6	7	8	9	10	11	12	13	14	15	16	17	18	19	20	21
Blades	7.8	.5	1.0	.1		.4	.4	.5														
Rods	12.0	1.4	2.3	.3	.2	.1	.9	.6	.1			.1	.2									
Spheres	17.9	2.4	3.6	.9	.2	.2	2.2	1.1	.1			.3	.4				.2					.1
Discs	23.8	2.7	4.6	.9	.2		3.2	1.5	.5			.2	.6									

Miles from source

mi. Note that the order of shape stability is the same as that indicated by the previous method. Although this plot is a crude tool, some useful approximations can be made. The distance 13.5 mi is selected as a reference distance because both rods and discs terminate here (note that discs are more abundant than rods). Blades only persist to about 8.5 mi. Spheres, on the other hand, persist to 20.5 mi. When mileage units with large numbers of pebbles are compared, for example 0 to 1 and 6 to 7 mi, spheres and discs have increased and rods and blades have decreased. When miles with small amounts of pebbles are compared to the 0- to 1-mi proportions, the changes are more irregular, possibly due to small sample size.

REASONS FOR PREFERRED SHAPE PRODUCTION

I suggested previously that crushing and abrasion are the dominant factors controlling final shape of the pebbles. The pebble-shape data also suggest that these factors dominate. With regard to crushing, the elongate pebbles (blades and rods) would clearly be most susceptible to fracture and the spheres the least susceptible. Experimental abrasion of the same lithologies has resulted in increasing of the percentages of spheres at the expense of blades (Drake, 1970). Presumably the results would be the same in subglacial abrasion.

Hence, blades are more susceptible to crushing and abrasion and are destroyed most rapidly. Spheres are least susceptible to crushing and are actually produced by abrasion and so are the most durable shape. Rods are susceptible to crushing but their proportions are little affected by abrasion (Drake, 1970) so they decrease much less rapidly than blades. Discs are slightly favored by crushing and their proportions are little affected by abrasion (Drake, 1970), so their numbers increase slightly. Discs, then, are slightly favored by combined crushing and abrasion because if the elongate pebbles (rods and blades) are most easily destroyed, then this can increase the relative proportions of discs as well as spheres.

From the above interpretation, crushing and abrasion appear to be about equally effective in destroying pebbles. If the cycling hypothesis is valid, then pebble-shape evolution is a relatively slow process compared to rates of crushing and abrasion, and many generations of crushing and abrasion are necessary to alter the net proportions of pebble shapes. The shaping effect is most noticeable after most of the pebbles have been destroyed and the survivors have the most stable shape for the subglacial environment. Therefore, it is possible that the boulders in New England till are well rounded and highly spherical because they may simply be the result of the preferential production of rounded spherical shapes by subglacial crushing and abrasion.

DISCUSSION

One factor not investigated is the durability of different lithologies in the above-mentioned processes. A very durable rock type, for example, quartzite, might travel for long distances in basal till with little change. Ideally, studies should be carried out separately for each lithology. However, 1,852 pebbles divided among the 28 recognizable rock types leave an average of only 65 pebbles per rock type, and as most of these pebbles are within a mile of their source, only a few of each were collected at greater distances from their source.

The holes in the distribution of values on Tables 1 through 4 are the result of a lack of sample sites located so that more than 3 pebbles are at that particular distance from their source (the computer program was designed to round off to zero those data points of less than 4 pebbles). The data points seem to cluster at 6- to 7-mi intervals (0 to 1, 6 to 7, 13 to 14, 20 to 21 mi). No other evidence was found to suggest whether this clustering was a fortuitous result of sample site location (such that most pebbles happened to be at these intervals) or a real distribution, perhaps resulting from a harmonic variation between erosion and deposition by the basal ice.

A plot of degree of pebble weathering versus

108

distance from source (not illustrated) indicated that highly weathered pebbles disappeared most rapidly and unweathered pebbles survived the longest.

ACKNOWLEDGMENTS

The author is indebted to Richard P. Goldthwait, Alonzo Quinn, Charles Corabato and Stephen Forster for their assistance. This project was part of a study conducted as partial fulfillment of the Ph.D. program at Ohio State University and was funded by National Science Foundation Grant GA-945.

REFERENCES CITED

Billings, M. P., 1956, The Geology of New Hampshire, pt. II, Bedrock geology: New Hampshire Dept. Res. and Econ. Devel., 203 p., maps.

Drake, L. D., 1970, Rock texture: An important factor for clast shape studies: Jour. Sed. Petrology, v. 40, no. 4, p. 1356–1361.

—— 1971, Evidence for ablation and basal till in New Hampshire, in Goldthwait, R. P., ed., Till Symposium Volume: The Ohio State Univ. Press (in press).

Goldthwait, R. P., 1968, Surficial geology of the Wolfeboro-Winnipesaukee area, New Hampshire: New Hampshire Dept. Res. and Econ. Devel., 60 p.

Holmes, C.D.S., 1960, Evolution of till stone shapes, central New York: Geol. Soc. America Bull., v. 71, p. 1645–1660.

Kingsley, L., 1931, Cauldron-subsidence of the Ossippee Mountains: Am. Jour. Science, 5th ser., v. 22, p. 139–168.

Krumbein, W. C., 1941, Measurement and geological significance of shape and roundness of sedimentary particles: Jour. Sed. Petrology, p. 64–72.

Moke, C. B., 1946, Geology of the Plymouth Quadrangle: New Hampshire Div. Econ. Devel., 21 p.

Quinn, A., 1937, Petrology of the alkaline rocks at Red Hill, New Hampshire: Geol. Soc. America Bull., v. 48, p. 373–402.

—— 1941, Geology of Winnipesaukee Quadrangle: New Hampshire Div. Econ. Devel., 22 p.

—— 1953, Geology of the Wolfeboro quadrangle: New Hampshire Div. Econ. Devel., 24 p.

Smith, A. P., Kingsley, L., Quinn, A., 1939, Geology of Mt. Chocorua quadrangle: New Hampshire Div. Econ. Devel., 14 p.

Von Engeln, O. D., 1932, Type form of faceted and straited glacial pebbles: Am. Jour. Science, 5th ser., v. 19, p. 9–16.

Wentworth, C. K., 1936, An analysis of the shapes of glacial pebbles: Jour. Sed. Petrology, v. 6, p. 85–96.

Zingg, Th., 1935, Beitrag zur Schotteranalyse: Schweiz. Min. u. Pet. Mitt., v. 15, p. 38–140.

MANUSCRIPT RECEIVED BY THE SOCIETY NOVEMBER 29, 1971

5

Reprinted from *Arctic Alpine Res.*, **5**(1), 17–27 (1973)

NATURE AND RATE OF BASAL TILL DEPOSITION IN A STAGNATING ICE MASS, BURROUGHS GLACIER, ALASKA*

DAVID M. MICKELSON[†]

Institute of Polar Studies
The Ohio State University
Columbus, Ohio 43210

ABSTRACT

Rates of basal till deposition ranging from 0.5 to 2.5 cm year^{-1} have been obtained from the Burroughs Glacier in southeast Alaska. Unlike minimum rates of till deposition which are derived by dividing the thickness of till deposited by the length of time the area was ice covered, these values are believed to more closely represent an actual rate of deposition.

Because of the emergence of hills during deglaciation, a change in ice flow direction of up to 90° has occurred near the southeast terminus. Since this change is recorded by maps and photographs dating to 1892, a rate of change of ice flow direction can be estimated.

Till fabric measurements and till composition at two or three depths in the till at seven localities reflect this change. Estimates of the rate of till deposition are obtained by assuming that the fabric azimuth represents the ice flow direction at the time the till was deposited.

Most till deposition seems to have taken place during late stages of deglaciation since at two locations fabric of till just above bedrock or a paleosol records a post-1892 flow direction. The rate obtained by dividing the average thickness of till by the number of years the area was ice covered is an order of magnitude lower than those calculated above.

INTRODUCTION

It is nearly impossible to obtain a measure of the rate of till deposition directly. Rates of till deposition, calculated by dividing the average till thickness in an area by the length of time the area was ice covered, provide only minimum rates because one must assume that till deposition took place at a constant rate throughout the whole time the area was ice covered. This obviously may not have been the case.

Maps and photographs dating from 1892 to 1960 show a change in ice flow direction of up to 90° near the southeast terminus of Burroughs Glacier. Changes in fabric and composition of basal till accompanied this change and these are described in the following sections. Because the rate of change of ice-flow direction is known, till fabrics at different depths provide more accurate estimates of the rate of basal till deposition.

---

*Contribution No. 234 of the Institute of Polar Studies, The Ohio State University, Columbus, Ohio 43210.

[†]Present address: Department of Geology and Geophysics, University of Wisconsin, Madison, Wisconsin 53706.

HISTORY OF DEGLACIATION

The Burroughs Glacier is located in the north central part of Glacier Bay National Monument in southeastern Alaska. It is a stagnating remnant of a large, active Neoglacial ice mass which retreated rapidly since its maximum advance about 1700 A.D. (Field, 1959; Goldthwait, 1966). A topographic map of the area was constructed between 1890 and 1892 by Reid (1896) which shows ice over the Bruce Hills (Figure 1) at an elevation between 610 and 685 m. The Wachusett Inlet area has been completely deglaciated and now lies at sea level (Figure 1). The elevation of the crest of Burroughs Glacier was between 500 and 525 m in 1960 (Taylor, 1963). Retreat of the calving ice margin in Wachusett Inlet was rapid and

FIGURE 1. Photographs taken (A) in 1907 (by the International Boundary Commission) and (B) in 1965 (by A. Post).

111

no doubt accounts for the rapid deglaciation of the area.

As deglaciation proceeded, two factors played an important role in changing ice flow directions near the southeast terminus of Burroughs Glacier. The Bruce Hills emerged and prevented ice flow directly over their summits. Because ice loss from Plateau Glacier was greater than at Burroughs Glacier due to calving into the Wachusett Inlet, the ice surface gradient shifted toward the inlet. By 1960, ice near the southeast end of the Bruce Hills flowed directly toward Wachusett Inlet.

The change in ice flow direction between 1892 and 1948 for three points in this area is given in Figure 2. These were obtained by drawing lines normal to contours on the ice surface on topographic maps constructed in 1892, 1907, and 1948 and by estimating the ice surface slope in 1929 from oblique aerial photographs.

TILL FABRIC MEASUREMENTS

The orientation of pebbles with *a:b* axes ratios of greater than 1.7:1 and long axes between 1 and 10 cm long was measured in the field. Strike and plunge of the long axis was measured on 50 pebbles at most locations. Data were plotted and contoured on an equal area

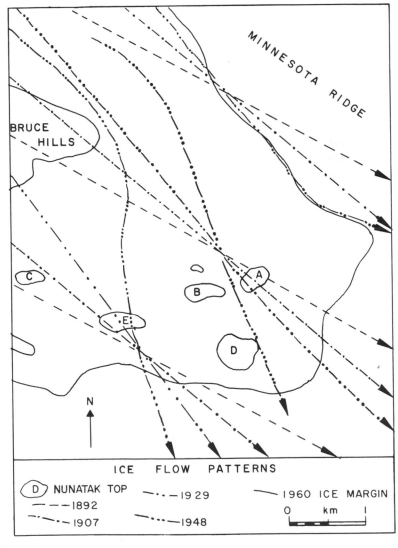

FIGURE 2. Changes in the direction of ice flow between 1892 and 1948.

projection using a computer program developed by C. E. Corbato (pers. comm., 1971) after Kamb (1959). The computer program is given by McKenzie (1968).

Fabric diagrams for each location are given in Figures 3 and 4. The lowest contour is one standard deviation and the contour interval above this is given. All diagrams are nearly symmetrical about the modal azimuth which plunges up-glacier. They are similar to many other till fabrics reported in the literature. Azimuth modes, three dimensional vector means and their standard deviations are given in Table 1.

It can be seen (Figures 3 and 4) that there is a change in till fabric azimuth with depth in the till at each location. Different fabric azimuths at different depths have been reported by numerous authors including fairly detailed studies by Young (1969) and Andrews and Smith (1970). The reasons for these changes

have not been established. It seems likely that the vertical changes of till fabric azimuth at the Burroughs Glacier are due to changes in ice-flow direction. At seven localities from 0.5 to 4 km apart in the southeastern terminus area the changes of fabric mode are consistent with ice flow direction changes observed on maps. Because of the thin glacial ice the effect of glacier-bed topography on basal ice flow direction seems to be reflected by ice surface slope observed on maps. All but one of the sites is on the up-ice side of a nunatak and there seem to be no large scale topographic features to affect the flow direction of basal ice. Crossing striations with azimuths in the same range as flow directions recorded on maps occur on Nunataks A, D, and E. This, plus the fact that till fabric measurements at all localities are consistent with the known changes in surface ice flow direction, suggests that the till fabrics are actually recording the change in ice flow direction.

TILL DEPOSITION DURING THE CHANGE OF ICE-FLOW DIRECTION

Although fabrics are related to ice-movement direction and the fabrics show a change in ice-flow direction parallel to that determined from early maps, it must be shown that till was actually being deposited while the flow direction change was taking place if a rate of till deposition is to be calculated. Tills reoriented by flowing ice have been reported by MacClintock and Dreimanis (1964) in the St. Lawrence Valley and they no doubt occur elsewhere. Therefore, changes in till composition in vertical sections must be examined to demonstrate changes in till provenance.

In the seven localities, tills were sampled when measurements of fabric were made and composition of samples is given in Tables 1 and 2. Near its southeast terminus, the Burroughs Glacier flows from bedrock which is predominantly diorite, to metasedimentary rocks. Acidic and basic dikes cut both lithologies. Although changes in till composition cannot be related to specific areas of bedrock up-ice stream, significant changes do take place. There is not, however, a progressive change in composition that parallels the change in fabric direction suggesting that bedrock contacts beneath the ice and up-ice from this location are complex.

Standard hydrometer procedures were used to determine the percentage less than 2μ of till samples (American Society for Testing and

Materials, 1958). Sand fractions were segregated by dry sieving after the total sand fraction had been separated by wet sieving. The percentage of metasiltstone fragments in the heavy mineral fraction of the fine and very fine sand fraction of a number of till samples was examined. By comparing these results with pebble lithology counts, it is clear that the percentage of metasiltstone fragments is a good indicator of the amount of locally derived metasiltstone in the till (Mickelson, 1971).

Although limestone was found in one small outcrop, small percentages of carbonates were found in all till samples. The Chittick apparatus (Dreimanis, 1962) was used to analyze carbonate content in samples of the <2 mm fraction of the till. Leaching is probably negligible except right at the surface because of the recent age of the tills (Ugolini, 1966); therefore, differences in carbonate content probably represent differences in till provenance. The highest percentages found are located down-ice from Nunatak E, where the small limestone outcrop was observed.

Clay minerals were identified by X-ray diffraction of (1) air-dried, (2) ethylene-glycolated, (3) heated (at 400°C), and (4) heated (at 550°C) samples of the <2-μ fraction. The clays were first saturated with magnesium, resuspended, and then plated in <1-mm-thick

113

TABLE 1
Composition and fabric data from seven sections of till

Location[a]	Sample number	Depth of till (cm)	% Total sample <2mm	<2mm fraction % Sand	% Silt	% Clay	% Metasiltstone Rock fragments	% Calcite	% Dolomite	% Total carbonate	Depth of fabric (cm)	Mode of azimuths (degrees)	Vector mean (azimuth)	Standard deviation of vector mean
North Side Nunatak A	124	0-25	83	46	44	9	33	1.6	0.6	2.2	0-25	330	330	5.1
	123	25-65	70	40	45	15	46	1.1	1.2	2.3	25-65	321	319	3.9
	122	65-80	83	40	50	10	41	0.8	0.2	1.0	65-80	301	302	4.4
North side Nunatak E	140	0-20	67	57	35	8	25	1.6	0.5	2.1	5-10	9	18	8.1
											10-20	9	15	8.9
	141	20- ?	28	48	43	9	24	1.4	0.4	1.8	25-40	319	315	10.4
West side Nunatak C	108	0-30	61	55	39	7	20	0.9	0.3	1.2	0-10	45	240	6.7
											10-30	320	326	6.4
	107	30- ?	87	52	42	6	13	0.9	0.3	1.2	45-70	290	289	5.6
West side Nunatak B	46	0-35	88	41	51	9	40	2.0	0.3	2.3	0-35	322	329	6.0
	48	35- ?	74	46	45	9	38	2.0	0.8	2.8	50-65	292	293	4.2
East Side Nunatak D	144	0-15	70	53	43	5	29	0.9	0.3	1.2	0-10	7	3	4.4
	143	15-40	68	60	34	6	44	2.3	0.8	3.1	10-25	350	350	4.6
West side Nunatak D	50	0-40	61	57	37	6	33	1.8	0.8	2.6	0-40	335	341	4.3
	51	40- ?	74	56	39	5	22	2.6	0.3	2.9	40-60	320	324	5.4
Between Nunataks D and E	101	0-10	58	63	31	6		5.5	0.8	6.3	0-10	0	350	6.0
	102	10-80	40	73	23	5		7.5	1.4	8.9	10-100	330	328	7.6

[a]See Figure 2 for locations.

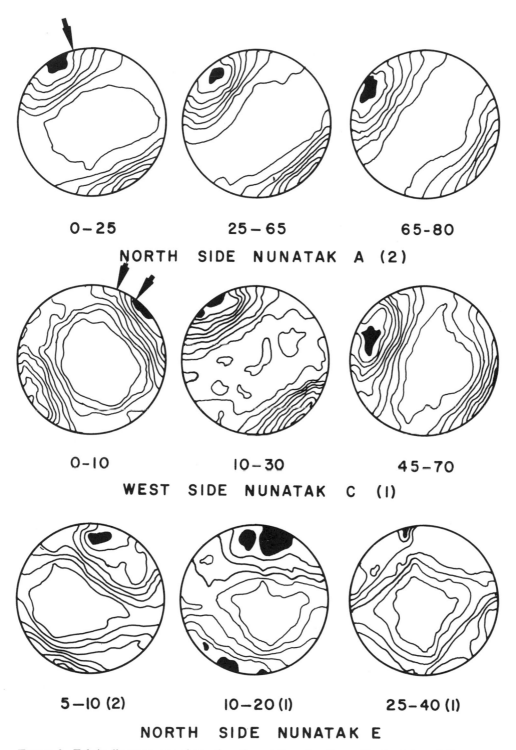

0-25   25-65   65-80

NORTH SIDE NUNATAK A (2)

0-10   10-30   45-70

WEST SIDE NUNATAK C (1)

5-10 (2)   10-20 (1)   25-40 (1)

NORTH SIDE NUNATAK E

FIGURE 3. Fabric diagrams one above the other at three locations. North at top of diagrams. Arrows show orientation of crag and tail feature at surface. Contour interval (number of standard deviations) is given in parentheses. Depth below surface (cm) is given below each diagram.

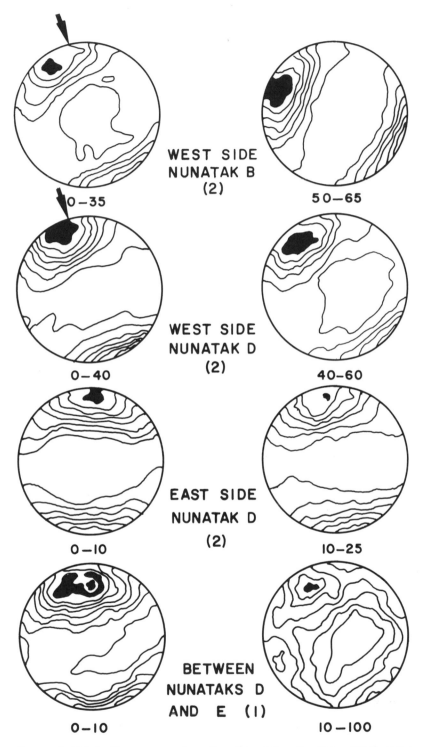

FIGURE 4. Fabric diagrams one above the other at four locations. North at top of diagrams. Arrows show orientation of crag and tail feature at surface. Contour interval (number of standard deviations) is given in parentheses. Depth below surface (cm) is given below each diagram.

TABLE 2

Relative clay mineral abundances in three sections of Glacier Bay till[a]

Locatiton	Sample Number	% Illite	% Vermiculite	% Quartz	% Chlorite	% Expandable
North side	124	T[b]	40	25	35	T
Number A	123	T	T	10	65	20
	122	10	35	15	35	T
West side	46	25	10	20	35	10
Nunatak B	48	T	T	40	60	T
West side	50	15	15	30	35	5
Nunatak D	51	10	T	5	75	10

[a]For depth and relation to fabric diagrams see Table 1.
[b]Trace.

layers on ceramic plates by applying a vacuum to the base of the plate and allowing water from the suspension to be drawn through. Semi-quantitative estimates of abundances are based on peak areas after the method of Johns *et al.* (1954) modified by Wilding and Drees (pers. comm., 1970).

In calculating abundances, montmorillonite was included with other expandable clays. Since glycolation produced a broadening of the 14 A peak toward 16 A, the montmorillonite may be interlayered with chlorite. Fairly large feldspar peaks were present for the clay fraction; however, the feldspars, calcite, dolomite, and amphiboles were not considered in the calculations of relative abundance.

The samples are all characterized by large amounts of chlorite and in some cases, quartz, relative to the other minerals considered. Relative abundances vary from those of McKenzie (1970) in that tills in Adams Inlet average 50% illite. O'Brien and Burrell (1970) report similar differences between the Muir Inlet area and areas to the east in relative amounts of illite and chlorite.

One sample of metasiltstone, the predominant rock type at the southeast end of the Burroughs Glacier, was ground and analyzed by X-ray diffraction. A relatively high mica peak (10 A) was present but the amount of illite in the till samples shows no relation to the amount of metasediment in pebble counts from the same localities. This suggests that in the short distance the till was transported, little of the mica was ground to the clay fraction, but a large number of bedrock samples would have to be analyzed before this could be demonstrated conclusively.

On the east side of Nunatak D a 15% decrease in metasiltstone fragments occurs in the upper till (Table 1). This may reflect a more eastward origin (over Nunatak A which is composed of diorite) of the upper till than the lower till, which would have been carried over metasiltstone after leaving the Bruce Hills. Till west of Nunatak C shows a decrease in metasiltstone fragments with depth (Table 1). This might be expected because the ice that deposited the lower till presumably flowed over less metasiltstone than the ice that deposited the upper till. Changes in the amounts of clay minerals in other samples are also significant but these have not been related to bedrock. Although the changes in compositional parameters are not progressive with depth in the till, they must indicate that till was being actively carried and deposited while the change in ice flow direction was taking place.

RATE AND MODE OF TILL DEPOSITION

If it is accepted that the till compositions listed in Tables 1 and 2 do represent real changes in till provenance accompanying the change in ice flow direction, then rates of basal till deposition can be estimated. Ice flow direc-

tions for 1960, 1948, 1941, 1907, and 1892 were estimated by measuring the direction normal to the ice surface contour over each location on published maps. Oblique aerial photographs were used to estimate the direction of ice-surface slope in 1929. These values were then plotted against time for each location (Figure 5). The directions are estimated to be correct to within ±10°.

Since fabrics were measured on 5-to-40-cm-thick till units, the fabric direction (mode in Table 1) is taken to be the fabric direction (thus the ice-flow direction) at the midpoint of each till unit. Since the ice-movement direction for any year is known, the "year" of deposition can be plotted against depth for each location (Figure 5). For example, on Nunatak A the upper till fabric was measured between 0 and 25 cm depth in the till. It's azimuth mode is 330° (Table 1). In constructing diagrams shown in Figure 5 it was assumed that ice flowed from azimuth 330° when the till 12.5 cm beneath the surface was being deposited. It follows (Figure 5) that this till was deposited in about 1948. The same procedure was used for other till fabric measurements. The slopes of straight lines through the points obtained by the above procedure yield rates of deposition ranging from 0.5 to 3.2 cm year^{-1}. The number of points is probably not sufficient to conclusively demonstrate changes in rate during this short time span but the curves suggest that till deposition began quite late in Neoglacial time and had ceased or slowed down when the ice was about 70 to 100 m thick. Since the tills can in some cases be differentiated as distinct units, deposition of the till may not have been continuous and some subglacial erosion may have taken place.

The significance of the rate of deposition is not readily apparent. Basal melting due to geothermal heat is sufficient to melt about 0.5 cm of ice a year. Heat added by friction is probably negligible at flow rates measured in 1960 (< 5 m year^{-1} in this area) but if higher flow rates are assumed to have been prevalent in the past, the total basal melting of ice might be on the order of 1 cm year^{-1}. If the ice-till mixture at the base of the ice was 25% ice, then 2 to 4 cm of ice-till mixture or 1.5 to 3 cm of till would be expected to have melted out each year. Thus basal melting might account for the till deposition.

Depending on the amount of water present in the till at the time of deposition, some reorientation of the till which had melted out might be expected. This does not seem to have happened except on the north side of Nunatak E, where the rate of deposition over one interval is very high (Figure 5).

A second possibility is that the calculated rate of till deposition represents the stagnation of a basal till-ice mixture under flowing ice as suggested by Boulton (1970). In this case a thin, stagnant ice-till mass would separate the overlying active ice from the till-water mixture being formed by basal melting, preventing reorientation of fabric.

Although till was observed in only one location under the Burroughs Glacier it tends to support the above hypothesis. One meter of frozen till (estimated 25 to 40% ice) was examined between Nunataks D and E under 5 m of relatively clear ice. Fabric diagrams indicate that the top 20 cm of till which is lighter gray does show a different flow direction than the till below. Thus the lower till is considered to be in place when determining the rate of deposition even though it is still frozen. If the frozen till is 25% ice, then the actual rate of deposition calculated after melting would be 2.3 cm year^{-1}.

If this stagnation of frozen till took place earlier or at a greater rate on the stoss side of bedrock knobs as described by Boulton (1970), then it might explain the apparently thick Neoglacial till in these positions at both ends of the Burroughs Glacier. At the southeast end, these till deposits indicate flow from the west-northwest. A similar flow direction is indicated by the till build-up on one hill at the terminus of the Cushing Glacier (Figure 1). However, at the western terminus of the Burroughs Glacier, in the same valley only 1 km southeast of Cushing Glacier, the till deposits suggest flow from the east. Thus, most till deposition must have taken place after the Burroughs reversed its flow direction shortly before 1892 (Mickelson, 1971).

That most basal till deposition took place in the late stages of deglaciation is supported by comparison of the calculated rates with that obtained by dividing the average till thickness in the area by the number of years the area was ice covered. Ice advanced over trees on the north side of Nunatak A about 2,520 ± 87 BP (OWU-489). Thus the area was ice covered for approximately 2,500 years. Although measurements were not taken for this purpose, the average till thickness in the area of the present terminus is estimated to be 5 m. The apparent rate of till deposition is thus about 0.2 cm year^{-1}, an order of magnitude lower than that calculated above.

118

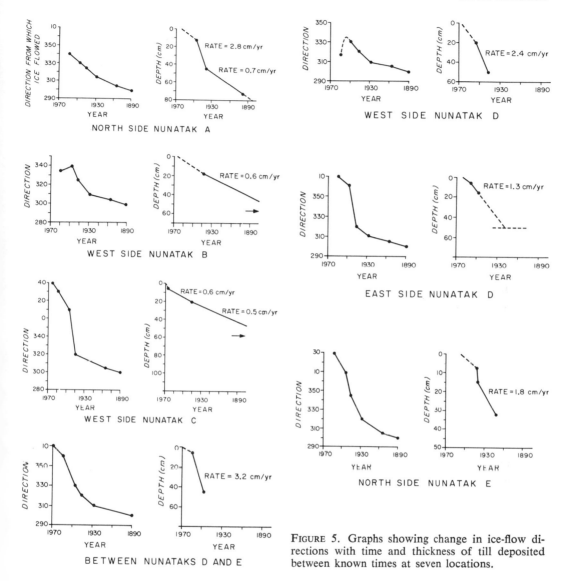

FIGURE 5. Graphs showing change in ice-flow directions with time and thickness of till deposited between known times at seven locations.

SUMMARY

Measured rates of basal till deposition range from 0.5 to 2.5 cm year^{-1} near the southeast terminus of Burroughs Glacier. This value is an order of magnitude higher than the rate which is obtained by estimating average till thickness in the area and dividing by the number of years the area was ice covered.

At one location the fabric mode azimuths of clasts in a till-ice mixture beneath the ice change with depth. This suggests that till deposition may take place by basal stagnation of a till-ice mixture as the ice thins.

ACKNOWLEDGMENTS

The author wishes to thank R. P. Goldthwait for valuable ideas and criticism. The study was supported by National Science Foundation Grant GA-12300, administered by the Institute of Polar Studies at Ohio State University.

119

REFERENCES

American Society for Testing and Materials
1958 : Grain-size analysis of soils. *In: Procedures for Testing Soils,* 83-94.

Andrew, J. T. and Smith, I. D.
1970 : Till fabric analysis: Methodology and local and regional variability (with particular reference to the North Yorkshire till cliffs). *Quart. J. Geol. Soc. London,* 125: 503-542.

Boulton, G. S.
1970 : On the deposition of subglacial and melt-out tills at the margins of certain Svalbard Glaciers. *J. Glaciol.,* 9: 231-245.

Dreimanis, A.
1962 : Quantitative gasometric determinations of calcite and dolomite by using Chittick apparatus. *J. Sed. Petrol.,* 29: 459-463.

Field, W. O.
1959 : Notes on the recession of Plateau and Burroughs Glaciers, Glacier Bay, Alaska. Unpublished Ms. and photographs on file with the American Geographical Society, New York.

Goldthwait, R. P.
1966 : Glacial history. *In* Mirsky, A., (ed.) *Soil Development and Ecological Succession in a Deglaciated Area of Muir Inlet, Southeast Alaska. The Ohio State Univ. Inst. Polar Stud. Rep.,* 20: 1-18.

Johns, W. D., Grim, R. E., and Bradley, W. F.
1954 : Quantitative estimations of clay minerals by diffraction methods. *J. Sed. Petrol.,* 24: 242-251.

Kamb, W. D.
1959 : Ice petrofabric observations from Blue Glacier, Washington, in relation to theory and experiment. *J. Geophys. Res.,* 64: 1891-1909.

MacClintock, P. and Dreimanis, A.
1964 : Reorientation of till fabric by overriding glacier in the St. Lawrence Valley. *Amer. J. Sci.,* 262: 133-142.

McKenzie, G. D.
1968 : Glacial history of Adams Inlet, southeast Alaska. The Ohio State University, Ph.D. dissertation. 200 pp.
1970 : Glacial geology of Adams Inlet, southeastern Alaska. *The Ohio State Univ. Inst. Polar Stud. Rep.* 25. 121 pp.

Mickelson, D. M.
1971 : Glacial geology of the Burroughs Glacier area. *The Ohio State Univ. Inst. Polar Stud. Rep.* 40. 149 pp.

O'Brien, N. R. and Burrell, D. C.
1970 : Mineralogy and distribution of clay size sediment in Glacier Bay. *J. Sed. Petrol.,* 40: 650-655.

Reid, H. F.
1896 : Glacier Bay and its glaciers. *U.S. Geol. Surv. Ann. Rep.,* 16, 1894-1895, pt. 1: 415-461.

Taylor, L. D.
1963 : Structure and fabric on the Burroughs Glacier, southeast Alaska. *J. Glaciol.,* 4: 731-751.

Ugolini, F.
1966 : Soils. *In* Mirsky, A., (ed.) *Soil Development and Ecological Succession in a Deglaciated Area of Muir Inlet, Southeast Alaska. The Ohio State Univ. Inst. Polar Stud. Rep.,* 20: 29-72.

Young, J. A. T.
1969 : Variations in till macrofabric over very short distances. *Bull. Geol. Soc. Amer.,* 80: 2343-2352.

Ms submitted May 1972

6

Glaciotectonic Structures in Drift

Stephen R. Moran

ABSTRACT

Glaciotectonic structures of three types are present throughout the glaciated portion of North America.

"Simple *in situ* deformation" involves small-scale folds and faults produced by ice push and bed shear in both bedrock and drift. These structures have long been used to reconstruct direction of ice advances.

"Large-scale block inclusion" is the incorporation of large intact blocks of bedrock and drift into younger drift. The incorporated blocks were sheared from bed material having low shear strength as a result of groundwater discharge in response to high heads produced by the glacier. Thick permafrost may have helped to reduce shear strength locally. Shearing occurred most readily where the compressive-flow regime was favored, as near ice margins or where the ice flowed up scarps. Failure to recognize repetition of glacial sequences and the occurrence of erratic blocks of older drift or bedrock in younger drift will lead to major errors in interpreting glacial sequences. Ridges produced by large-scale block inclusion may be erroneously interpreted as end moraines.

"Transportational stacking within single till sheets" by sporadic differential movement along shear planes in the debris-charged basal zone of the ice may produce disturbed sequences. Continual deposition and re-erosion along different planes of failure cause the original sequence within the till to be shuffled, so that slices from the base of the drift-rich zone may be deposited anywhere throughout the till sheet. This process may produce, by faulting and stacking in one till sheet during deposition, complex patterns resembling normal sequences of several till sheets.

INTRODUCTION

Observations in three areas that are widely separated, both geographically and in type of glacial processes and features, indicate that structural deformation of both bedrock and glacial drift is common. Deformed drift and bedrock have been observed in many places in exposures along strip-mine highwalls in the glaciated Allegheny Plateau of eastern Ohio and northwestern Pennsylvania. Numerous examples of such deformation have been seen in exposures in Illinois and Indiana. Repeated examination of active strip mines near Danville, Illinois, as stripping proceeded from 1965 to 1969, has revealed several examples of thrust faulting in both bedrock and drift. Studies of surface exposures in the Hudson Bay Area (63 D and C) in east-central Saskatchewan have revealed other examples of deformed bedrock and drift (Moran, 1969). Test drilling by the Saskatchewan Research Council in the Hudson Bay Area encountered two major glacio-tectonic structures.

For purposes of discussion, glaciotectonic structures have been subdivided into three classes on the basis of the type and magnitude of deformation: (1) "simple *in situ* deformation," (2) "large-scale block inclusion," and (3) "transportational stacking within single till sheets."

SIMPLE IN SITU DEFORMATION

Description

The first class of deformation involves bedrock or glacial drift which has been contorted by ice push or bed shear into folds and faults with only minor displacements. Structures in this class, which are generally small, consist of simple folds in till, stratified drift, and bedrock, and faults of visibly small displacement. The attitude of beds prior to deformation can be readily reconstructed because of the simplicity of the structures or proximity to the site of *décollement*. Deformation of this type is generally of small enough scale that the entire structure is visible in a single outcrop. This type of deformation has been recognized for many years, because of its generally small-scale proportions and because it is common near the bedrock-drift contact where contrasting lithologies are present to delineate structures. Most of the common reported occurrences of glacial deformation are included in this class.

Around the turn of the century, F. W. Sardeson (1905, 1906) described several folds and faults in dolomite and shale in and around Minneapolis which he attributed to glacial disturbance. He demonstrated that these structures were not the result of frost heave, as had been claimed by some, but were the result of horizontal compression associated with Pleistocene glaciation.

Bluemle (1966) described blocks of shale sheared up into till along thrust faults in Cavalier County, North Dakota. T. C. Brown (1933) reported minor structures associated with glacial deformation of an over-ridden esker in Massachusetts. Deformed Pennsylvanian bedrock in Iowa and Kansas had been considered to be the result of ice push by Lammerson and Dellwig (1957) and by Dellwig and Baldwin (1965).

In the quarry of the Carbon Limestone Company about five miles southeast of Youngstown, Ohio, near Hillsville, Pennsylvania, minor folding of shale just below the bedrock surface was observed by the author. These folds, which die out a few feet below the base of the drift, are the result of simple *in situ* deformation by subglacial shear. In a strip mine along the Pennsylvania Turnpike one mile northeast of New Galilee, Beaver County, Pennsylvania, a coal bed up to one foot thick and approximately one hundred feet long occurs at the base of the Titusville Till overlying strongly oxidized glacial gravel. Other occurrences of coal beds sheared up into till were observed elsewhere in the region. Along the east wall of the quarry of the New Castle Limestone Company, two miles northeast of Lowellville, Mahoning County, Ohio, a four-foot-thick bed of Vanport Limestone (Pennsylvanian) overlies six inches of Titusville Till for a horizontal distance of several hundred feet. The limestone was incorporated into the basal drift of the advancing glacier and transported intact along a horizontal shear plane. Two well-developed folds exposed in the upper part of the Titusville Till on the same wall were formed during transportation of the till. On the west wall of the north pit of the same quarry, the upper part of the Titusville Till is repeated along two low-angle thrust faults. The two- to four-foot-thick thrust slices are clearly marked by the threefold repetition of brown oxidized till overlying gray unoxidized till.

Additional evidence of this type of simple *in situ* deformation was seen

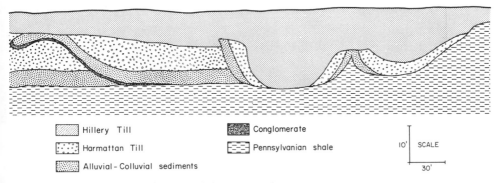

Figure 1. Glaciotectonic structures exposed in Harmatton strip mine near Danville, Illinois, June, 1969.

in strip mines west of Danville, Illinois. Evidence of movement along bedding planes in Pennsylvanian shales below the lowermost till was commonly observed. As stripping proceeded from 1965 to 1969, continually exposing new highwall, thrust faults carrying gray shale up into reddish-pink till or carrying the reddish-pink till up into the gray till overlying it have been observed. Along a fault exposed in 1966, red till could be traced for a horizontal distance of 100 feet and for a vertical distance of up to 15 feet above the base of the overlying gray till. Figure 1 shows a structure exposed in the mine during the spring of 1969.

Mechanism of Formation

Many writers have attributed glacial structures of the simple *in situ* deformation type to ice push or shearing along the bed of the ice. The author agrees that these small-scale structures were in fact probably caused by the normal stresses exerted on the upstream faces of protuberances from the bed (ice push), shear exerted by moving ice along its bed, or a combination of both.

Significance

Small-scale *in situ* deformation of both bedrock and drift has long been used as evidence of post-depositional overriding by glaciers. The direction of overturning of folds and of movement of thrust blocks has been used as an indicator of the direction of glacial flow. Because structures in this class are small and usually marked by beds of distinctive lithology, they are readily recognized and therefore generally do not pose significant problems in interpretation. A small mass of bedrock whose base is not exposed in shallow exposures might be interpreted as being *in situ*, while in fact it overlies drift, thereby causing an erroneous concept of drift thickness and makeup. Other than local cases such as this, these structures are not significant for stratigraphic interpretation.

LARGE-SCALE BLOCK INCLUSION

Description

The second class of glacially derived structures involves incorporation of large masses of bedrock or pre-existing glacial drift into the ice and transportation of the intact block away from the site of *décollement*. It is evident that any boundary between simple *in situ* deformation and large-scale block inclusion is placed arbitrarily, because both the scale of deformation and distance of transport are gradational from one class to the other. In general, if the original location and attitude of the block involved cannot be reconstructed, the deformation is considered to belong in the second class. Because of their very different stratigraphic implications, structures in this class of deformation are divided on the basis of lithology into two subgroups: bedrock structures and glacial drift structures.

Like simple *in situ* deformation, bedrock structures in this group have been recognized for many years. Sardeson interpreted several large anomalous outliers of Cretaceous rock in Minnesota as being blocks transported intact by glacier ice (Sardeson, 1898). Wolford (1932, p. 362-67) reported a large erratic block of limestone, several acres in area, buried in the drift

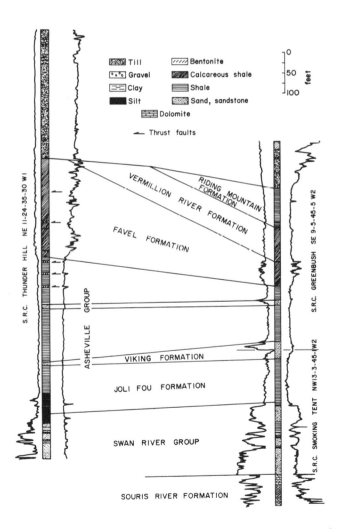

Figure 2. Repetition of sequence in Upper Cretaceous sediments resulting from large-scale block inclusion, Thunder Hill, eastern Saskatchewan.

125

in central Ohio. Wickenden (1945, p. 75-76) interpreted anomalous structures down to a depth of several hundred feet in a well on Thunder Hill in
east-central Saskatchewan as the result of glacial deformation. Rutten
(1960) and Brinkmann (1953) discuss the origin of glacially derived bedrock structures in northern Europe. Kupsch (1962) and Byers (1960) described contorted bedrock along the Missouri Coteau of Saskatchewan and
attributed the deformation to glacial action. Hopkins (1923) concluded
that the faults, folds, and steeply dipping rocks exposed in the Mud Buttes
and Tit Hills of Alberta were the result of glacial deformation. Slater
recorded numerous ice-thrust bedrock structures in Britain, Europe, and
North America (Slater, 1926, 1927a, 1927b, 1927c, 1927d, 1927e, and 1929).

In the Hudson Bay Area of east-central Saskatchewan, a testhole drilled
by the Saskatchewan Research Council on Thunder Hill (NE 11-24-35-30
W1)[1] encountered a complexly faulted sequence consisting of at least six
overthrust slices of Cretaceous shales with Pleistocene gravels and till (?)
included along at least one of the faults (Figs. 2 and 3). Another SRC testhole near Steen, Saskatchewan, encountered 18 feet of glacial drift over

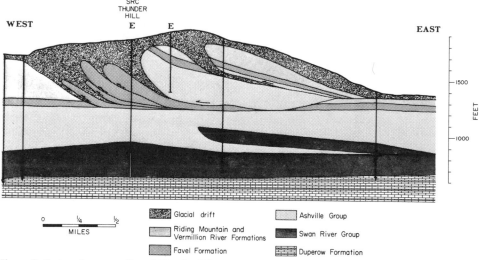

Figure 3. East-west cross section of Thunder Hill, eastern Saskatchewan, showing large-scale block inclusion
of Cretaceous sediments. "E" denotes interpretation based on electric logs.

113 feet of clayey silt of the Riding Mountain Formation (Cretaceous),
which in turn overlay five feet of till resting on Riding Mountain Formation
in situ (Fig. 4). This large block of shale incorporated into the drift is
expressed at the surface as a series of isolated hills rising above a flat to
gently rolling lake plain.

1. For an explanation of the National Topographic System of land location used in Canada,
see Christiansen, 1971; this volume, p. 169, Fig. 2.

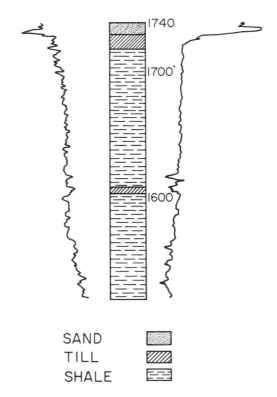

SAND

TILL

SHALE

Figure 4. 110-foot-thick block of shale underlain by till, encountered in a Saskatchewan Research Council testhole near Steen, Saskatchewan (NW 15-31-42-11 W2).

Large-scale block inclusion of glacial drift has not been generally recognized, but as more detailed stratigraphic information becomes available, more and more instances of this type of structural deformation are being reported. Repetition of the same peat bed and of till sequences under end moraines has been observed in borings in central Illinois (J. P. Kempton, personal communication, 1968).

In Saskatchewan, testholes drilled in areas of hummocky moraine often encounter four or five oxidized zones, whereas testholes in adjacent areas encounter only one, or at most two, such weathered zones (E. A. Christiansen, personal communication, 1969). Large-scale deformation involving large blocks of drift has been encountered in several locations throughout Saskatchewan, particularly along the Missouri Coteau (E. A. Christiansen, W. A. Meneley, S. H. Whitaker, personal communications, 1969). These blocks of drift commonly rest on younger drift, causing repetition of sequence, and, in a few cases, are completely inverted so that progressively younger drift is encountered with depth.

Mechanism of Formation

The origin of large-scale block inclusion structures has been the subject of considerable debate. Most of those who have discussed the problem have considered it necessary that the material be frozen in order to be incorporated (e.g., Mathews and MacKay, 1960; Kupsch, 1962; Rutten, 1960). Undoubtedly, in some instances, the included blocks have been frozen at the time of their incorporation, but this is not a necessary prerequisite for incorporation if the porewater pressure in the subjacent beds was sufficiently high.

The shear strength of a soil material is directly related to the effective normal stress on the soil. The effective normal stress is defined as the total normal stress minus the porewater pressure (Terzaghi and Peck, 1967).

$$\bar{\sigma} = \sigma - U, \qquad \text{where: } \sigma = \text{total normal stress}$$
$$\bar{\sigma} = \text{effective normal stress}$$
$$U = \text{porewater pressure}$$

It follows that if the porewater pressure is increased an amount greater than any accompanying increase in the total normal stress (the total load), then the effective normal stress will be decreased. This decrease in effective normal stress produces a corresponding decrease in the shear strength of the material. Under certain conditions, it is possible for the porewater pressure to be sufficiently increased to permit shear failures in materials overridden by glaciers. Two general sets of conditions must be met to permit the incorporation of large intact blocks into an advancing glacier. First, the soil or rocks which are overridden must contain beds of permeable material confined by less permeable material. In most instances, the permeability discontinuities also coincide with changes in strength of the materials. Secondly, the ground-water flow must be modified so that water moves into these confined beds, generating elevated porewater pressures.

In addition to the geohydraulic conditions, the incorporation of large blocks of soil or rock into glaciers requires conditions of compressive flow within the glacier itself. Under the compressive-flow regime, the ice is moving, either by plastic deformation or by shear, upward from the base. This flow regime is favored in regions where ablation is in excess of accumulation and where the bed of the glacier is concave upward, as where the ice flows up over an obstruction on its bed. Under such conditions, large blocks of material sheared into the ice as a result of the decreased effective stress are carried up into the glacier and may be transported some distance essentially intact.

One or more geologic situations may have existed to generate the necessary confined conditions to produce large-scale block inclusion in any given instance. The common interbedding of sands or sandstones with clays, silts, or shales in bedrock and glacial sequences produced the necessary stratigraphic situation for this type of failure. Where the permeable beds in such a sequence were confined laterally, the migration of ground water was prevented, and porewater pressures built up. Facies changes from sand to clay in either bedrock or glacial sequences would cause such confinement. Truncation of a permeable bed by an erosional unconformity over which an impermeable material was deposited would also generate such a condition. Where the permeable beds were cut into less permeable material, as in a gravel valley-fill cut into older shales or clays, the valley sides would act as the obstruction to lateral ground-water migration. The presence of ground ice or permafrost in the outcrop area of a permeable bed might prove sufficiently confining to permit the development of high porewater pressures.

Several different mechanisms probably operated to produce the ground-water flow necessary to create the elevated porewater pressures. As was the case with the geologic controls, a given case of deformation may have resulted from the operation of one or more of these mechanisms.

The occurrence of continental glaciers in low areas caused the discharge areas of large, regional-scale, ground-water flow systems to be blocked. The directions of ground-water flow were modified or reversed, producing new flow systems which discharged beyond the limits of the glacier. Where such a modified flow system interacted with one of the geologic situations described above, abnormally high porewater pressures may have resulted. The operation of this mechanism has not yet been fully investigated and therefore little is known of the magnitude of porewater pressure which can be developed.

Generation of elevated porewater pressures by the rapid formation of a permafrost layer at the time of glaciation is discussed by Mathews and MacKay (1960). It is not clear whether such a permafrost layer can form sufficiently rapidly to prevent porewater pressure conditions from coming to equilibrium with each successive increase in permafrost thickness. If such rapid formation of permafrost is in fact possible, then this mechanism may have played a significant role in the formation of large-scale block-inclusion structures.

Where a glacier overrode unfrozen cohesive sediments, the load of the glacier would tend to produce consolidation of the sediments. If drainage from the clay or silts into sands or gravels was possible, the cohesive sediments would consolidate and the water forced into the permeable beds would flow toward areas of lower head, near the margin of the ice sheet. Where the permeable beds were confined as described above, such consolidation of cohesive sediments back under the ice could produce sufficiently elevated porewater pressure nearer the margin of the glacier to permit the incorporation of large blocks of material (Fig. 5).

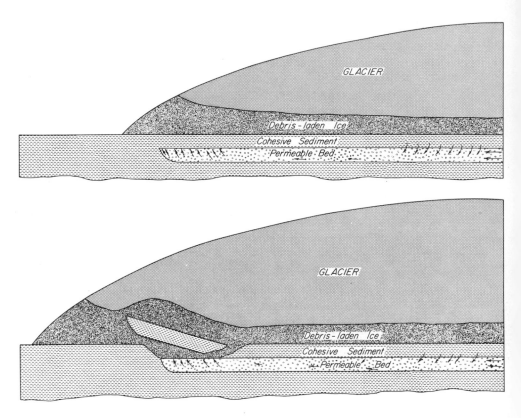

Figure 5. Schematic drawing of large-scale block inclusion resulting from elevated pore-water pressures.

High porewater pressures would also be generated where a glacier advanced over debris containing buried ice blocks remaining from a former glacier. The ice blocks would serve as source layers of water as they melted, maintaining anomalously high porewater pressures in the manner described by Hanshaw and Bredehoeft (1968). The high fluid pressures thus generated would cause a zone of weakness at the interface between the buried ice and the overlying debris. The shear stress resulting from the overriding glacier could cause failure along this surface, thereby incorporating the entire debris layer into the younger glacier.

Significance

Large-scale block inclusions of both bedrock and drift material can have profound effects on the interpretation of glacial stratigraphy. Failure to recognize the presence of large blocks of bedrock, some of which may be over one hundred feet thick, as in SRC Steen (NW 15-31-42-11 W2), within a drift sequence can lead to very great errors in determination of drift thickness, as well as of the configuration of the bedrock surface. For this reason, anomolously high bedrock elevations should be suspect.

Large-scale block inclusion of drift poses especially serious problems, because it is usually very difficult to detect. In the absence of detailed knowledge of the regional stratigraphy it may be impossible to recognize the presence of such blocks. Because blocks of drift carried upward in this manner can be redeposited above younger drift, it is immediately obvious that considerable confusion may occur as the result of the presence of older drift and of weathering zones occurring above younger drift. In the absence of knowledge of regional lithostratigraphy, studies of small areas must be undertaken with great care, and correlations based solely on the occurrence of a paleosol without knowledge of the stratigraphic framework involved are questionable.

Large-scale block inclusion has probably played a major role in the formation of many if not most of the classical end moraines of midwestern United States. Because many of the natural and man-made exposures throughout the midwest are located in morainic belts, the recognition of the presence of large-scale block inclusion of drift becomes very important in the correct interpretation of these outcrops. The land forms mapped throughout the midwest as end moraines can be separated into three types: true end moraines, moraines resulting from bed configuration, and moraines resulting from hydrodynamics in the bed.

True End Moraines

The true end moraines include both terminal and recessional moraines which formed at or very near the margin of an ice sheet. Large-scale block inclusion probably played a major role in the formation of many true end moraines. Rapid changes in porewater pressure resulting from the rapid expansion of a glacier over ice-free marginal areas would favor the incorporation of large blocks by one of the mechanisms described above. In areas where the stress environment in the ice favored compressive flow (Nye, 1952, p. 89), large blocks incorporated into the ice by any of these three processes would be carried upward in the ice and stacked. Optimum conditions favoring compressive flow occur near the ice margin in the zone of ablation where the bed is concave upward.

The combination of these two processes at or near glacial margins — the common occurrence of large-scale block inclusion and the compressive flow regime to lift and stack the blocks — probably accounts for a large proportion of the volume of most true end moraines. Because this block stacking may have involved glacial drift and/or bedrock material of varying ages, great care must be taken in interpreting glacial sequence within true end moraines.

Moraines Resulting from Bed Configuration

In areas where the bed of the glacier was concave upward, such as where the ice flowed up a cuesta scarp, the compressive-flow regime was favored, and large-scale block inclusion would be expected. In such areas, a thicker deposit consisting of large blocks as well as smaller debris would be formed by the resultant stacking. Because the state of stress over this slope or obstruction continued to favor compressive flow as long as the ice continued to advance, this greater accumulation of material persisted after the margin had passed the area. When ablation occurred, the thicker debris would be let down to form a ridge or moraine which was the result of bed configuration (Fig. 6). The coincidence of several moraines in Illinois with bedrock

A. Compressive Flow

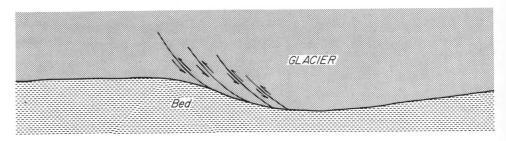

B. Imbrication of Slabs of Bed Material

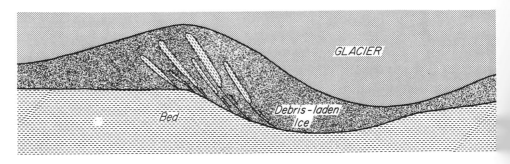

C. Ridge after Ablation of the Glacier

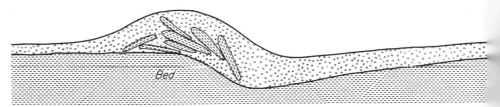

Figure 6. Formation of a "moraine" ridge by thrust faulting in a zone of compressive flow.

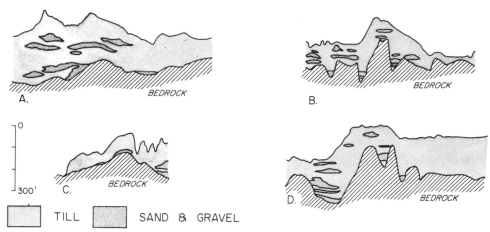

TILL   SAND & GRAVEL

Figure 7. Examples of "end moraines" in Illinois, associated with large-scale bedrock obstructions. (Modified from Piskin and Bergstrom, 1967, Plate 2).

uplands or scarps suggests that they may have formed in this manner (Fig. 7).

Moraines Resulting from Hydrodynamics of Bed

If, for one of the reasons discussed above, the hydraulic uplift in an area under a glacier should become greater than the superincumbant load of ice and rock debris, the material would be lifted up into the ice. Such a condition would continue to produce large-scale blocks until the uplift pressure

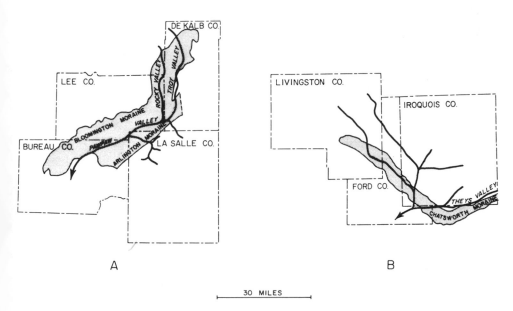

30 MILES

Figure 8. Examples of moraines in Illinois associated with buried bedrock valleys.

133

became sufficiently reduced. When ablation occurred, a ridge would result in the same manner as in the case above. The coincidence of a number of moraines in Illinois with large bedrock valleys suggests the operation of such a mechanism (Fig. 8).

TRANSPORTATIONAL STACKING WITHIN A SINGLE TILL SHEET

Description

The third class of structural deformation involves differential movement and overthrusting of elements within a single till sheet while it is being transported and deposited. Although the existence of this type of deformation has not been previously demonstrated, two authors have suggested such a mechanism to explain observed characteristics of tills.

From his studies of the fabric of clay till, Harrison (1957, p. 300-301) concluded the debris-laden zone at the base of a glacier advanced by often sporadic but generally continuous differential movement along more or less horizontal shear planes. He stressed that when friction along a particular plane became too great because of the abundance of debris, movement ceased and temporary deposition occurred. Renewed movement could begin on any of the shear planes within the stationary debris-laden ice mass when friction along that plane was overcome by greater shear stress exerted on it.

Virkkala (1952) reported the presence of thin partings within till sequences in Finland which he referred to as bed limits. The bed limits were finer textured than the surrounding till and were generally contorted. In all sections containing more than one bed limit, "the deformation of the bed limits increases downward" (Virkkala, 1952, p. 105). Virkkala (1952, p. 107-9) concluded that the bed limits reflected the shear planes within englacial drift along which movement had occurred. He suggested a model of the base of a glacier consisting of beds of drift-free ice interbedded with drift-rich ice. Continued concentration of debris in the dirty ice decreased the ability of the ice to flow until movement ceased. Re-erosion of deposited debris would occur, Virkkala suggested, by movement along the already-present shear planes, causing further distortion of the planes (bed limits) each time they were utilized.

In both of these models, the base of the ice was considered to consist of a series of discrete slabs of debris-laden ice moving sporadically along horizontal shear planes. Through the continual process of deposition and re-erosion of material along different shear planes, the sequence of these slices could become thoroughly disordered. Material originally located at the base of the debris-laden zone could be brought fairly readily to the top of the zone by a very few episodes of deposition and re-erosion. The operation of this mechanism should, therefore, result in till deposits in which the original sequence of slices is nearly everywhere out of order.

The author's work in the Allegheny Plateau of Ohio lends support to this hypothesis by demonstrating the presence of stacking of elements within a single till sheet. Gross and Moran (1971; this volume, p. 251) reported a fairly regular decrease in feldspar content with increased depth below the top of the Titusville Till throughout the Plateau. However, only four of the seventeen sections near Youngstown, Ohio, studied by the author (Moran, 1967) display this trend. Most of the remaining sections display the expected gradation, except for major breaks in the sequence at which feldspar content decreased abruptly only to continue to increase above the break (Fig. 9). Although these abrupt changes in feldspar content were originally interpreted as representing contacts between separate till sheets (Moran, 1968), it is now suggested that these breaks indicate the presence of thrust faults along which the sequence has been duplicated. This interpretation is strongly supported by the thickness of the Titusville Till in the two groups of sections. In the four sections which display the normal trend of increasing feldspar content upward, the mean thickness of the Titusville Till is 17.1 feet (13.4 feet, excluding associated stratified drift). In the thirteen sections which are believed to contain one or more thrust faults, the mean thickness of the Titusville Till is 26.25 feet (20.6 feet, excluding associated stratified drift).

The textural composition and/or potassium feldspar content of the till in two sections in the Allegheny Plateau corroborates the interpretation of the breaks in feldspar content trend as thrust faults. In section 80, the lowermost part of the sequence appears to be repeated once (Table 1). Marked similarities occur between samples 83 and 85, and between 84 and 86. Except for the sand content in the second pair, all differences appear to be the result of random variation. Table 2 gives the mean values of the reconstructed sequence, with the influence of the shear plane between samples 84 and 85 removed. Section 71 does not contain any apparent trend in

TABLE 1

COMPOSITION OF TITUSVILLE TILL IN SECTION 80[*]

Sample	% Sand	% Clay	% Feldspar	% K Feldspar
80	33	22	16	42
81	37	24	12	42
82	40	20	15	42
83	39	24	14	56
84	62	13	10	46
85	38	23	14	56
86	47	15	10	44
87	28	28	8	28

[*]Located in an old Bessemer Limestone Company quarry (½ mile south of Ohio Rt. 630, 150 yds. east of State Line Rd., in northern Beaver Twp., Lawrence Co., Pennsylvania).

TABLE 2

RECONSTRUCTION OF SECTION 80*

Sample	% Sand	% Clay	% Feldspar	% K Feldspar
80-82	37	22	14	42
83,85	38	24	14	56
84,86	55	14	10	45
87	28	28	8	28

*For location see note to Table 1.

TABLE 3

COMPOSITION OF TITUSVILLE TILL IN SECTION 71*

Sample	% Sand	% Clay	% Feldspar
71	39	15	16
72	38	15	15
73	39	19	10
74	37	21	13
75	31	22	15
76	39	19	13
77	36	21	12
78	32	22	14

*Located in American Fire Clay Company quarry (NW¼ SE½, Sec. 6, Beaver Twp., Mahoning Co., Ohio).

feldspar composition, but investigation of textural data (Table 3) indicates the possibility of a repetition of the base of the section on a thrust fault between samples 75 and 76. Removal of the effect of this duplication of section produces the reconstructed sequence shown in Table 4. The apparent reversal of trend in feldspar is believed to be the result of error variation and is not significant though it may reflect a reversal of sequence by thrust faulting.

The position of postulated thrust planes in a number of the remaining sections is shown in Figure 9. None of these sections contain additional proof of repetition of particular beds, but the existence of the trends in feldspar content is considered a sufficiently valid basis to interpret these sections in this manner. If the model proposed by Virkkala (1952) and Harrison (1957) is considered, it seems obvious that clear-cut sequences containing only a single shear plane should be the exception rather than the rule. The data themselves are by no means conclusive and could be subjected to alternate interpretations, but in view of the clear evidence of

TABLE 4

RECONSTRUCTION OF SECTION 71*

Sample	% Sand	% Clay	% Feldspar
71-72	38.5	15	15.0
73,76	39.0	19	11.5
74,77	36.5	21	12.5
75,78	31.5	22	14.5

*For location see note to Table 3.

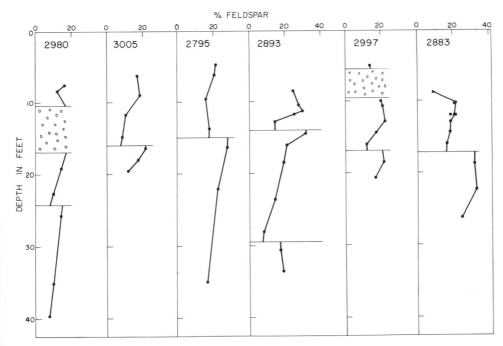

Figure 9. Plot of feldspar content of the Titusville Till (Youngstown, Ohio, area) against depth. Discontinuities in the trend of upward-increasing feldspar content resulting from the presence of thrust faults.

stacking shown in some of the data, the author feels justfiied in interpreting them in this manner.

Evidence for a number of thrust planes within a single till was observed in a section in the strip-mining area west of Danville, Illinois, in SW¼ SW¼ Section 33, T. 20 N., R. 12 W. Table 5 contains textural data from the

TABLE 5

COMPOSITION OF BLOOMINGTON-TYPE TILL, EMERALD POND SECTION*

Sample	% Sand	% Clay	Element
11	40	26	A
12	40	25	A
13	40	21	B
14	26	20	C
15	39	21	B†
16	29	20	C
17	39	20	B
18	39	21	B
19	41	20	B
20	40	21	B
21	27	20	C
22	40	17	B
23	38	19	B
24	26	19	C
25	37	19	B

*Located west of Danville, Ill. (in SW¼ SW¼, Sec. 33, T.20N, R.12W).
†Lines indicate position of thrust faults.

Bloomington-type till (Johnson and others, 1971) in this section, called the Emerald Pond Section. Examination of these data reveals a series of three distinct textural groupings within this till unit. These three groups have sand, silt, and clay percentages of approximately 40-35-25, 40-40-20, and 27-53-20, respectively. The position of the thrust faults necessary to produce the reconstructed sequence shown on Table 6 is indicated on Table 5. Two of these faults were clearly marked by lenticular masses of sand located along the fault plane. As was the case in the sections through the Titusville Till, this section is considerably thicker (18 feet) than are nearby sections of the same unit which exhibit no evidence of stacking (mean of 8 feet for three sections).

TABLE 6

RECONSTRUCTION SEQUENCE OF BLOOMINGTON-TYPE TILL,
EMERALD POND SECTION*

Element	% Sand	% Clay
A(2)	40.0	25.5
B(9)	39.2	19.9
C(4)	27.0	19.8

*For location see note to Table 5.

Mechanism of Formation

Transportational thrust faulting is believed simply to be the product of differential movement along discrete shear planes within the debris-laden basal zone of ice, such as was postulated by Harrison (1957) and by Virkkala (1952). It is generally agreed that the majority of the differential movement occurring within an ice sheet occurs near the base, and it is reasonable to assume that this movement occurs along discrete shear planes in this zone, because the large concentration of debris should tend to decrease other forms of movement within the ice.

Significance

Where vertical compositional changes occur in till sheets, transportational thrust faulting becomes quite significant. As described above, the studies of Gross (1967) and of Gross and Moran (1971) indicate that feldspar content progressively increases toward the top of tills in the Allegheny Plateau of Ohio and Pennsylvania. Gross (personal communication, 1969) found vertical gradations in clay content within single till sheets in northeast Illinois. Elson (1961) suggested that a single ice sheet should deposit four types of till, each of which differs from the others in texture and/or mineralogical composition, as well as in origin: superglacial ablation till, subglacial abla tion till; deformation till, and comminution till. Although he indicated that it is not likely for all four types of till to occur in any single section, several of them may occur together. The four types of till should occur in the same order as listed above if they all occurred together. It is evident, in the cases of both of the above types of compositional change within a single till sheet, that the operation of the Harrison-Virkkala transportation-deposition model may produce complex patterns resembling normal sequences in several till sheets by faulting and stacking during one ice advance. In this case, as in the case of large-scale block inclusion, confusion and errors in interpretation can be prevented by a knowledge of the regional stratigraphic pattern and of the potential of the deformational mechanism.

SUMMARY

Glacial tectonic structures of three types occur generally throughout Pleistocene glacial sequences.

1. Simple *in situ* deformation involves small-scale folds and faults produced by ice push on upstream faces of protuberances and by shear exerted on the bed by an advancing glacier.

2. Large-scale block inclusion of bedrock and drift includes shearing of large masses of material up into the ice and transporting them more or less intact away from the area of *décollement*. This process involves a combination of the proper hydrogeologic and hydrologic conditions in the bed below the glacier to produce low shear strength and compressive flow in the basal zone of the glacier.

3. Transportational thrust faulting involves the deposition and re-erosion of slices of debris-laden ice in the basal zone of the glacier. The use of different shear planes during each re-erosion causes the original pattern of slices to be shuffled. As a result, slices which originally occurred at the base of the drift zone may occur anywhere throughout the till sheet and may overlie slices from the top of the original sequence.

Failure to recognize large-scale block inclusion and transportational thrust faulting can result in the creation of nonexistent local units which are actually the result of repetition of other beds in the sequence. It can also lead to very erroneous models of the thickness and make-up of glacial-drift sequences where older beds are encountered near the surface.

Structures of all three types are believed to be ubiquitous throughout the glaciated part of North America, but can be expected to be most prevalent where the ice flowed up over scarps, over confined aquifers which were continuous back under the ice, and at ice-marginal positions. End moraines, in particular, were the sites of extreme glaciotectonic activity. Many if not most of the moraine-like ridges throughout the midwest may have been formed by glaciotectonic activity and thus never represented, during a glaciation, ice-marginal positions.

ACKNOWLEDGMENTS

The ideas presented here have been formulated over the past four years from observations by the author in Ohio, Illinois, and Saskatchewan. S. H. Whitaker and W. A. Meneley of the Saskatchewan Research Council, Frank Patton of the University of Illinois, and J. A. Cherry of the University of Manitoba, who have been working on similar problems, have been instrumental in clarifying and modifying the ideas presented. Discussion with G. W. White, W. H. Johnson, J. P. Kempton, P. B. DuMontelle, Lee Clayton, E. A. Christiansen, D. L. Gross, and others have also been very helpful to the author.

REFERENCES

Bluemle, J. P., 1966, Ice thrust bedrock in northeast Cavalier County, North Dakota: Proc. North Dakota Acad. Sci., v. 20, p. 112-18 (North Dakota Geol. Survey Misc., Series, No. 33).

Brinkmann, R., 1953, Uber die diluvialen Storumgen aus Rugen: Geol. Rudsdrau, v. 41, Sanderband, p. 231-41.

Brown, T. C., 1933, The waning of the last ice sheet in central Massachusetts: Jour. Geology, v. 41, no. 2, p. 144-58.

Byers, A. R., 1960, Deformation of the Whitemud and Eastend Formations near Claybank, Saskatchewan: Royal Soc. Canada, Trans., v. 53, ser. 3, sec. 4, p. 1-16.

Christiansen, E. A., 1971, Tills in southern Saskatchewan, Canada, in this volume.

Dellwig, L. F., and Baldwin, A. D., 1965, Ice-push deformation in northeastern Kansas: Kansas Geol. Survey Bull. 175, part 2, 16 p.

Elson, J. A., 1961, The geology of tills: in Proc. 14th Canadian Soil Mechanics Conf. 13 and 14 October 1960, Nat. Res. Council, Canada, Assoc. Comm. on Soil and Snow Mechanics, Tech. Memorandum No. 69, p. 7-13.

Gross, D. L., 1967, Mineralogical gradations within Titusville Till and associated tills in northwestern Pennsylvania: M.Sc. thesis, University of Illinois, 77 p.

Gross, D. L., and Moran, S. R., 1971, Textural and Mineralogical gradations within tills of the Allegheny Plateau: in this volume.

Hanshaw, B. B., and Bredehoeft, J. D., 1968, On the maintenance of anomalous fluid pressures: II source layer at depth: Geol. Soc. America Bull., v. 79, no. 9, p. 1107-22.

Harrison, P. W., 1967, A clay-till fabric: its character and origin: Jour. Geology, v. 65, no. 3, p. 275-308.

Hopkins, O. B., 1923, Some structural features of the plains area of Alberta caused by Pleistocene glaciation: Geol. Soc. America Bull., v. 34, p. 419-30.

Johnson, W. H., Gross, D. L., and Moran, S. R., 1971, Till stratigraphy of the Danville region, east-central Illinois: *in* this volume.

Kupsch, W. O., 1962, Ice-thrust ridges in western Canada: Jour. Geology, v. 70, p. 582-94.

Lammerson, P. R., and Dellwig, L. F., 1957, Deformation by ice push of lithified sediments in south-central Iowa: Jour. Geology, v. 65, no. 5, p. 546-50.

Mathews, W. H., and MacKay, J. R., 1960, Deformation of soils by glacier ice and the influence of pore pressure and permafrost: Royal Soc. Canada Trans., v. 54, ser. 3, sec. 4, p. 27-36.

Moran, S. R., 1967, Stratigraphy of Titusville Till in the Youngstown region, eastern Ohio: M.Sc. thesis, University of Illinois, 73 p.

————, 1968, Stratigraphic divisions of the Titusville Till near Youngstown, Ohio (abs.): Geol. Soc. America, Abstracts for 1967, p. 393-94.

————, 1969, Geology of the Hudson Bay area, Saskatchewan: Ph.D. dissertation, University of Illinois, 194 p.

Nye, J. F., 1952, The mechanics of glacier flow: Jour. Glaciology, v. 2, p. 82-93.

Piskin, K., and Bergstrom, R. E., 1967, Glacial drift in Illinois: thickness and character: Illinois Geol. Surv. Circ. 416, 33 p.

Rutten, M. G., 1960, Ice-pushed ridges, permafrost and drainage: Amer. Jour. Sci., v. 258, no. 4, p. 293-97.

Sardeson, F. W., 1898, The so-called Cretaceous deposits in southeastern Minnesota: Jour. Geology, v. 6, p. 679-91.

————, 1905, A peculiar case of glacial erosion: Jour. Geology, no. 105, no. 4, p. 351-57.

————, 1906, The folding of subjacent strata by glacial action: Jour. Geology, v. 14, p. 226-32.

Slater, George, 1926, Glacial tectonics, as reflected in disturbed drift deposits: Geologists' Assoc. Proc., v. 37, 392-400 p.

————, 1927a, The structure of the disturbed deposits of the Hadleigh Road area, Ipswich: Geologists' Assoc. Proc., v. 38, p. 183-261.

————, 1927b, The structure of the disturbed deposits of Moens Klint, Denmark: Proc. Roy. Soc. Edinburgh, v. 60, part 2, no. 12, p. 289-302.

————, 1927c, The disturbed glacial deposits in the neighborhood of Lonstrip, near Horving, North Denmark: Proc. Roy. Soc. Edinburgh, v. 60, part 2, no. 13, p. 303-315.

————, 1927d, The structure of the disturbed deposits in the lower part of the Dipping Valley near Ipswich: Proc. Geol. Assoc., v. 28, p. 157-182.

————, 1927e, Structure of the Mud Buttes and Tit Hills in Alberta: Geol. Soc America Bull. v. 38, p. 721-30.

————, 1929, The structure of the drumlins exposed on the south shore of Lake Ontario: New York State Mus. Bull. 281, p. 3-19.

Terzaghi, K., and Peck, R. B., 1967, Soil mechanics in engineering practice: New York, John Wiley and Sons, Inc., 729 p.

Virkkala, K., 1952, On the bed structure of till in eastern Finland: Komm. Geol. de Finlande, Bull. no. 57, p. 97-109.

Wickenden, R. T. D., 1945, Mesozoic stratigraphy of the eastern plains, Manitoba and Saskatchewan: Geol. Survey Canada Mem. 239, 87 p.

Wolford, J. J., 1932, A record size glacial erratic: Amer. Jour. Sci., v. 224, p. 362-367

II
Drumlins and Fluted Moraine

Editor's Comments on Papers 7 Through 10

The origin of streamlined glacial forms, especially drumlins, has been a source of controversy among glacial geologists for more than a century. Drumlins appear "ready made" in ground moraine as it is exposed from under receding glaciers (Goldthwait, 1974). Typical elliptical drumlins occur only in large clusters of 10 to 4000 in certain restricted areas, so the long arguments about their origin have been focused in regions of drumlin concentration such as upper New York State (Tarr, 1894; Fairchild, 1929; Slater, 1929) and central England–Ireland (Hollingworth, 1931). The classic descriptive work was done by Upham (1896) and Alden (1905) in southeast Wisconsin. Because collectively these articles are too long to include here, the Gravenor article (Paper 8), which gives a good summary of the erosional versus depositional controversy, was chosen. Another recent summary is found in the general reference, Embleton and King, 1968, or Muller, 1974.

The genetic relationship between "typical" drumlins (length to breadth ratio between 2 to 1 and 5 to 1) and the extremely elongated forms (10 to 1 and perhaps 100 to 1) is still another controversy. The extreme end form, till flutings, also occurs today in front of certain, but not all, retreating glaciers. Gilbert (1904) first related fluting to mountain glaciers in Alaska. More drumlin-like are the cigar-shaped elongate forms reported by Lemke (1958). In 1953 two leading Swedish glacial scientists, Hoppe and Schytt, presented an adequate explanation of fluting, so their article is chosen (Paper 7). They would now modify their original article reproduced here. A note by Schytt (1963) explains further examples and adds that regular spacing of the furrows and ridges (p. 152) has been abandoned.

Recently attention has been paid to the statistical treatment of drumlins and the circumstances under which they form (Chorley, 1959; Smalley, 1966). Here, two of the first studied are chosen: Reed and others (Paper 9) and Smalley and Unwin (Paper 10).

One critical aspect in all these articles concerns the internal content of the drumlins and the "till fabric" (alignment of the stones). Since some fabrics are so unusual (such as jointed boulders) as to merit special explanations (e.g., Kupsch, 1955) articles concerning these fabrics are not included. Several others which give details of fabrics (e.g., Wright, 1957) are omitted. Most fabrics in the outer till shell nearly parallel the long axis of the drumlin and the presumed ice motion. Well-exposed drumlins contain cores. These cores may be just a harder till with pebbles oriented at an angle to the long axis of the ellipse or they may consist of stratified material (kame, esker, or lacustrine varves). Where these core materials are clearly older than the outer till carapace (Goldthwait, 1974) they give credance to the erosional reshaping of earlier deposits as the early or first phase of drumlin making. Alden (1905) and Crosby (1934) noticed bedrock cores sticking up within some drumlins. Their argument indicated that such knobs initiated the clotting of till at a critical pressure-water-load condition on the sole of the moving ice sheet. This raises the question of whether "rock drumlins" are not just well-eroded preglacial hills that received minimal additions of till. As such they are an erosional product of glaciers and will be considered in a companion volume.

References Cited

Alden, W. C., (1905) The drumlins of southeastern Wisconsin: U.S. Geol. Survey Bull. 273, 46 p.

Chorley, R. J. (1959) The shape of drumlins: Jour. Glaciol. v. 3, no. 25, p. 339–344.

Crosby, I. B. (1934) Evidence from drumlins concerning the glacial history of Boston Basin: Geol. Soc. America Bull., v. 45, p. 135–158.

Embleton, C., and C. A. M. King (1968) *Glacial and Periglacial Geomorphology:* Edward Arnold, London, 608 p. (also St. Martin's Press, New York).

Fairchild, H. L. (1929) New York drumlins: Rochester Acad. Sci. Proc., v. 7, p. 1–37.

Gilbert, G. K. (1904) Glaciers and glaciation of Alaska: in *Harriman Alaska Expedition,* New York, v. 3, 231 p.

Goldthwait, R. P. (1974) Rates of formation of glacial features in Glacier Bay, Alaska: in *Glacial Gemorphology,* ed. D. R. Coates, State University of New York, Binghamton, N.Y., p. 163–185.

Hollingworth, S. E. (1931) The glaciation of western Edenside and adjoining areas and the drumlins of Edenside and the Solway Basin: Geol. Soc. London Quart. Jour., v. 87, p. 281–359.

Kupsch, W. O. (1955) Drumlins with jointed boulders near Dollard, Saskatchewan: Geol. Soc. America Bull., v. 66, no. 3, p. 327–337.

Lemke, R. W. (1958) Narrow linear drumlins near Valva, North Dakota: Amer. Jour. Sci., v. 256, p. 270–283.

Müller, E. H. (1974) Origin of drumlins: in *Glacial Geomorphology,* ed. D. R. Coates, State University of New York, Binghamton, N.Y., p. 187–204.

Schytt, V. (1963) Fluted moraine surfaces: a letter in Jour. Glaciol., v. 4, no. 36, p. 825–827.

Slater, G. (1929) The structure of drumlins exposed on the south shore of Lake Ontario: N.Y. State Mus. Bull. 281, p. 3–19.

Smalley, I. J. (1966) Drumlin formation: A rheological model: Science, v. 151, no. 3716, p. 1379–1380.

Tarr, R. S. (1894) The origin of drumlins: Amer. Geologist, v. 13, p. 393–407.

Wright, H. E. (1957) Stone orientation in the Wadena drumlin field, Minnesota: Geografiska Annaler, v. 39, p. 19–31.

7

Copyright © 1951 by the Swedish Geographical Society

Reprinted from *Geo. Ann.*, **35**, 105–115 (1953)

SOME OBSERVATIONS ON FLUTED MORAINE SURFACES

By *GUNNAR HOPPE and VALTER SCHYTT*[1]

Geographical Institutions, University of Uppsala and University of Stockholm

Introduction

Observations from a large number of glaciers in various parts of the world show that ground moraine surfaces recently exposed by the retreating ice are often "fluted" in a striking manner. The first record of this seems to be from the 1899 Harriman expedition to the Columbia Glacier in Alaska (Gilbert 1904, pp. 77 f.). Several other observations of the phenomenon also originate in Alaska: from the Petrof Glacier (Grant and Higgins 1913, p. 66), the Heney Glacier (Tarr and Martin 1914, pp. 448 f.), the Mendelhall Glacier (Ray 1935, pp. 310 f.), and the Woodworth and Schwan Glaciers (Washburn, acc. to Dyson 1952, p. 205). In a paper devoted to this parallel fluting, Dyson describes examples from 3 small glaciers in the Glacier National Park, Montana—the Sperry, Grinell and Jackson Glaciers. Another very recent account is from the Morveno Glacier in Patagonia (Nichols and Miller 1952, p. 46). The first observation of such fluting in Europe is from the Biferten Gletscher in Switzerland, where it occurs on the surface of a recent drumlin (de Quervain and Schnitter 1920, Heim 1919, p. 440).

Fluted ground moraine is thus not at all uncommon, and the instances described here are confirmation of this; indeed, it is not improbable that a fuller survey of terrain exposed by the general glacial recession might reveal that the phenomenon is normal for level moraine surfaces.

It has often been suggested that such parallel fluting is due to low ridges formed on the moraine surface with the same direction as the ice. Their height is said not to exceed one metre, the breadth seems to vary between one half and four metres, while the length may be as much as some hundred metres. Another explanation of the parallel pattern is that it is due to furrows in the moraine surface; the depth of the furrows is given as several centimetres, the breadth up to 30 cm (Nichols and Miller, p. 46). To judge from the descriptions, two different phenomena appear to be involved. On occasion, however, there seem to be grounds for suspecting that what is on the one hand regarded as primary and most significant (either ridges or furrows) is on the other considered secondary and less significant (and interpreted as erosion remnants or spaces between the furrows, as the case may be); in other words, that it is really a question of identical formations.

---

[1] The "Introduction" and "Observations from Icelandic glaciers" are by G. Hoppe Uppsala; "Fluted moraine in front of the Isfall Glacier, Kebnekajse" by V. Schytt Stockholm.

Fig. 1. Fluted moraine in front of Fláajökull.

This latter remark anticipates to some extent the account of modern ideas on the matter. It is possible to distinguish between erosion and accumulation theories, and it should be stressed, with reference to what has just been said, that both may be justifiable. According to the first, the surface of the ground moraine was grooved, either by the uneven lower surface of the ice (e. g. Ray, p. 310), or by stones and boulders embedded in the ice (de Quervain and Schnitter, p. 146). According to the second, the ridges were formed by debris pressed up into channel cavities under the ice (de Quervain and Schnitter, Dyson). Important evidence for the latter view is that the ridges often terminate at a boulder fixed in the surface of the moraine; there are a number of observations to show that if the ice is not very thick, long channels in the ice appear in the lee of such boulders, owing to the relatively slight plasticity of the ice. It should be particularly easy for debris to be pressed up into such channels when it is very wet and of almost fluid consistency—conditions which Dyson states to be normal for the glaciers he investigated. When ridges occur in the lee of boulders in this way, they bear a close resemblance to crag an tail phenomena (Schuttfahnen)—a resemblance pointed out by de Quervain.

Observations from Icelandic glaciers

The observations here put forward were made during a journey in Iceland in the summer of 1952, when some twelve lobes and tongues running out from Vatnajökull were visited. Fluted ground moraine was found adjacent to three of them—Brúarjökull, Svínafellsjökull

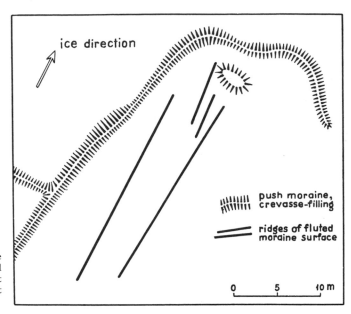

ice direction

push moraine,
crevasse-filling

ridges of fluted
moraine surface

0   5   10 m

Fig. 2. Detail of the moraine
plain in front of Svínafellsjökull
in Hornafjördur. Notice that
the ridges in this case are not
quite parallel.

in Hornafjördur and Fláajökull. The fact that fluted surfaces were not discernible near the
other glaciers is not surprising, since the ice front in several cases remains at the 19th
century end moraines, and in other cases the surface exposed by regression is covered by
wide stretches of water.

The largest area of fluted moraine was by Brúarjökull. Between the two rivers Kverká
and Kringilsá is an area about 5 km wide, which has been uncovered since a considerable
advance in 1890 (Todtmann 1952). As far as I could discover, the characteristic pattern
occurred only on relatively elevated terrain, within an area of about a square km near the
present ice margin. It occurred even where the moraine was studded with boulders. But
it did not occur just inside the 1890 end moraine, nor on lower terrain near the glacier.
These gaps might have been caused by later erosion, not least that due to melt-water, or
obliterated by glaciofluvial deposits on the ground moraine; though in some cases they seem
to be definitely primary. The fluted areas near Svínafellsjökull and Fláajökull were not
quite so large, but they were nonetheless characteristic. The phenomenon was everywhere
confined to level terrain.

The fluting was in the form of small parallel ridges at a distance varying from some
dm to several metres. The distance was often fairly constant within a particular region,
without there being any real regularity. The dimensions were the same as those commonly
given in the literature: the height up to a metre or so, though usually about 10 cm or less;
the length usually some tens of metres. The direction coincided with that of the ice movement. This could be seen most convincingly by Brúarjökull, where the direction of the
ridges was exactly the same as that of striae on nearby basalt outcrops and on numerous
boulders fast embedded in the moraine, as well as the orientation of elongated pebbles

Fig. 3. Fluted moraine in front of Brúarjökull. In the foreground a ridge in the lee of a fast embedded
boulder.

and cobbles at a depth of some dm in the moraine. The directions of adjacent ridges were
found to diverge up to 10° in some places (Fig. 2).

As regards the way in which the ridges were formed, an important observation is that
at Brúarjökull many of them were in the lee of rather large boulders—though there was no
general rule about this. These boulders were fixed in the underlying material, very often
with their longest axes in the direction of the ice movement, and often with distinct uniform
striae. The height of the ridges gradually decreased away from the boulders. Boulders not
fixed prior to the recession did not exhibit such orientation and striae, nor were there
longitudinal ridges in the lee of them (though small transverse "push" ridges were quite often
seen). There seems to be no other possible explanation of such parallel ridges in the lee of
fixed boulders than that advanced by de Quervain and Dyson: that debris was pressed up
into channels in the lee of the boulders.

In many cases, however, there was no terminal upstream boulder, and the height of the
ridges then scarcely decreased in the downstream direction. An interesting fact is that all
stages between the parallel ridges and basal crevasse-fillings could be observed. There
were many crevasses in the marginal parts of the glaciers visited, and the ground moraine
had sometimes been squeezed up several metres into these crevasses (Fig. 4). This showed
that the ground moraine was well saturated with water, not only at the ice front—where
one often sank several dm in it—but also far in under the glacier; it also showed that all

Fig. 4. Crevasse-filling at the front of Hoffellsjökull. Height of the formation 1.8 m. Photo: Jón Jónsson.

hollows underneath the ice were soon filled. In several cases my notes record that the finest material—sand and finer—was found on the crests of the ridges, where there were also numerous desiccation cracks. This finer material must have been especially mobile, and this observation therefore supports the idea of accumulation in subglacial hollows; as a matter of fact, a similar sort of material was also found in the upper parts of basal crevasse-fillings. Another point is that though the elongated stones in the level ground moraine exhibit a clear orientation in the direction of the ice movement—particularly noticeable at Brúarjökull and Skaftafellsjökull—the parallel ridges lack every preferred orientation of the stones. This is perhaps a further argument against the erosion remnant explanation.

Everything mentioned so far indicates that the ridges were fillings in low subglacial channels. But it seems that their origin is in general probably more complicated. It was observed at Fláajökull that an ice-boulder several m³ in size had been pushed along by the ice front, giving rise to a miniature striation of the moraine beneath. The moraine seemed to have been scraped up. Push moraines occurring in many places, seldom more than a metre high, also indicated that there had been erosion of the ground moraine.

The formation of parallel ridges by debris pressed up into subglacial channels may be regarded as definitely established; erosion as a collaborating factor in the formation of fluted surfaces is a possibility in certain cases. Other alternatives might be put forward, but their validity could only be tested by following the changes at the ice front over a longer period, and preferably by investigation ьnder the glacier as well—desirable measures for the further verification of the first alternative as well, incidentally. It was unfortunately quite impossible to carry out such measurements during the period of my visit.

When the ice had receded from the ridges, many of them seem to have been soon obliterated, because of their looseness and lowness; though Dyson says that he found ridges as much as 50 years old. There seems, thus, to be little likelihood of finding such small formations from the Ice Age; but the way in which they are formed may be relevant when considering larger formations, and they may therefore well be ascribed an interest greater than the merely local.

Fluted moraine in front of the Isfall Glacier, Kebnekajse

The rapid recession of Scandinavian glaciers at the present time uncovers large areas of moraine every year, providing excellent opportunities for the study of glacio-geological formations in the actual process of development.

During glaciological investigations in the Kebnekajse region of northern Lappland in recent years, the emergence of a fluted moraine of the same kind as Hoppe describes from Iceland and such as has been described from several places in Europe and America has been followed with considerable interest.

The phenomenon has been followed mainly in front of the Isfall Glacier, but it also occurs at two other glaciers in the Kebnekajse region, at the Kårsa Glacier and two other small glaciers near Riksgränsen, at a glacier in the Tärna Mountains, and by at least one glacier in the Sulitelma massif. Erik Bergström, who has studied the moraines adjacent to these glaciers, has made similar observations in Norway, where Olav Liestöl and Alf Bergersen of the Norwegian Polar Institute have also observed the same type of formation adjacent to several glaciers. According to an as yet unpublished paper by H. Holmen and S. A. Svensson, such fluting of the surface occurs in front of one of the Storstein Glacier's tongues some tens of km south of Narvik.

The Isfall Glacier is a small glacier (2.3 km²) with a rather broken profile. Forty years ago the glacier extended up to and even over the crest of the innermost end moraine. Since then, the front has retreated 200 m, and during the whole of this time the receding tongue of ice has left a large number of narrow parallel ridges, all in the direction of the ice movement. The ridges can often be followed from the crest of the 15 m high end moraine right up to the present ice front. They disappear in one place into a small tarn and under a delta in this, but they reappear on the other side just as clearly formed. The distance between the centres of two adjacent ridges is normally 1—1.5 m and the most well-developed are 30—40 cm high.

A detailed investigation of the ridge-formations was made in July 1949, and further comparative measurements in August 1952. There was a great deal of snow in 1949, and the summer was cold, so the ridges could be studied only within a 90 m broad belt, i. e., within about a third of the area they covered.

Fig. 5. Fluted moraine in front of Isfall Glacier.

The length of the moraine ridges varies considerably—partly because of rock sliding and solifluction on the slope of the end moraine—but, as already mentioned, many of the ridges could be traced from the crest of the end moraine to the ice front. All 54 measured were orientated exactly in the direction of the glacier's movement. Their relative positions are given in table I below, which also gives their height.

The profile was perpendicular to the ridges and about 40 m from the ice front, with the origin (o m) at the northern edge of the belt (the right in Fig. 5).

d — distance between ridges (from centre to centre)
l — height of ridge, 5—15 cm
ml —   »   »   »   15—25 cm
mh —   »   »   »   25—35 cm
h —   »   »   »   35—45 cm

Notes at the time of measurement also contain the following remarks:

1. Traces of further ridges near no. 8.
2. The direction of striae on several boulders near no. 18 is parallel to the direction of the ridges.
3. Traces of some very low ridges near no. 24.
4. After no. 54 several fragmentary ridges up to 87 m.

It will be seen from the above table that most of the ridges were relatively low, but all those noted were well defined.

It is very striking that all the ridges are at a distance of a little more than a metre from one another. Analysis of the figures shows that 29 (55 %) lie within the interval 1.0—1.4 m. The notes indicate that where the d-value is large there are some traces of further ridges, so that the above percentage would probably have been still higher if erosion, especially solifluction, had not made them too indistinct to be included in the table.

The surface material at least is unsorted near the ice front, with all grades from fine clay to large boulders. The farther from the ice front, the more fine material is found between the ridges, while gravel, stones and boulders on the crests have been able to withstand solifluction and water erosion. Near to the ice front the moraine was frozen, and the shape of

Table 1

ridge no.	dist. from origin (m)	d (m)	height	ridge no.	dist. from origin (m)	d (m)	height
1	0		l	28	42.2	1.0	mh
2	1.5	1.5	l	29	43.5	1.3	l
3	2.5	1.0	l	30	44.5	1.0	l
4	3.5	1.0	ml	31	45.4	0.9	l
5	5.1	1.6	l	32	46.9	1.5	l
6	6.9	1.8	h	33	47.7	0.8	l
7	8.0	1.1	l	34	48.6	0.9	l
8	11.0	3.0	l	35	50.0	1.4	l
9	13.1	2.1	mh	36	50.8	0.8	h
10	14.4	1.3	h	37	51.9	1.1	ml
11	17.5	3.1	l	38	53.7	1.8	l
12	18.3	0.8	l	39	55.0	1.3	l
13	19.3	1.0	ml	40	57.1	2.1	h
14	20.6	1.3	l	41	58.4	1.3	l
15	21.7	1.1	mh	42	61.9	3.5	ml
16	24.6	2.9	l	43	63.0	1.1	l
17	26.0	1.4	l	44	64.2	1.2	l
18	28.0	2.0	mh	45	66.1	1.9	l
19	29.4	1.4	l	46	70.2	4.1	l
20	30.4	1.0	l	47	72.5	2.3	mh
21	31.5	1.1	l	48	73.8	1.3	ml
22	33.1	1.6	ml	49	74.9	1.1	l
23	34.1	1.0	l	50	76.3	1.4	mh
24	37.5	3.4	mh	51	78.3	2.0	l
25	39.2	1.7	mh	52	79.4	1.1	l
26	40.2	1.0	mh	53	80.5	1.1	l
27	41.2	1.0	ml	54	81.7	1.2	ml

the ridges well preserved. Farther away, the moraine had thawed out, and the surface had that loose, almost fluid consistency so often encountered just outside the ice front. It was only with difficulty that one could avoid sinking a foot or so into the porridge-like mass, and it seemed quite astonishing that the surface formations could persist in such plastic waterlogged moraine.

Yet farther away from the glacier, on the upstream side of the innermost end moraine, the ground was altogether thawed and the ridges dry and "solid".

The ridges seemed to continue straight in under the glacier, and thus consist of ground moraine. The internal moraine of many glaciers is at present thawing out, and it protects the underlying ice from ablation, but in the case of the Isfall Glacier the old ridges on the slope of the end moraine and the total absence of dead-ice hollows indicate that here the ridges do not consist of such material. Moreover, photographs show that boulders of the surface of the moraine were in exactly the same places in the summer of 1952 as they were in 1949 (comparison of photographs gives a very good stereoscopic effect).

Since the ridges run parallel and almost straight on both up and down slopes, and since digging reveals no stratification, they cannot be associated with erosion or accumulation caused by running water.

In order to examine some ridges under the glacier, a trench was dug in July 1949 about 5 m above the point where one of the most distinct ridges (no. 10) disappeared under the ice.

Fig. 6. The inner wall of the trench, showing a cross section of ridges nos 9 and 10.

The upper part of no. 10 was found 35 cm below the ice surface, and three ridges were soon exposed. The appearence of the inner wall of the trench is shown in Fig. 6. Ridge no. 10 is seen furthest to the left, between the ice axes, and was of considerable dimensions here inside the glacier. The height and breadth were 90 cm, and the sides more or less vertical. The edges of the stones were rounded off and the material was quite unsorted, containing all grades from clay to boulders. There was a remarkably high ice content, consisting not of glacier ice but of frozen melt-water. The crystals were columnar, and several layers of ice were often parallel with the sides of the ridge, which were quite distinct from the adjacent glacier ice.

Ridge no. 9 was also 90 cm wide, but only 45 cm high, and its cross-section was more rounded. The distance between the centres of nos 9 and 10 was 135 cm, which agrees well with the 1.3 m given in the table. Another ridge was uncovered 130 cm further to the north. It was only 20 cm high, and is probably missing from the table, where the distance between nos 8 and 9 is given as 2.1 m.

Five metres further up in the glacier the upper surface of no. 10 was found beneath 90 cm of ice, and the profile which has now become apparent is shown in Fig. 7.

The 90 cm high ridge sank to about half the height just outside the edge of the glacier, because the ice content had there melted away. The ratio between the frozen parts of the

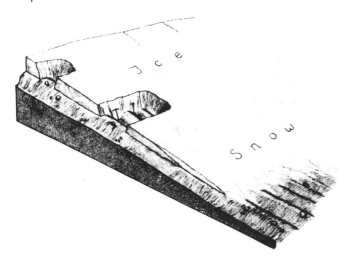

Fig. 7. Structure of bottom moraine as revealed by trenching.

ridge within the glacier and those already exposed and thawed out was about the same in both ridges.

Another ridge was examined in August 1952, this time about 15 m to the north of no. 10 4.5 m from the ice front, the upper surface of the ridge was 130 cm below the surface of the glacier. It was not so symmetrical as the previous ones, as there was a slight overhang on its northern side. Its dimensions are shown most clearly in Fig. 7 (the rule shown is 100 cm long). The debris in this ridge was also quite unsorted, and of all sizes. The crest of the ridge sloped a few degrees downwards into the ice.

A common feature of all the ridges examined was the abundance of fine material. As a matter of fact, we have never observed similar ridges in coarse moraine. Their fine grain size, high ce content, and shape—especial y as illustrated by the ridge in fig. 8.—suggest that the debris was pressed up into cavities in the ice while more or less fluid. It seems probable that this might have happened in the lee of a boulder, but no such boulders have been exposed by the recession of the glacier. It is also worthy of note that the ground moraine under this glacier is frozen all the year round from the ice front to a point some distance in under the glacier. Moreover temperature measurements in glacier ice in the Kebnekajse region have shown that the glaciers there do not really belong to the temperate type, as defined by Ahlmann, since the ice temperature is negative, at least in the outer layers, even at the end of summer.

The ridges may therefore have been formed by the great weight of ice pressing water-soaked ground moraine at the pressure melting point up into hollows formed in the lee of boulders. The release of pressure might then cause some of the ground moraine to freeze to the basal ice, so that it is carried along with the ice, while fresh debris continues to be pressed up in the lee of the boulder. In this way the ridge is carried forward and built up from behind until it reaches the zone where the low winter temperatures extend through the ice into the ground moraine. There, the ridge freezes to the substratum and can no longer

Fig. 8 Stereo-photographs of a moraine ridge dug out in 1952. Note its very smooth and even walls and the slight overhang to the right. Big crystals of glacier ice are visible along the right side of the ridge. The black line in the background shows a shear plane probably separating the moving upper layers from the stagnant bottom ice.

move with the ice. This means that there is a solid frozen moraine ridge between the initiating boulder and the "stranded" outer end. It is thereafter impossible for any hollow space to persist in the lee of the boulder, and the formation of the ridge is at an end.

Such an explanation accounts for the parallelism with the direction of motion, as well as the constant height and considerable length of individual ridges.

It is not clear how the relatively constant distance between ridges is to be accounted for. It could be a coincidence for this particular glacier, and only further observation can help to decide what significance may be assigned to it.

References

DYSON, J. L., 1952. Ice-ridged moraines and their relation to glaciers. American Journal of Science, 250.

GILBERT, G. K., 1904. Glaciers and glaciation. Alaska, vol. III.

GRANT, U. S. and HIGGINS, D. F., 1913. Coastal glaciers of Prince William Sound and Kenai Peninsula, Alaska. U. S. Geological Survey, Bull. 526.

HEIM, A., 1919. Geologie der Schweiz, I.

NICHOLS, R. L. and MILLER, M. M., 1952. The Moreno Glacier, Lago Argentino, Patagonia. Journal of Glaciology, 11.

DE QUERVAIN, A. and SCHNITTER, E., 1920. Das Zungenbecken des Bifertengletschers. Denkschriften der Schweizerischen Naturforschenden Gesellschaft, 55:2.

RAY, L. L., 1935. Some minor features of valley glaciers and valley glaciation. Journal of Geology, 43.

TARR, R. S. and MARTIN, L., 1914. Alaskan glacier studies of the National Geographical Society in the Yakutat Bay, Prince William Sound and Lower Copper River regions.

TODTMANN, E. M., 1952. Im Gletscherrückzugsgebiet des Vatna Jökull auf Island, 1951. Neues Jahrbuch für Geologie und Paläontologie, 1952, Abt. B.

8

Reprinted from *Amer. Jour. Sci.*, **251**(9), 674–681 (1953)

THE ORIGIN OF DRUMLINS

CONRAD P. GRAVENOR

ABSTRACT. Theories on the origin of drumlins are grouped as either depositional or erosional, and the merits of these theories are discussed. The fact that many drumlins have a small amount of clay and contain stratified materials makes the depositional theory untenable. It is believed that a modification of the erosional theory will fit the known facts on drumlins.

INTRODUCTION

ALTHOUGH several theories have been advanced to account for drumlin formation, there is still difference of opinion on their origin. Essentially there are two ideas: first, drumlins are formed by the erosion of pre-existing drift; second, they are depositional in origin. Recently the latter theory has gained much favor even though it does not explain fully drumlins of extreme composition (Flint, 1947, p. 121-126). The depositional theory has been applied to only those drumlins which are made up almost entirely of till, and the erosional theory is left to explain those drumlins which clearly have been formed from pre-existing drift or rock.

The purpose of this paper is to review briefly some of the better-known theories and to sum up the facts known about drumlins. It is considered that a modification of the erosional theory best fits the known facts.

DEPOSITIONAL THEORY

In general the depositional theory states that drumlins are formed by the progressive deposition of drift. However, there is a lack of agreement on the factors that determine the start of accumulation. Russell (1895) suggested that since ice under pressure behaves as a plastic solid the introduction of debris into this plastic mass will decrease the rate of flow. At certain points excessive amounts of debris stop ice flowage, and hence nuclei are formed. Ice, containing smaller amounts of debris, which passes over these nuclei will deposit material and form drumlins. An examination of the excellent drumlin sections on the south shore of Lake Ontario led Slater (1929) to believe that in this case a till core initiated the drumlin accumulation.

674

Chamberlin (1883, p. 306) found that some drumlins have rock cores and from this intimates that a deeply hidden rock boss is usually and perhaps universally the determining cause of the accumulation. Crosby (1934, p. 150-151) concluded that some of the drumlins in the Boston basin area have rock cores. From a study of the drumlins of central British Columbia Armstrong (1949, p. 14) inferred that the nuclei were knobs of frozen till and the drumlins were built up from these knobs. Alden (1918, p. 253-256) found that few of the drumlins of Wisconsin have rock cores and consequently suggested that the radial spreading of ice developed transverse stresses which, although not actually creating longitudinal crevasses, may have induced localized piles or ridges of drift which were later shaped into drumlins.

The concentric layering of material found in a few drumlins has led many geologists to believe that drumlins have been built up by successive additions of clayey till (Alden, 1918; Goldthwait, 1924, p. 91-96; Flint, 1947; and Fairchild, 1929). Fairchild suggested that "the accretion was because of the greater friction between clay and clay than between clay and ice."

The theories of Millis (1911) and Upham (1892) differ from the accretion theory, but fall into the depositional class. Millis suggested that material accumulated in crevasses which were enlarged by melting. Then as the remaining ice melted, drumlins were left. Upham believed that as the ice melted by ablation, englacial drift would appear at the surface in depression areas. When the ice "re-livened," this superglacial material again would become englacial as a stratum of drift. This englacial drift would be shaped into lenticular masses by ice movement and then let down either as a completed drumlin or as an accumulation point.

EROSIONAL THEORY

From an examination of the materials which form drumlins it is evident that many of them were formed from materials which were present prior to the ice advance. This has led to the belief that all drumlins were formed by erosion. Shaler (1889, p. 550-551) thought that the drumlins of New England were formed by two glaciations. The first glaciation provided an irregular till surface and the second scoured this surface leaving the drumlins. Tarr (1894) found that rock drumlins and till drumlins have the same shape, and consequently he concluded that the same erosive process produced both types.

OBJECTIONS TO THE DEPOSITIONAL THEORY

Although drumlins with cores of bedrock or pre-existing drift are found, more often a central core is absent. While this fact demonstrates that for a drumlin to accumulate no pre-existing core is necessary, it does not disprove the depositional theory. Contrary to Alden's theory, Hollingworth (1931) found that drumlins were formed in regions where there was no radial spreading of the ice.

The presence of stratified materials in drumlins constitutes a more serious objection to the depositional hypothesis. Drumlins containing stratified materials have been recorded in most drumlin fields both in North America and Europe (Ebers, 1937; Deane, 1950; Alden, 1918; Hollingworth, 1931; Tarr, 1894; Upham, 1894; Slater, 1929). Although drumlins containing stratified materials have been described by many geologists, for the most part their origin has been ignored. Deane (1950, p. 12-14) suggested that stratified materials in drumlins indicate a readvance of the ice after each layer of stratified material was deposited. Tarr (1894) pointed out that flowing water could not exist under the great thickness of ice necessary to form drumlins. It seems reasonable that deep within an active glacier any openings which would give access to meltwater would be closed by plastic flow. If eskers and associated deposits are evidence of stagnation, then the time when meltwaters are flowing at the base of the ice is not the time of drumlin formation.

Although it has been suggested by Fairchild that accretion takes place because of the greater friction between clay and clay than between clay and ice, the writer has been unable to find quantitative data which would substantiate this conclusion. Many drumlins have been described which contain little or no clay. The drumlins of northern Saskatchewan, described by Sproule (1939, p. 102-103), are made up largely of sand. Mechanical analyses made by Goldthwait (1948, p. 9-11) of drumlin materials in New Hampshire show that the drumlins of that region contain an average of about 10 per cent clay and in one case the clay content was as low as 5 per cent. In southern

Ontario, Chapman (1951) has found that the drumlins are more numerous in loamy till than in clay till. Mechanical analyses of drumlin materials of southern Ontario give an average of about 12 per cent clay (Chapman, 1951). Indeed most published analyses indicate that drumlins are sand-rich rather than clay-rich. In the states of Indiana and Illinois, where the tills generally contain more clay than those of southern Ontario and New England, no drumlins are found. Therefore it appears that clay is not necessary for drumlin formation.

Concentric banding, found in some drumlins, has been an important factor in the formulation of the depositional theory. However, it seems that this concentric banding is a rarity. Fairchild thought that the banding in the drumlins on the south shore of Lake Ontario was evidence of accretion, but Slater (1929) found that these bands consist largely of stratified materials. In many areas where the internal structure of drumlins has been investigated banding is absent (Hollingworth, 1931; Deane, 1950; Ebers, 1937). Alden (1918) suggested that a definite cleavage found in certain drumlins in Wisconsin could be accounted for either by accretion or by pressure effects.

The alignment of drumlins and their streamlined shape seems sufficient to discount the theory proposed by Millis. Tarr (1894) has objected to Upham's theory on the grounds that no evidence of shearing is found in drumlins. Another objection is the presence of so much englacial material in an ice sheet.

OBJECTIONS TO THE EROSIONAL THEORY

Objections to this theory have been outlined by Thwaites (1941, p. 43-45), but certain of these objections do not appear valid. Thwaites suggested that drumlins of the erosional type should be shaped like roches moutonnées; yet it is well known that drumlins formed from pre-existing materials have the same shape as till drumlins. Other objections advanced by Thwaites include the following:

(1) The stratified materials found in drumlins is unlike that found in kames.
(2) The width of drumlin belts (10-20 miles) exceeds that of most moraines.
(3) The drift in drumlins is apparently the same age as the surrounding drift.

While the second objection holds for most moraines of the Mississippi basin area it is not true of the wide morainic belts found over much of western Canada. The third objection is perhaps the most valid since in many areas where there are till drumlins there is no evidence of readvance of the ice.

MODIFIED EROSION THEORY

It is known that rock drumlins and drumlins carved from pre-existing materials are found in the same drumlin fields as those made of till (Tarr, 1894; Deane, 1950; Hollingworth, 1931; Shaler, 1889; Armstrong, 1949). Consequently it is reasonable to suppose that all these drumlins were formed at the same time and by the same process. During the past 80 years many facts have been learned about drumlins and any one theory should satisfy the following conditions:

(1) Drumlins may consist of (a) clay till, (b) sandy or loamy till, (c) rock, (d) pre-existing drift.

(2) They frequently have lenses and layers of stratified materials which sometimes are faulted and folded.

(3) Rock drumlins are found side by side with other varieties and have the same shape.

(4) Many glaciated regions do not support drumlins.

(5) They exist in fields wider than most moraines and rarely occur singly.

(6) They have a streamlined shape with the stoss end usually pointing upstream.

(7) Lamination may or may not be present.

(8) Some drumlins have cores but most do not.

(9) They are found behind terminal moraines which mark approximately the outer limit of the ice advance.

(10) Their long axes parallel the direction of ice movement.

Generally it is agreed that drumlins are formed under actively flowing ice and their form is one which offers the least resistance to ice movement. This assumption seems valid since it accounts for the position of the drumlins with respect to moraines, their alignment and streamlined shape.

The usual interpretation of the erosional theory is that during a retreat of the ice moraines are formed which are shaped by a later advance. It is the writer's belief that this retreat is not necessary. Upon the retreat of an ice sheet moraines are formed which mark stationary positions of the ice. Is it not possible that during the advance of an ice sheet the rate of movement would vary? If this were the case there should be evidence of "moraines of advance." However, such surface irregularities would be shaped by the over-riding ice and would not be recognizable as moraines.

Where the ice advance is more rapid and the material distributed homogeneously throughout the ice a relatively flat till plain is formed. Chapman (1951) found that the flat till surfaces of southern Ontario are scored with shallow grooves or "flutings." These flutings may result from the same process as that which formed the drumlins.

The complete process of drumlin formation can be outlined as follows:

(1) Masses of till and stratified materials would be deposited at the front of an advancing glacier if there was a temporary halt during the ice advance.

(2) Ice riding over this drift would erode and shape it and thereby produce drumlins.

Debris derived from this erosion and shaping eventually could move to the front of the ice and be redeposited either as "moraine of advance" or as terminal moraine. It is known that the second part of this theory is feasible since drumlins of erosional origin are found. Consequently all that is needed is an irregular surface of till with or without stratified materials. It is believed that such a surface can be formed during an ice advance.

Obviously, one other prerequisite to drumlin formation is a supply of drift. Consequently drumlins form best in the softer materials, such as shales and limestones, which are removed readily by glacial erosion.

Objections to the previously outlined erosional and depositional theories can be explained by this modification of the erosional theory.

(1) The absence of drumlins from certain regions of clay till can be explained by a relatively rapid ice advance.

(2) The lamination found in a few drumlins could be the result of ice pressure. Any clay minerals which are present would have preferred orientation as a consequence of the applied pressure, and this would give rise to a rough fissility in the till.

(3) Meltwater action at the front of the advancing ice would give rise to stratified materials.

(4) Although Thwaites pointed out that drumlin belts are much wider than most moraines, it should be realized that the nature of retreat is quite different from that of advance. During a retreat, wide intermorainal areas, covered by eskers, superglacial till, and related deposits, are left by the downwasting process. An advance would not produce these ablation areas since wasting would take place mainly at the front of the ice, and wider morainic belts would result.

(5) Thwaites also stated that the stratified material found in drumlins is unlike that found in kames. It may be pointed out that the bedded material in drumlins is quite similar to the stratified material found in ground moraine. It is thought that the origin of the stratified material found in ground moraine is similar to that postulated for drumlins.

(6) Where drumlins have been found with cores it is possible that during a slow advance any pre-existing irregularities would have a layer of till deposited on top of the obstruction and thus present the type of surface necessary for drumlin formation.

SUMMARY

Since some drumlins are made of pre-existing materials, it is known that erosion can produce a drumlin. It is believed that halts or a slow advance during the forward movement of a glacier can give rise to a wide irregular surface of drift which would be shaped into drumlins by the advancing ice.

This modification of the erosion theory is less intricate in its mechanics than those previously described. It avoids the necessity for two theories and offers adequate explanations for the known facts on drumlins.

The erosional theory is not acceptable in its present form because of the following objections:

(1) In many areas where there are till drumlins there is no evidence of readvance of the ice.

(2) Morainic belts usually are not as wide as drumlin fields.

(3) The stratified materials found in drumlins is unlike that found in kames.

The currently accepted depositional theory is untenable because of the following objections:

(1) Stratified materials found in drumlins could not have been deposited under a thick ice sheet.

(2) The accretion theory hinges on the idea that drumlins are clay-rich; however, descriptions and mechanical analyses show that many drumlins contain little or no clay.

(3) Many areas of clay-rich till do not support drumlins.

(4) The absence of nuclei and concentric banding from most drumlins.

REFERENCES

Alden, W. C., 1918, The Quaternary geology of southeastern Wisconsin, with a chapter on the older rock formations: U. S. Geol. Survey Prof. Paper 106.

Armstrong, J. E., 1949, Fort St. James map-area, Cassiar and Coast districts, British Columbia: Canada Geol. Survey Mem. 252.

Chamberlin, T. C., 1883, Terminal moraine of the second glacial epoch: U. S. Geol. Survey 3rd Ann. Rept., p. 291-402.

Chapman, L. J., and Putman, D. F., 1951, The physiography of southern Ontario, University of Toronto Press, Toronto.

Crosby, I. B., 1934, Evidence from drumlins concerning the glacial history of Boston basin: Geol. Soc. America Bull., vol. 45, p. 135-158.

Deane, R. E., 1950, Pleistocene geology of the Lake Simcoe district, Ontario: Canada Geol. Survey Mem. 256.

Ebers, Edith, 1937, Zur Entstehung der Drumlins als Stromlineinkorper; Zehn weitere Jahre Drumlinforschung (1926-1936): Neues Jahrb., Beilage Band 78, Abt. B, p. 200-240.

Fairchild, H. L., 1929, New York drumlins: Rochester Acad. Sci. Proc., vol. 7, p. 1-37.

Flint, R. F., 1947, Glacial geology and the Pleistocene epoch, John Wiley and Sons, Inc., New York.

Goldthwait, J. W., 1924, Physiography of Nova Scotia: Canada Geol. Survey Mem. 140.

Goldthwait, Lawrence, 1948, Glacial till in New Hampshire: New Hampshire State Plan. and Devel. Comm. Min. Res. Survey, part 10, p. 3-11.

Hollingworth, S. E., 1931, The glaciation of western Edenside and adjoining areas and the drumlins of Edenside and the Solway basin: Geol. Soc. London Quart. Jour., vol. 87, p. 281-359.

Millis, John, 1911, What caused the drumlins? Science, vol. 34, p. 60-62.

Russell, I. C., 1895, The influence of débris on the flow of glaciers: Jour. Geology, vol. 3, p. 823-832.

Shaler, N. S., 1889, The geology of Cape Ann, Massachusetts: U. S. Geol. Survey 9th Ann. Rept., p. 529-611.

Slater, George, 1929, The structure of the drumlins exposed on the south shore of Lake Ontario: New York State Mus. Bull. 281, p. 3-19.

Sproule, J. C., 1939, The Pleistocene geology of the Cree Lake region, Saskatchewan: Royal Soc. Canada Trans. 3d ser., vol. 33, sec. 4, p. 101-109.

Tarr, R. S., 1894, The origin of drumlins: Am. Geologist, vol. 13, p. 393-407.

Thwaites, F. T., 1941, Outline of glacial geology, Edwards Bros., Ann Arbor, Michigan.

Upham, Warren, 1892, Conditions of accumulation of drumlins: Am. Geologist, vol. 10, p. 339-362.

————, 1894, The Madison type of drumlins: Am. Geologist, vol. 14, p. 69-83.

UNIVERSITY OF ALBERTA
EDMONTON, ALBERTA, CANADA

9

Reprinted from *Amer. Jour. Sci.*, **260**(3), 200–210 (1962)

SOME ASPECTS OF DRUMLIN GEOMETRY

BRUCE REED,[*] CYRIL J. GALVIN, JR.,[**] and JOHN P. MILLER[†]

ABSTRACT. Measurements of form, orientation, and spacing of drumlins were made from topographic maps and aerial photographs of areas (1) near Boston, Massachusetts, (2) near Weedsport, New York, and (3) in Blaine County, Montana, and Pierre County, North Dakota. It was found that individual drumlins depart slightly from the ideal ellipsoidal form. Orientations of drumlins within small areas are characterized by normal distributions. Spacing between drumlins is characterized by multimodal distributions with some indications of periodicity.

INTRODUCTION

Current uncertainties about genesis of specific landforms serve to emphasize the need for (1) more detailed knowledge of mechanics and dynamics of surficial processes, and (2) more precise description of landform characteristics, including size, shape, and orientation. For most landscape features the details of form are so complex that efforts at quantitative description have been limited and only partially successful. Preoccupation with these difficult problems has tended to inhibit quantitative description of the less abundant relatively simple and symmetrical landforms characteristically associated with specific processes and commonly restricted in areal distribution.

It is generally recognized that of all topographic forms none possesses greater symmetry than drumlins. This, combined with the fact that each drumlin is a distinct isolated feature in a "field" which may include dozens or hundreds of drumlins, greatly facilitates precise description. The primary objective of this paper is to compare certain geometrical properties of drumlins located in three different regions of the United States. All the data to be presented here were obtained from topographic maps and aerial photographs. The areas chosen for study were those for which recent, photogrametrically constructed topographic maps on a scale of 1:24,000 and with contour intervals of 10 feet are available. These restrictions limited the study to three areas: (1) near Boston, Massachusetts, (2) central-western New York, and (3) the Voltaire area, North Dakota. Supplementary information was obtained from aerial photographs of drumlin fields in Montana and North Dakota. Other extensive and important drumlin fields were eliminated because maps with suitable scale and contour interval were not yet available.

The problem of recognizing drumlins from topographic maps and aerial photographs without field examination contributes some uncertainty to the results of this investigation. For our purposes, drumlins were indicated by parallel groups of elliptical contour patterns on topographic maps of areas known to be underlain by glacial deposits and described in the literature as possessing drumlins. Personal acquaintance with New England, and to a lesser extent with New York, makes the results more dependable for these cases than for the North Dakota and Montana areas, where none of the authors have appreciable first-hand experience.

[*] Department of Geological Sciences, Harvard University, Cambridge, Massachusetts.

[**] Department of Geology and Geophysics, Massachusetts Institute of Technology, Cambridge, Massachusetts.

[†] Department of Geological Sciences, Harvard University, Cambridge, Massachusetts. Deceased July 29, 1961.

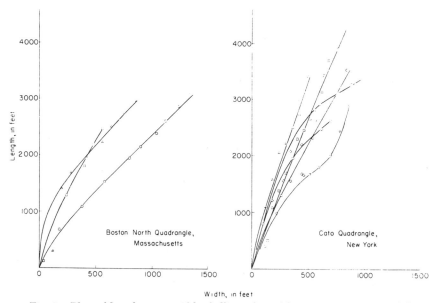

Fig. 1. Plots of length versus width of ellipses formed by successive contours of three drumlins in Massachusetts and six drumlins in New York. Base of drumlin corresponds to uppermost part of curve.

The form of individual drumlins and their orientation and spacing within drumlin fields are the geometrical characteristics most susceptible to quantitative study. The results presented in this paper should be viewed as preliminary only, because the samples are relatively small, and in some cases the procedure and measuring technique may be subject to future revision.

DRUMLIN FORM

Drumlin form has been described in terms of such picturesque expressions as "half-torpedo", "inverted bowl of a spoon", and "egg-shaped". Chorley (1959) related the plan form of drumlins to lemniscate loops and showed the comparison is a reasonably accurate one.

In this paper, drumlin form is compared with the ideal ellipsoid. The standard equation of an ellipsoid with its center at the origin and its semi-axes, a, b, and c coinciding with the rectangular coordinate directions x, y, and z is as follows:

$$\frac{x^2}{a} + \frac{y^2}{b} + \frac{z^2}{c} = 1$$

By generalizing this equation, it can be shown that any topographic contour about an ellipsoidal landform must be an ellipse. Furthermore, the major and minor axes of the ellipse formed by successive contours must have a constant ratio.

The axes of ellipses formed by successive contours of an individual drumlin are readily measured from maps. Results of such measurements from drumlins in the vicinity of Boston, Massachusetts and Cato, New York are

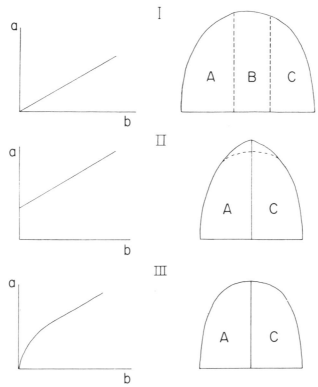

Fig. 2. Idealized relations between drumlin shape and ellipsoids. Sections are perpendicular to long axis *a*. Part I: ellipsoid. Part II: ellipsoid with middle segment B removed. Part III: typical drumlin; upper surface is equivalent to dotted line in section of Part II.

plotted in figure 1. If drumlins were truly ellipsoidal, the curves of figure 1 would be straight lines passing through the origin. A few of the examples shown approximate this condition. In most cases, however, the straight line relation applies only for larger values of length and width (lower drumlin contours) with departure from linearity toward the origin of the curve (higher drumlin contours). As shown in the idealized examples of figure 2, the typical drumlin resembles an ellipsoid from which a middle segment parallel to the long axis has been removed and the resulting sharp crest rounded off.

It appears, then, from figure 1 that drumlins closely approximate the ellipsoidal form, but depart slightly from it in the vicinity of their summits. Furthermore, the steeper curves (fig. 1) for the Weedsport area indicate drumlins are more elongate than those near Boston, a difference not immediately apparent from the topographic maps.

DRUMLIN ORIENTATION

Drumlin orientation as used here refers to the azimuth, in degrees east or west of north, of a line drawn through the long axis of each drumlin. Data ob-

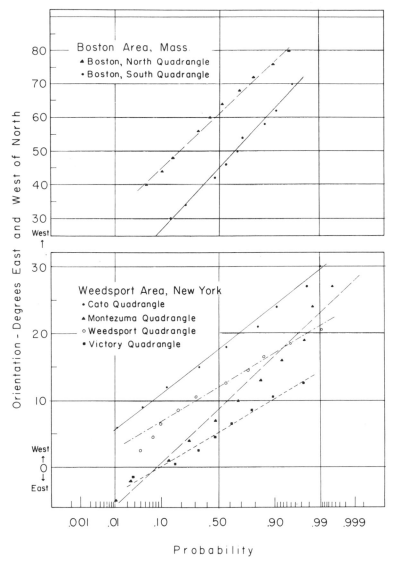

Fig. 3. Frequency distribution of drumlin orientations.

tained from measurements in 7½-minute quadrangles near Boston, Massachusetts and Weedsport, New York are summarized in table 1 and plotted as frequency distributions in figure 3. The data plot as nearly straight lines on arithmetic probability paper, indicating that the distributions of orientation are approximately normal. Furthermore, dispersion about the mean values is quite small, especially in the Weedsport area. The larger spread in orientation of drumlins near Boston may be the result of greater bedrock control. Accord-

TABLE 1

Summary of data on drumlin orientation

Location	Area Measured (sq. mi.)	Number of Drumlins	Length: Width ratio and Drumlin Shape	Range of orientation	Mean Orientation	Standard Deviation	Drumlin Composition
Cato, New York Quadrangle	36	85	3-4:1 Lenticular	5°W-36°W	18°W	4.0	Hard, compact cores of till with abundant stones and cobbles grade upward into concentric banding of till with some sandy clay.
Montezuma, New York Quadrangle	24	87	3-4:1 Lenticular	6°E-27°W	10°W	5.7	
Victory, New York Quadrangle	18	36	3-4:1 Lenticular	2°E-14°W	6°W	4.1	
Weedsport, New York Quadrangle	16	71	3-4:1 Lenticular	2°W-27°W	12°W	4.5	
Boston North, Massachusetts Quadrangle	24	19	2-4:1 Mammilary to Lenticular	38°W-84°W	64°W	13.8	Bedrock cores common; drumlins composed of compact till.
Boston South, Massachusetts Quadrangle	48	22	2-4:1 Mammilary to Lenticular	0°W-88°W	46°W	17.4	

ing to Crosby (1934), at least 25 percent of the drumlins in the Boston Basin have rock cores and a majority of the others rest upon rock hills or ridges.

SPACING OF DRUMLINS

Parallel spacing of drumlins is defined here as the distance between centers of any two adjacent or in-line drumlins measured in the direction of mean local orientation. The center is located midway along the long axis drawn through the lowest enclosing contour. No significant difference in the results would be obtained by using the highest elevation of each drumlin instead of the center point as a reference for measurement, because drumlins within the small areas measured have fairly uniformly asymmetric profiles.

In order to measure drumlin spacing parallel to the direction of long axis orientation, it is necessary to adopt certain conventions which are illustrated in figure 4. In this example, Drumlin A is measured only with reference to Drumlin B, namely distance A′B′. Drumlin C is not adjacent to Drumlin A. and therefore the center point distance A′C′ is not used. Drumlin B can be measured with reference to three drumlins which are adjacent—A, C, and D. Drumlin D can be measured with reference to adjacent Drumlin B and in-line

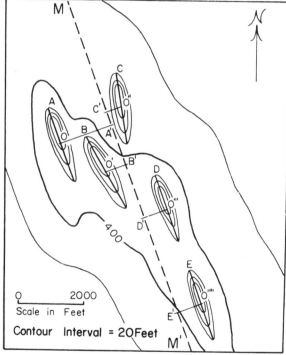

Fig. 4. Procedure used for measurement of drumlin spacing. Center points of drumlins A, B, C, D, and E are indicated by the symbol O. Line M-M′ is the mean orientation of the drumlins. Perpendiculars drawn from the center points to line M-M′ define points A′, B′, C′, D′, and E′. The distance between any two of these points on line M-M′ is taken as the distance between center points and is defined as parallel spacing.

Drumlin E. Parallel spacing was not measured for cases in which (1) the distance perpendicular to the line of mean orientation between adjacent drumlin axes was greater than 2000 feet, or (2) the distance between ends of the lowest enclosing contour of the in-line drumlins was greater than 1000 feet. It is apparent that there is duplication in the measurements. However, in taking these data only one of any duplicate measurements was recorded.

Perpendicular spacing, defined as the distance between drumlin centers measured at right angles to the line of mean orientation, was also determined. Measurements were made on the same drumlins used to obtain the data on parallel spacing.

The data on drumlin spacing are plotted as histograms in figures 5 through 7. As shown by these diagrams, and also by table 2, which summarizes data for the New York area only, drumlin spacing varies through a wide range, even showing large differences between adjacent quadrangles.

The histograms of parallel spacing (fig. 5) for the New York area are multimodal, with some suggestion of a 500- to 600-foot interval between modes.

Perpendicular spacings of drumlins (fig. 6 and 7) also show multimodal distributions with tendencies toward equal intervals between modes. Results obtained from the Montana-North Dakota area seem to be in general agreement with the finding of Gravenor and Meneley (1958) who showed that fluting wavelengths in Alberta generally have a strong mode at about 300 feet and in some cases a lesser mode at about 600 feet.

DISCUSSION OF RESULTS

Basically, there are two principal hypotheses of drumlin origin: (1) formation by erosion of pre-existing glacial drift, and (2) formation by localized deposition of drift from the ice. At present, there is no comprehensive

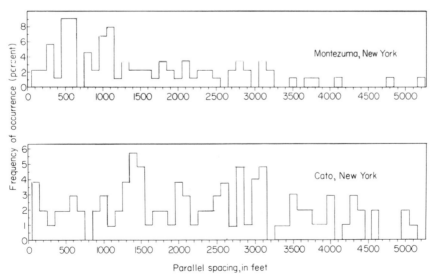

Fig. 5.   Parallel spacing of drumlins in the Weedsport area, New York.

171

TABLE 2

Comparison of drumlin spacing in adjacent quadrangles

Location	Area Measured (square miles)	Number of Drumlins	Number of Measurements	Mean Drumlin Size (feet)		Parallel Spacing (feet)		Perpendicular Spacing (feet)	
				Length	Width	Range	Mean	Range	Mean
Cato, New York Quadrangle	36	85	105	3910	1080	100-5100	2280	100-2800	1140
Montezuma, New York Quadrangle	24	87	89	3500	920	100-5200	1540	100-3000	1060

172

theory to account for the interaction between terrain and ice to produce the streamlined drumlin form. The data presented in this paper do not permit elaboration of a theory, but it is believed that information of this kind is necessary before specified mechanisms of drumlin origin can be fully understood.

It appears that too much emphasis has been placed on drumlin origin's being entirely the result of either erosion or deposition, because both must be in-

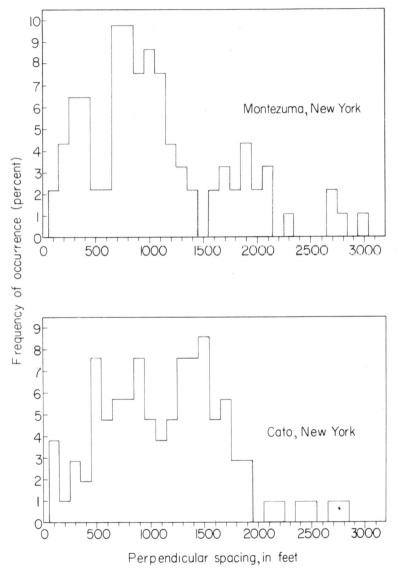

Fig. 6. Perpendicular spacing of drumlins in the Weedsport area, New York.

volved. Rather than overemphasizing the role of either deposition or erosion, it seems preferable to consider drumlins as ice-molded features.

Any theory for drumlin formation must take into account both the dynamic conditions within the ice, and variations in physical properties of till and bedrock. However, it appears that the behavior of the ice may be of greater significance than the properties of rock and unconsolidated material affected by it. Glacial flutings in Alberta described by Gravenor and Meneley (1958) are developed under a wide variety of bedrock and topographic conditions. Aronow (1959) concluded that the occurrence of drumlins and other streamline glacial features in North Dakota cannot be correlated with either bedrock lithology or topographic conditions. This implies that locations of drumlin fields and spacing of drumlins within fields are controlled by characteristics of the moving ice.

Dispersion of drumlin orientation within small areas emphasizes the diverging character of glacier flow, and segregation of the ice margin into lobes

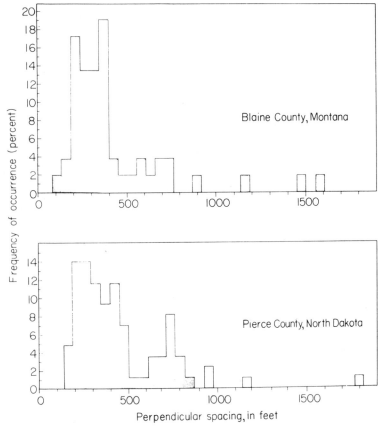

Fig. 7. Perpendicular spacing of drumlins in Montana and North Dakota. Measurements, which were made on air photos, are identical in all respects with those made on topographic maps of other areas.

of small dimensions. Although uniform spacing is not as well defined for the drumlins discussed here as for the flutings in the western Great Plains, there is considerable indication of preferred spacing. Some property of the ice, possibly a wave phenomenon, may be the critical factor involved. Thickness and velocity of the ice and variations in the physical properties of unconsolidated materials under the glacier are among the factors that would affect wave-motion. One way to relate geometrical properties of drumlins to thickness of former glaciers is through preconsolidation-pressure studies of overridden silts in drumlin areas (Harrison, 1958).

SUMMARY AND CONCLUSIONS

Measurements of geometrical properties of drumlins in three areas of the United States indicate that:

1) individual drumlins are approximately ellipsoidal in form; greatest departures from the ideal ellipsoids occur in the summit area.

2) orientations of drumlins within small areas are characterized by normal distributions with fairly small dispersions.

3) spacing between drumlins, both parallel and perpendicular to the axis of elongation, is characterized by multimodal distributions with some indication of periodicity. In the areas considered here, and also those described by other workers, it seems likely that spacing between drumlins is more closely related to characteristics of moving ice than to bedrock lithology or topography.

Development of a satisfactory theory of drumlin origin requires collection of additional data on drumlin properties and also further development of the theory of glacier flow. The present investigation emphasizes that the difficulty of precise description of even simple landforms is a formidable obstacle to progress toward their genetic explanation.

ACKNOWLEDGMENTS

We are grateful for assistance at various stages of this study from Arnold Lent, Department of Metallurgy, Massachusetts Institute of Technology, and Joseph H. Hartshorn and Roger B. Colton, U. S. Geological Survey.

REFERENCES

Aronow, Saul, 1959, Drumlins and related streamline features in the Warwick-Tokio area, North Dakota: Am. Jour. Sci., v. 257, p. 191-203.
Chorley, R. J., 1959, The shape of drumlins: Jour. Glaciology, v. 3, p. 339-344.
Crosby, I. B., 1934, Evidence from drumlins concerning the glacial history of Boston Basin: Geol. Soc. America Bull., v. 45, p. 135-158.
Gravenor, C. P., and Meneley, W. A., 1958, Glacial flutings in central and northern Alberta: Am. Jour. Sci., v. 256, p. 715-728.
Harrison, P. W., 1958, Marginal zones of vanished glaciers reconstructed from the pre-consolidation-pressure values of overridden silts: Jour. Geology, v. 66, p. 72-95.

10

Reprinted from *Jour. Galciol.*, **7**(51), 377–390 (1968)

THE FORMATION AND SHAPE OF DRUMLINS AND THEIR DISTRIBUTION AND ORIENTATION IN DRUMLIN FIELDS

By IAN J. SMALLEY and DAVID J. UNWIN

(University College, London W.C.1, England)

ABSTRACT. If glacial till contains more than a certain minimum boulder content, it is dilatant and requires a much larger stress to initiate shear deformation than to sustain it. If the stress level at the glacier-terrain interface drops below a certain critical level, or the till reaches its critical boulder-content density, then the till beneath the glacier packs into stable obstructions. These are shaped into streamlined forms by the glacier and are found distributed at random in drumlin fields. Due to drumlin coalescence there is a normal distribution of drumlin axes about the direction of ice movement.

RÉSUMÉ. *La formation et morphologie de drumlins et leur répartition et orientation en champs de drumlins.* Si un argile à blocaux contient plus d'une certaine quantité minimum de blocaux, il est dilatable et exige une plus grande tension pour débuter une déformation par cissaillement que pour le soutenir. Si le niveau de tension entre le glacier et le terrain descend au-dessous d'un certain niveau critique, ou si l'argile à blocaux atteint sa densité critique de blocaux, alors l'argile à blocaux au-dessous du glacier s'agglomère et forme des obstacles stables de formes carénées et distribués au hasard en champs de drumlins. À cause de leur coalescences, les axes des drumlins s'alignent avec la direction du mouvement glaciaire en une répartition normale.

ZUSAMMENFASSUNG. *Über Entstehung und Gestalt von Drumlins sowie deren Verteilung und Orientierung in Drumlinfeldern.* Wenn glazialer Geschiebemergel mehr als ein bestimmtes Minimum von Geröll enthält, ist er dilatant und benötigt eine weit höhere Spannung zur Einleitung von Scherformationen als zu deren Fortführung. Sinkt das Spannungsniveau zwischen Gletscher und Untergrund unter einen bestimmten kritischen Wert oder erreicht der Geschiebemergel seinen kritischen Geröllgehalt, so häuft sich das Geschiebe unter dem Gletscher zu stabilen Hindernissen auf. Diese werden vom Gletscher stromlinienförmig gestaltet und erscheinen in zufälliger Verteilung als Drumlinfelder. Infolge der Verschmelzung von Drumlins haben die Drumlinachsen eine Normalverteilung um die Richtung der Eisbewegung.

INTRODUCTION

The Pleistocene glaciers left many signs of their passage; some of the most remarkable are the low, smooth hills known as drumlins. Drumlins have been observed and investigated for a considerable time but to date no completely satisfactory theory has been evolved to explain their formation. Several factors appear to be involved and the problem of choosing the significant and avoiding the irrelevant is very difficult. This paper represents the development of the theory, already published in outline (Smalley, 1966[b]), that the formation of glacial-till drumlins is a consequence of the dilatancy of the material of which they are composed. Most drumlins are formed from till, a material with very complex rheological properties and this paper is mainly concerned with these, but any other formation which may be called a drumlin (e.g. a rock drumlin) is also considered.

The drumlins we see now were formed in the Pleistocene and like most aspects of Pleistocene geology there is a vast literature relating to them. Literature up to the mid-1950s has been surveyed by Charlesworth (1957) in his great compendium. He does, however, tend to concentrate on an assessment of theory and opinion rather than on the collection of available drumlin facts. The best collection of drumlin-shape and size data is probably still that by Ebers (1926). There have been several notable papers since the Charlesworth survey and these are discussed in later sections of this paper.

Two questions need to be answered to provide an explanation of the observed nature of drumlins. These concern (1) the nature of the geomorphic force which shaped the drumlins, and (2) the way in which the force accomplished the shaping. The necessary geomorphic force is generally ascribed to the action of glaciers, so only one problem remains, that of the mode of interaction of the glacier and the terrain which leads to the formation of drumlins. The consequences of this interaction are low streamlined hills which tend to occur in groups or fields and within these fields they have a certain distribution of orientations and certain positional relationships.

176

In this paper the formation of the drumlins is explained by invoking the dilatancy mechanism, the shape by the requirements for streamline flow, and the orientation is simulated by a random model. The distribution of drumlins in a drumlin field has been investigated in two ways. Random models have been produced and the spacings of drumlins in these compared with the measured spacings of real drumlins, and drumlin-spacing data derived from maps have been analysed for indications of random spacing. Both approaches indicate that drumlins do occur at random within drumlin fields.

FORMATION

The relative scarcity of drumlin forms suggests that the conditions necessary for their formation were rigorous, and infrequently achieved. It is proposed that the basic conditions for the formation of glacial-till drumlins were:

   (i) The glacier–terrain relationship was such that at the base of the glacier the terrain material was being continuously deformed. Some of this terrain material was carried along by the glacier so that shear deformation occurred within the terrain material.

   (ii) The deformed layer was composed of a concentrated dispersion of boulders and large rock particles in a dense clay–water system, the material usually called boulder clay or glacial till. For drumlins to form, the large particles in this till layer had to form a dilatant system.

Dilatancy is a property of granular masses. When a granular mass, for example some dry sand, is at rest, it forms a stable heap and the particles in the heap are relatively closely packed together. When the granular material is deformed, it expands; this is the phenomenon of dilatancy, first observed by Reynolds (1885), related to geology by Mead (1925), and fully investigated by Andrade and Fox (1949). There is no completely satisfactory definition of dilatancy. Boswell ([1961], p. 73) stated that dilatant systems are those in which the anomalous viscosity increases with increase of shear. This is just an elaborate way of saying that dilatant materials are more resistant to shear stresses than might be expected. It is the high resistance to initial deformation of till which leads to the formation of drumlins.

The dilatancy of granular masses under compressive and shear stresses has been demonstrated using the simple apparatus illustrated in Figure 1a. A cylindrical ram of diameter 7.5 cm is forced down at a constant rate by a hydraulic press into a cylindrical container of diameter 15 cm which contains dry sand of particle size about 0.5 mm. A graph showing load vs deformation (i.e. ram travel) is produced automatically by the machine recorder as shown in Figure 1b. When the load is first applied, the sand aggregate starts to expand because its natural close packing is being disturbed and a more open packing is developing. The material continues to expand and to resist deformation until it reaches the state of maximum expansion, at point A in Figure 1b. Further deformation causes a collapse and the required deforming load drops; point B on the graph is reached and further deformation only requires loads of about this magnitude.

The curve shown in Figure 1b is an *ideal* example produced with dry sand as the granular medium. A curve of similar form is produced when dilatant glacial till is deformed from rest. The constraints produced by the container are more noticeable when the till is deformed and because of these constraints quantitative measurements were not attempted; unfortunately an uncomfortably large container would be required to eliminate the constraints. The very large particles in the till cause irregular deformation but the general shape of the load-deformation curve (Fig. 1c) is similar to the ideal model case shown in Figure 1b.

In the suggested drumlin-forming mechanism the glacial till is being continuously deformed by the movement of the glacier and a stress level in the general range indicated by c in Figure 1b and c is involved. Within the thin deformed layer of till there is a certain variation in stress level. If the stresses drop below level B then the expanded material collapses into the static stable form and there are no stresses of magnitude A available to cause sufficient dilation to get the compacted material moving again so the flowing till flows around it, shaping it so that it causes the minimum of disturbance in the flowing stream of till.

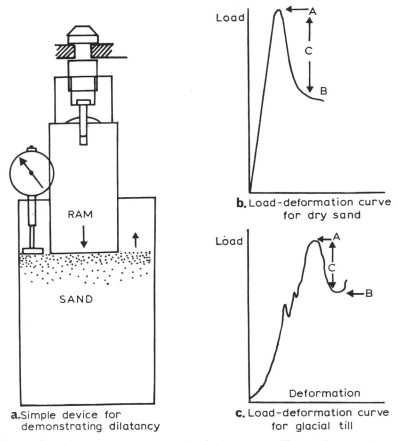

a. Simple device for demonstrating dilatancy

b. Load-deformation curve for dry sand

c. Load-deformation curve for glacial till

Fig. 1. The dilatancy of granular materials. a. Simple device for demonstrating dilatancy; b. Load-deformation curve for dry sand; c. Load-deformation curve for a glacial till.

If the glacier–terrain relationship is such that the general stress level is greater than A, then no drumlins can form; the glacier sweeps all before it. If the general stress level is below B, no drumlins can form because continuous deformation of the till is impossible. In most large continental glacier systems one would expect the mean stress level to be greater than A so that most tracts of glaciated land have no drumlins. Towards the periphery of the ice sheets the stress levels drop and there is a region in which the stress levels are such that drumlins form and farther towards the periphery the stress levels drop too low to allow drumlin formation and end-moraine structures form. The drumlins form when the general stress level beneath the ice sheet is in the region represented by range C in Figure 1b and c, i.e. below A but above B; this is illustrated diagrammatically in the ice-sheet section shown in Figure 2. Fairly rapid thinning might be expected to occur at the edges of large ice sheets, giving rise to a relatively narrow drumlin belt. Small local glaciers, such as covered the northern half of Ireland during the late Weichsel glaciation, mostly produced stress levels in the C range and thus very large and extensive drumlin fields exist in Ireland.

It is the large rock fragments in glacial till which make it dilatant; it may be that the clay part of the material is thixotropic and thus aids the shaping process. When part of the till layer packs into an obstruction, a local high-pressure zone is formed as the rest of the till layer flows past the obstruction. Under the influence of this local high pressure the clay part of the till becomes more fluid. The more fluid clay flows more easily around the obstruction carrying the large particles with it and giving the obstruction a smooth streamlined shape. The phenomenon of thixotropy is even more difficult to define satisfactorily than that of dilatancy. The two phenomena are effectively opposites in that thixotropic materials are less resistant to shear stresses than might be expected. A well-known example of a thixotropic material is non-drip paint; in the can this appears almost solid but under the pressure of the brush it flows quite smoothly and easily. In the same way the till clay flows more easily at the drumlin–till flow interface and thus facilitates the formation of a smooth streamlined shape.

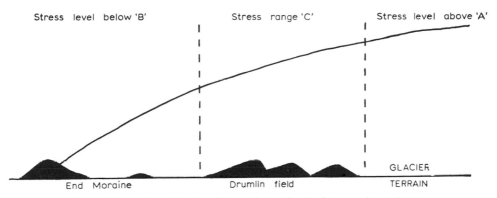

Fig. 2. Cross-section at the edge of the ice sheet with critical stress regions indicated.

Thus the model for the drumlin-forming mechanism depends on two postulates: that the glacier is separated from the terrain by a continuously deformed layer of till, and that within this till layer there is a range of stress levels. The drumlins formed are basically accretional forms, although the material is gathered by local erosion; this falls in fairly neatly with the views of Thornbury ([1954], p. 391). He described drumlins as characteristically lying several miles to the back of end moraines and suggested that their streamlined forms indicated lodgement of till, with subsequent over-riding and reshaping by the ice. He also suggested that drumlin development just behind end moraines may possibly be related to rapid thinning of the ice in this zone; this is the state of affairs illustrated in Figure 2.

If the continuously deformed till-layer postulate is discarded (the action being justified by reference to Occam's razor), a related erosional mechanism suggests itself. It can be imagined operating in Ireland where the late Weichselian glaciers advanced over morainic material resulting from an earlier glacial phase. Within this earlier till there was a certain distribution of relatively large rock fragments, in some places more tightly packed together than in others. The glacier advanced with stress level c in operation at the working interface and, while the till with the loose packing of boulders was smeared easily across the landscape, the till patches with the high boulder content were obstinately dilatant. For the closely packed parts to be eroded they had first to expand but the weight of the glacier pressing down prevented them from doing so. The glacier provides the eroding force but, paradoxically, this very force prevents erosion. The obstruction, *very* closely packed by the passage of some of the ice, was sculptured into the most convenient shape as the glacier forged on. The most convenient shape was the one which caused the least interference with the flow pattern; this was a streamlined shape (Chorley, 1959) and thus drumlins achieved their characteristic form.

179

Drumlins formed from older, stratified drift material are presumably formed by this simple mechanism and drumlins formed from the latest drift are formed by the more complex method. Actually, both can operate at the same time, and must in fact necessarily do so. There is, as Gravenor (1953) noted, both an erosional and a depositional aspect to the formation of drumlins. Aronow (1959) suggested that to explain the phenomenon of drumlin formation recourse must be had to "something" in the now vanished ice. He was only very slightly off target, and it is suggested that this "something" was the dilatancy of the till.

Flint ([1957], p. 68) stated that ". . . there is a complete gradation, independent of outward form and within a single field, from rock to drift. This suggests that any one group was molded contemporaneously under a single set of conditions". Within the various forms there will be basic material requirements; solid obstructions will require stresses far above the A level in order to be eroded away completely, these gain an accretionary smoothing of clay to give streamline flow contours. The drumlins with high boulder contents also represent zones of resistance because the material was dilatant and required stresses consistently at the A level for complete erosion. As the rock content decreases a minimum point is reached where the mean c level stresses at the glacier–terrain interface cause erosion so effectively that no drumlins can form. Thus drumlins can have assorted contents, provided the minimum rock or boulder content is reached.

It will be seen from Figure 2 that due to the rapid thinning at the periphery of the ice sheet the critical stress range at the A level is passed through much more slowly than the critical stress at the B level. In other words, the area where the drumlin-forming mean stress level c is operating is bounded at the up-stream side by a fairly diffuse boundary and on the down-stream side by a more precise boundary. If a drumlin field could be examined in its entirety, it should exhibit a concentration of drumlins near the fringe of the ice sheet, behind the end moraine, and the frequency of drumlins should decrease in the parts of the field which were farther under the ice. When the general stress level drops below the A range, drumlins may begin to form but the general stress level is too high for large-scale formation; an occasional drumlin forms, possibly with an elongated shape due to the relatively high ice pressure. The stress level drops as the ice thins and more drumlins form. As the edge of the ice sheet is approached the ice thins rapidly and the stress level drops rapidly below the critical B range. At this level the glacier is still an efficient debris transporter and the carried till load is eventually dumped as end moraine.

SHAPE

Two methods have been advocated to describe the shape of drumlins. Chorley (1959) has proposed that the plan form of the drumlin can be described by a polar equation of the form $\rho = l \cos k\theta$. In this equation l is the length of the drumlin, k is a dimensionless number which effectively indicates the width of the drumlin, ρ and θ are the two variables. Reed and others (1962) preferred to use an equation of the form

$$(x^2/a)+(y^2/b)+(z^2/c) = 1.$$

This produces an ellipsoid with its centre at the origin; a, b and c are the semi-axes, and x, y and z are the rectangular coordinate directions.

Each method has advantages and disadvantages, but the Chorley method appears to be the more useful. The disadvantages of the Chorley method are that it only operates in two dimensions and the equation is in polar coordinate form which is perhaps more difficult to manipulate than the more common rectangular coordinate system which is used by Reed and others. The great advantage of the Chorley method is that the value k serves to classify the shapes of drumlins; if the value of k is known then the shape of the drumlin is known. The other desirable thing about the Chorley equation is that it is obviously related to the mode of formation of the body it describes. It is accepted that drumlins are a consequence of glacier flow and flowing systems flow most easily around a streamlined object. Chorley's equation is that for a streamlined form; actually it is a slight simplification but in terms of goodness of fit and simplicity it is the best available.

Reed and others claimed no particular advantage for their representational method but it appears to have two: it gives a three-dimensional model and it works in rectangular co-ordinates. Actually, these advantages are relatively trivial and are completely overshadowed by the method's overwhelming disadvantage—that it lacks the physical meaning which Chorley's model has. Reed and others decided that drumlins have ellipsoidal shapes and stated the equation for an ellipsoid without really considering the consequences or implications of either the observation or the equation.

The value k has been calculated by Chorley (1959) for some drumlins described by Alden (1905). He gave values for 23 drumlins; the mean value was just under 6 and the mode value was between 3 and 4. If comparisons are to be made with these pioneering values of Chorley, then it is necessary to facilitate the production of k values for observed drumlins. Figure 3 is a

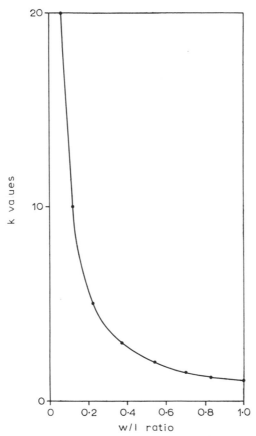

Fig. 3. Variation of Chorley value k with width/length ratio for lemniscate-type curve.

nomogram devised to convert the width–length ratio (w/l) into the k value. The curve of $\rho = l \cos k\theta$ was drawn for different values of k and the widths were measured, the length l remained constant. The w/l ratio was then plotted against k to produce the graph shown in Figure 3. The w/l ratio is preferred to the l/w ratio because the limits are 0 — 1 (real) rather than 1 — ∞ (unreal).

Charlesworth (1957, p. 389) has given some data for average dimensions of drumlins, indicating that the mean w/l ratio is of the order of 0.40; this gives a k value of just under 3. Some more recent observations by Heidenreich (1964) include values of l and w for Canadian drumlins. He found that in the four drumlin fields investigated w/l ratios decreased fairly uniformly until a critical width was reached, after which the drumlins could increase in length while width remained constant. Drumlins appeared to be divisible into two distinct populations, those of k value around 3–4 and those of high k values which usually had reasonably constant width within the same field. Heidenreich stated that until constant width was reached, width and length increased fairly constantly at a ratio of 0.37 which gives a k value of 3. It appears that a k value of about 3 is the most usual.

DISTRIBUTION

If either the accretionary or the erosional or, as seems most likely, both, mechanisms are working to produce drumlins their distribution might be expected to be completely random, or in other words there is no reason to expect a non-random distribution. Within the boundary conditions for drumlin formation the most important variable is probably the variation of properties in the available glacial till and these can be expected to vary randomly. The distribution of drumlins has been investigated in two ways. A random model has been devised and the distribution of drumlins in the model field has been compared with the observations of Reed and others (1962) and Vernon (1966), and drumlins in real fields have been subjected to nearest-neighbour analysis.

Random-placement model

The random process used to produce the model was devised initially to give one-dimensional random packings (Smalley, 1962) for subsequent comparison with three-dimensional packings of sedimentary particles (Smalley, 1964). To produce a model of a drumlin field, the method must function adequately in two dimensions; it has been applied successfully in two dimensions in producing models of crack systems in lava flows (Smalley, 1966[a]).

The terrain is represented by a square 100 × 100 frame (the numbers have arbitrary units). The glacier flows over this chosen piece of terrain and drumlins are formed within the square demarcation. The drumlins are placed at random, the points at which they occur being indicated by coordinates taken from a set of random-number tables. A very suitable set of tables is that by Kendall and Smith (1951); this gives lists of two-digit random numbers between 00 and 99 which are ideal when taken in pairs to represent positions within the square field. Two adjacent sides of the square frame are considered as axes, the first two-digit number of a pair represents the x coordinate, the second number the y coordinate. The plotted point marks the stoss end of the drumlin. The number of points plotted depends on the density of drumlins in the field; the mode density according to Charlesworth (1957, p. 389) is 3 per square mile. Thus if the side of the square frame represents 2 miles, 12 random points are needed.

In this model some overlaps will probably occur, giving rise to some rather odd shapes and this also appears to occur in nature. It is possible, for comparison, to produce a model in which overlaps do not occur. Each drumlin is drawn in as soon as its position is known and any subsequent drumlins which would overlap are rejected.

To produce a model field containing 12 drumlins the following procedure would be adopted: 12 random points are plotted in the square field and a drumlin is drawn, using a template, for each point. Some of these drumlin shapes will overlap giving several larger, more complex drumlins. Each of these larger drumlins counts as one so more coordinates are plotted until 12 distinguishably separate drumlins are formed; this is the model distribution and spacing measurements can be made and compared with real fields. Reed and others (1962) made perpendicular and parallel measurements which are difficult to define and make.

182

The Vernon (1966) direct measurements are capable of a more rigorous definition; every drumlin has two measurements associated with it. These are the distances of the two nearest drumlins in an up-stream direction, one on each side of the long axis of the reference drumlin.

In order to produce a comprehensive set of measurements of the drumlin in the random-model field special boundary conditions have to be introduced for the field otherwise the drumlins near the edge have no adjacent drumlin. The edge effect is eliminated by introducing edge drumlins into special adjacent fields. If a drumlin is placed at 05 51 by the random number coordinate then one is also placed at 105 51; similarly, one at 62 02 has a corresponding placement at 62 102. This has the effect of producing a closed field so that every drumlin has the required neighbours. Overlaps at the top are introduced at the bottom and so on. A sample field is shown in Figure 4; the measurements taken are indicated. Four

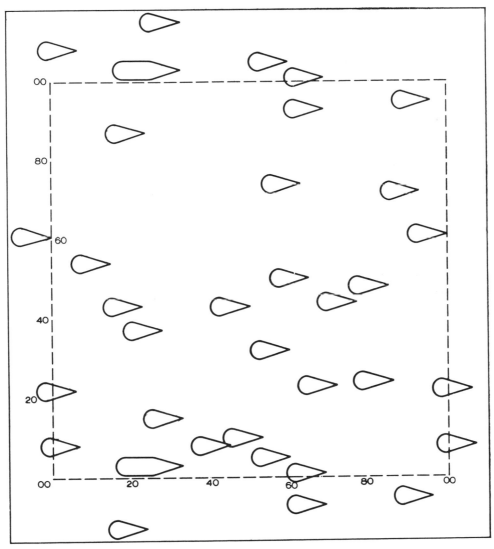

Fig. 4. Drumlin-field model produced by random placement of drumlins in a square field; edge of field represents a length of 4 572 m, each drumlin is 457.2 m long.

183

closed fields were produced, each containing 24 drumlins. Each field yielded 48 measurements and these were amalgamated to produce the histogram shown in Figure 5; this should be compared with the histogram produced by Vernon (1966) from similar measurements on the Ards Peninsula drumlins, a distinct similarity is apparent.

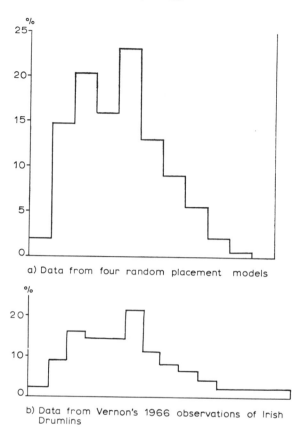

a) Data from four random placement models

b) Data from Vernon's 1966 observations of Irish Drumlins

Fig. 5. Histograms of direct spacings of drumlins. a. Data from four random-placement models; b. Data from Vernon (1966), observations of Irish drumlins.

Nearest-neighbour analysis

It is possible to test the fit between a natural field and a random model without recourse to direct simulation since the "nearest-neighbour" statistic tests the manner and degree to which the distribution of individuals in a population in a given area departs from that of a random distribution. The statistic, which was devised initially for use in the biological sciences by Clark and Evans (1954) and has since been used in settlement geography (Hagget, [1965]), is defined as:

$$R = D_{obs}/0.5(A/N)^{-1/2}$$

where D_{obs} is the linear distance between any one point in a specified area A and the nearest neighbouring point, and N is the number of points within the area. The statistic is such that values of R range from zero for maximum aggregation to 2.1491 for maximum (hexagonal) spacing. Random distributions give values of unity. However, it is important to note, as Clark and Evans pointed out, that a random distribution in this sense is defined as one in which any point has the same chance of occurring on any sub-area as any other point, and that any sub-area of specified size has the same chance of receiving a point as any other sub-area of that size. So defined, randomness is a spatial concept, depending upon the boundaries chosen. Thus a distribution may be random with respect to one area but non-random with respect to a larger area which includes it.

The method can be easily applied to a drumlin field. Natural patterns are taken off either

Fig. 6. Drumlin distribution in the Vale of Eden.

aerial photographs or maps of any scale consistent with identification of the features and the nearest-neighbour distances are found for a specified number of individuals within a sub-area. This is to eliminate side effects; a drumlin's nearest neighbour may fall outside the area A chosen. The distances are summed and a mean value D_{obs} calculated. Knowing the area, values of R can be calculated directly.

Table I presents the results of such an analysis performed using 1 : 25 000 topographic maps for small areas in Ireland, and one in England. Recognition of drumlins on the maps was done on the basis of contour pattern, drumlins being recognized where ellipsoidal patterns occur. This may result in the recognition of some forms which are not drumlins, but it is thought that the bias introduced is not great. Figure 6 shows the field which was used in England.

TABLE I. NEAREST-NEIGHBOUR ANALYSES FOR FOUR NATURAL DRUMLIN PATTERNS

Number	Location	Area km²	Density number/km²	Number	D_{obs} m	R
1	Co. Clare, Ireland	24.08	1.33	32	563.9	1.3096
2	Co. Clare, Ireland	24.08	1.53	37	554.2	1.3842
3	Co. Clare, Ireland	56.16	1.62	91	497.9	1.2784
4	Vale of Eden, England	40.04	0.92	37	583.7	1.1295

As the table shows, the R values derived for natural patterns range from 1.1295 to 1.3842, indicating that the distribution lies somewhere between "uniformly spaced" and "random", as the figures are usually interpreted. It should be noted, however, that the random-placement models gave R values of 1.1500, 1.0441, 1.2766 and 1.1367, indicating that the real distributions are actually truly random.

In general, both methods used indicate that the drumlin fields examined contain a random distribution of drumlins.

ORIENTATION

If a closely packed drumlin field is produced by the random-placement method, there will be considerable overlapping. Figure 7A shows a close packing of drumlins in which the

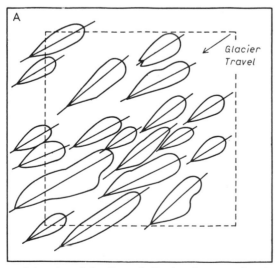
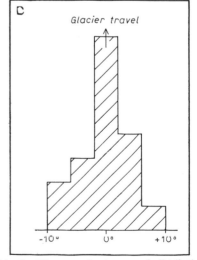

Fig. 7. Orientation of drumlins. A. Random-placement orientation model; B. Histogram showing normal distribution of orientations about chosen glacier-flow direction.

outlines have been slightly modified to produce coalescence and a smooth outline for the complex drumlins. This produces some drumlins of a more complicated shape and having orientations which do not lie exactly along the line of glacier travel. The direction of glacier travel is chosen at random after a suitable number of location points have been plotted. In Figure 7A the apparent drumlin axes are marked; the orientations have been measured and the results are shown as a histogram in Figure 7B. A normal symmetrical distribution of orientations is produced; this agrees exactly with the orientation measurements on real drumlins reported by Reed and others (1962).

DISCUSSION

Initially, the rigorous nature of the postulated boundary conditions suggests that drumlins should only rarely be observed in nature. Most authors who have written on the subject have agreed that this is the case. Fairchild (see Alden, 1911, p. 734) spoke of "The combination of several factors which do not commonly occur in nature", whilst Aronow (1959) considered drumlins to be related to "unknown conditions" which are not simply related to either the terrain or to the nature of the available materials. The dilatancy mechanism suggested requires both that the material be dilatant and that the stresses within the glacier system are within a certain critical range, and it is suggested that these are the "unknown conditions".

Available evidence indicates that drumlin fields are to be found in bands paralleling, and some distance to the rear of, end moraines. Vernon (1966) has noted that the drumlin fields of eastern Ireland are bounded to the south by the Carlingford Re-advance moraine, whilst Alden (1911) showed that in eastern Wisconsin the drumlin fields are confined to a zone within 30 to 50 miles (48 to 80 km) of the limits of an ice lobe where the ice was radiating towards its margin. He suggested that the height of ice cover over the drumlin field varied from 1 450 ft (442 m) over the initial part to 450–830 ft (137–253 m) where drumlins cease to be found, some 5 miles (8 km) from the limit of the advance. Taylor (1931) noted that drumlins occur in wide basins or lowland areas such as some of the Great Lakes basins and the lowlands of Ireland, Scotland and Scandinavia, suggesting that either dying or slowly moving ice is an essential pre-requisite to their formation. Similarly, Aronow (1959) has noted that there is a continuous gradation from perfect drumlin forms through drift patches to true end moraine; this is a critical observation since this is exactly what would be expected if the dilatancy mechanism were responsible for drumlin formation. As the stress level falls below the lower critical level B, drumlins cease to form and end-moraine material may begin to be deposited.

The observed distribution of drumlins relative to each other is as would be expected from the dilatancy mechanism. As Flint ([1957]) pointed out, single isolated drumlins are rare, possibly indicating that certain boundary conditions are involved and that a definable stress range is required for drumlin formation. This tends to occur over a relatively wide area when it does occur, although within a large ice sheet covering a large area of suitable terrain the occasional anomaly may occur to produce an isolated drumlin.

It has been the usual practice to divide theories of drumlin formation somewhat arbitrarily into erosional and accretional ones. Aronow (1959) placed in the erosional group suggestions by Gravenor (1953) and Flint ([1957]), and in the accretional group theories by Thwaites (1957), Thornbury ([1954]) and Charlesworth (1957). Reed and others (1962) observed the traditional dichotomy but also suggested that both erosion and deposition must be involved.

Few, if any, of the so-called theories are actually theories, if a theory is required to account for the formation of the streamlined hill of till or rock. Most "theories" are really suggestions for boundary conditions within which some unspecified process operates, and the stipulated conditions may be remarkably imprecise. They are perhaps none the worse for that and it is dangerous to be too precise with very little data to base judgements on and, although much has been written about drumlins, very little hard fact has emerged. Only recently, in accordance with the general trend in geomorphology, has the quantitative investigation of drumlins and drumlin fields been undertaken. Chorley (1959) has given a meaningful interpretation of the shape of drumlins; Reed and others (1962) have measured distributions and orientations, and Vernon (1966) has measured spacings and distribution; these three papers represent the basis of the new approach to the problem of drumlin formation.

Vernon (1966) has suggested that drumlins form when ice flow induces a pressure differential between the front and back of an obstacle. Pressure melting at the front of the obstacle creates a zone of greater mobility in the ice which moves to the zone of low pressure behind the obstacle, leaving the debris behind as till; presumably in front of the obstacle. The idea of pressure melting is attractive and obviously feasible but there are objections to incorporating it into the drumlin-forming mechanism. If the glacier is moving enough till to subsequently form a drumlin or an end-moraine deposit, the bedrock obstacle, unless it is drumlin-sized itself, will tend to be insulated from the actual ice by the till load. Also, the fact that pressure melting may be involved does not throw any light on the problem of drumlin distribution.

Gravenor (1953) has listed ten conditions which must be satisfied by a theory of drumlin formation and these can be summarized as follows:

1. Drumlins may consist of a variety of materials.
2. They may have layers of stratified materials which may be faulted or folded.
3. Rock and till drumlins have the same shapes and occur in the same fields.
4. Many glaciated areas do not have drumlins.
5. They occur in fields which are wider than most moraines and they rarely occur singly.
6. They have a streamlined shape with the blunt end pointing up-stream.
7. Lamination may be present.
8. Some drumlins may have cores but most do not.
9. Drumlins are found behind end moraines.
10. They are aligned parallel to the ice-flow direction.

It is suggested that the dilatancy theory provides a mechanism which satisfies these conditions, and especially those relating to distribution (numbers 4, 5 and 9). Perhaps one further condition could be added to Gravenor's ten:

11. They are formed beneath temperate glaciers.

The suggestion by Nye (1965) that temperate glaciers slip on their bed is to some extent supported by the observation of MacNeill (1965) that there appeared to be free water present when some of the drumlins in south-western Nova Scotia were formed. A glacier whose bottom ice is below the melting point probably does not slip on its bed and thus cannot form drumlins.

CONCLUSIONS

A comprehensive theory of drumlin formation can be derived from two basic precepts: (a) glacial-till drumlins are formed from dilatant material, and (b) they were formed when the stresses in the till–glacier interface zone were within certain critical limits. There must be deformation of the actual till, which is the dilatant material. The bulk of transported till may be carried within the ice sheet; for drumlins to form there must be shear deformation in till at the glacier–terrain interface. It may be that this requires the ice to be advancing over already deposited till.

MS. received 14 April 1967 and in revised form 23 February 1968

REFERENCES

Alden, W. C. 1905. The drumlins of south-eastern Wisconsin. *U.S. Geological Survey. Bulletin* No. 273, p. 9–46.

Alden, W. C. 1911. Radiation of glacial flow as a factor in drumlin formation. *Bulletin of the Geological Society of America*, Vol. 22, p. 733–34. [Abstract. Discussion by H. Fairchild, p. 734.]

Andrade, E. N. da C., *and* Fox, J. W. 1949. The mechanism of dilatancy. *Proceedings of the Physical Society*, Sect. B, Vol. 62, Pt. 8, p. 483–500.

Aronow, S. 1959. Drumlins and related streamline features in the Warwick–Tokio area, North Dakota. *American Journal of Science*, Vol. 257, No. 3, p. 191–203.

Boswell, P. G. H. [1961.] *Muddy sediments*. Cambridge, W. Heffer and Sons, Ltd.

Charlesworth, J. K. 1957. *The Quaternary era, with special reference to its glaciation*. London, Edward Arnold. 2 vols.

Chorley, R. J. 1959. The shape of drumlins. *Journal of Glaciology*, Vol. 3, No. 25, p. 339–44.

Clark, P. J., *and* Evans, F. C. 1954. Distance to nearest neighbour as a measure of spatial relationships in populations. *Ecology*, Vol. 35, No. 4, p. 445–53.

Ebers, E. 1926. Die bisherigen Ergebnisse der Drumlinforschung. Eine Monographie der Drumlins. *Neues Jahrbuch für Mineralogie, Geologie und Paläontologie. Beilagebände*, 53, Abt. B, p. 153–270.

Fairchild, H. 1911. Discussion. [See Alden, 1911, p. 734.]

Flint, R. F. [1957.] *Glacial and Pleistocene geology*. New York, John Wiley and Sons, Inc.

Gravenor, C. P. 1953. The origin of drumlins. *American Journal of Science*, Vol. 251, No. 9, p. 670–81.

Hagget, P. [1965.] *Locational analysis in human geography*. London, Edward Arnold.

Heidenreich, C. 1964. Some observations on the shape of drumlins. *Canadian Geographer*, Vol. 8, No. 2, p. 101–07

Kendall, M. G., *and* Smith, B. B. 1951. *Tables of random sampling numbers*. Cambridge, University Press. (Tracts for Computers, No. 24.)

MacNeill, R. H. 1965. Variation in content of some drumlins and tills in south-western Nova Scotia. *Maritime Sediments*, Vol. 1, No. 3, p. 16–19.

Mead, W. J. 1925. The geologic rôle of dilatancy. *Journal of Geology*, Vol. 33, No. 7, p. 685–98.

Nye, J. F. 1965. The flow of a glacier in a channel of rectangular, elliptic or parabolic cross-section. *Journal of Glaciology*, Vol. 5, No. 41, p. 661–90.

Reed, B., *and others*. 1962. Some aspects of drumlin geometry, by B. Reed, C. J. Galvin, Jr., and J. P. Miller. *American Journal of Science*, Vol. 260, No. 3, p. 200–10.

Reynolds, O. 1885. On the dilatancy of media composed of rigid particles in contact. *Philosophical Magazine*, Fifth Ser., Vol. 20, No. 127, p. 469–81.

Smalley, I. J. 1962. Packing of equal O-spheres. *Nature*, Vol. 194, No. 4835, p. 1271.

Smalley, I. J. 1964. Representation of packing in a clastic sediment. *American Journal of Science*, Vol. 262, No. 2, p. 242–48.

Smalley, I. J. 1966[a]. Contraction crack networks in basalt flows. *Geological Magazine*, Vol. 103, No. 2, p. 110–14.

Smalley, I. J. 1966[b]. Drumlin formation: a rheological model. *Science*, Vol. 151, No. 3716, p. 1379–80.

Taylor, F. B. 1931. Distribution of drumlins and its bearing on their origin. *Bulletin of the Geological Society of America*, Vol. 42, No. 1, p. 201. [Abstract.]

Thornbury, W. D. [1954.] *Principles of geomorphology*. New York, John Wiley and Sons, Inc.

Thwaites, F. T. 1957. *Outline of glacial geology*. Revised edition. Ann Arbor, Edwards Brothers Inc.

Vernon, P. 1966. Drumlins and Pleistocene ice flow over the Ards Peninsula/Strangford Lough area, County Down, Ireland. *Journal of Glaciology*, Vol. 6, No. 45, p. 401–09.

III
Shear and End Moraines

Editor's Comments on Papers 11 Through 14

End moraines of various sorts, large and small, are the third and last group of deposits to be composed mainly of till. Looped moraines around the ends of mountain glaciers have been mapped and described as they were made in the Alps during the last five to seven centuries. Clearly every moraine represents some readvance of ice with maximum stand for a decade or two before retreat, but the exact origin of the deposit is not deduced. Descriptions involve the three general processes of dumping, pushing, and thrusting of till into a ridge form. These are the hypotheses considered by Price (1973, p. 81–94) for all glaciers, especially continental ice sheets.

Dumping till involves a conveyor-belt mechanism to move the debris up to the marginal surface of an ice cap. Ice with new dirt is moving up continually on shear zones. This was first fully described and partly documented in 1950 on Barnes Ice Cap, Baffin Island, so that article comes first (Paper 11). But this article led to some disagreement among glaciologists; the most cited objection is by Weertman (Paper 12; also found in Weertman, 1962). Weertman questions the source of the "endless belt" debris. Most field studies have supported some facets of the original "shear moraine" hypothesis (Østrem, 1964; Souchez, 1966), but some, such as Hooke (1970), provide excellent measurements upon which to question the nature of the shearing motion in cold glaciers frozen to the ground. Still, most glacial geologists doing lithologic pebble counts would agree that 70 to 95 percent of the morainal material comes from within a kilometer or two of the ice edge in temperate continental moraines, so the emphasis turns again to basal gathering of debris on rough surfaces by regelation, as postulated by Weertman.

In many cases the deposition of an end moraine involves action at the base of the glacier as well. For this reason I include Kupsch's 1962 article (Paper 14) on "thrust stacking," which is one of the first clear demonstrations of basal processes. One could refer back also to Paper 6 by Moran in Part I; he applies these ground moraine features equally well to end moraine, as was demonstrated long ago in German "stauchmöranen" (Woldstedt, 1961, p. 101–110). Another more recent stratigraphic discovery is the existence of fossil older moraines which were later coated with thin tills from overriding ice (Totten, 1969).

Push moraine is a simple concept, but active ridges bulldozed by ice rarely exceed 5 meters in height. Except as they become ridges in larger moraine systems, even these are minor moraines. De Geer pioneered the idea that small detached moraines could be made by the annual "winter" push while (summer) ablation was curtailed (1940; Paper 23). Although formation of these minor moraines is limited to ice cliffs in water, and varves are indeed annual, very careful recent work by Strömberg shows that the ridges must have other explanations (Paper 13).

There is a very large literature on minor moraines. In North America, Gwynne of Iowa (1942 and later) discussed very minor broad "washboard" features on prairie till plains, adhering to the winter push explanation. Hoppe (1952) found evidence (fabrics) in Swedish Lappland that such material squeezed up into crevasses. Andrews (1963) and Andrews and Smithson (1963) measured fabrics on the striking "cross-valley moraines" of Baffin Island to show deposition below water level in crevasses parallel the ice cliffs of Barnes Ice Cap. Less known are squeeze-ups observed above water level in receding glaciers of Glacier Bay (Mickelson, 1971, 1974). Possibly Elson's summary (1969) is the best source for most of the early literature not reproduced here.

The most famous moraine system of all, the Salpausselka Moraines of Finland (Sauramo, 1929), are not genetically explained here because these moraines are not built of till (Virkkala, 1963). These are broad deltaic masses of washed sand which developed in one to three centuries of ice edge halted in Baltic water bodies. As yet there seems to be no detailed study of the formation of such a moraine at a present ice margin, although there are other ice-age examples, such as in eastern Maine.

References Cited

Andrews, J. T., (1963a) Cross-valley moraines of the Rimrock and Isortorq river valleys, north-central Baffin Island, N.W.T.; a descriptive analysis: Geograph. Bull., no. 19, p. 49–77.

———— and Smithson (1963) Cross-valley moraines of the Rimrock and Isortorq river valleys, north-central Baffin Island, N.W.T.; a quantitative analysis: Geograph. Bull., no. 20, p. 82–127.

De Geer, G. (1940) Geochrononlogia Suecica: Principles: K. Svenska Vetenskaps Akad. Handl., 3rd ser., v. 18, no. 6, 367 p.

Elson, J. A. (1969) Washboard moraines and other minor moraines: in *The Encyclopedia of Gemorphology*, ed. R. Fairbridge, Van Nostrand Reinhold, New York, p. 1213–1219.

Gwynne, C. S. (1942) Swell and swale pattern of the Mankato lobe of the Wisconsin drift plain in Iowa: Jour. Geol. v. 50, p. 200–208.

Hooke, R. L. (1970) Morphology of the ice-sheet margin near Thule, Greenland: Jour. Glaciol., v. 9, p. 303–324.

Hoppe, G. (1952) Hummocky moraine regions with special reference to the interior of Norrbotten: Geografiska Annaler, v. 34, p. 1–72.

Mickelson, D. M. (1971) Glacial geology of the Burroughs Glacier area: Ohio State University, Institute of Polar Studies, Rept. 40, 149 p.

———— (1974) Till ridges presently forming above and below sea level in Wachusett Inlet, Glacier Bay, Alaska: Geografiska Annaler, v. 56, ser. A, no. 1–2, p. 111–119.

Østrem, G. (1964) Ice-cored moraines in Scandinavia: Geografiska Annaler, v. 46, p. 282–337.

Price, R. J. (1973) *Glacial and Fluvioglacial Landforms:* Oliver & Boyd, Edinburgh, 242 p.

Sauramo, M. (1929) The Quaternary geology of Finland: Comm. Géol. Finlande Bull., v. 86, 110 p.

Souchez, R. A. (1966) The origin of morainic deposits and the characteristics of glacial erosion in the western Sør Rondane, Antarctica: Jour. Glaciol., v. 6, p. 249–254.

Totten, S. M. (1969) Overridden recessional moraines of north-central Ohio: Geol. Soc. America Bull., v. 80, p. 1931–1946.

Virkkala, K. (1963) On ice-marginal features in southwestern Finland: Comm. Géol. de Finlande Bull., v. 210, 76 p.

Weertman, J. (1962) Mechanism for the formation of inner moraines found near the edge of cold ice caps: U.S. Army Cold Regions Res. Eng. Lab. Res. Rept. 94, 11 p.

Woldstedt, P. (1961) *Das Eiszeitalter,* v. 1, Stuttgart, 374 p.

11

Reprinted from *Jour. Geol.*, **59**(6), 567–577 (1951)

DEVELOPMENT OF END MORAINES IN EAST-CENTRAL BAFFIN ISLAND[1]

RICHARD P. GOLDTHWAIT

Ohio State University

ABSTRACT

Five steps in the accumulation and deposition of end moraines are postulated from observations of profiles and ablation features around the south end of Barnes Ice Cap. This end of the icecap is generally more expanded than it was throughout the past century, but a retreat, producing end moraines, has begun on the southern and southwestern sides.

Dirt is raised to the surface on shear planes along a narrow fringe of black ice around the edge of the icecap. This is till, exposed in small amounts by 5–7 feet of ablation each summer. This film of debris slips and slides down the 10°–30° marginal slope and gathers over the lower ice slope. Where the ice becomes blanketed by 3 feet of ablation moraine and dirt-filmed ice above melts back rapidly, a trough develops parallel to the margin, isolating a steep-sided moraine on an ice core. After many years this material is let down irregularly to solid ground by flow of the till cover, cuts by transecting streams, and lateral undermining of the exposed core ice.

INTRODUCTION

The complete accumulation and emplacement of an end moraine involves more than one summer season. It was obvious at the outset of the Baffin Island Expedition of 1950[2] that the whole event could be neither witnessed nor measured at any one spot. The best substitute consisted of (1) careful observation of processes operating along a retreating ice edge and (2) comparison of different sectors of the ice edge, in order that stages might be deduced. Some 40 miles of the southeastern margin of Barnes Ice Cap were inspected on foot or ski; another 75 miles were observed from the air (fig. 1). From these observations the production of the end moraine is pieced together.

It is not proposed that all end moraines form this way. The Barnes Ice Cap is 90 miles long and 45 miles wide; hence its climatic regime cannot be identical with that of a large Pleistocene ice sheet. Furthermore, it is a cold, arctic icecap: the deep ice proved to be about 13° F., which is undoubtedly near mean annual temperature.[3] Such ice is more viscous than ice near 32° F. (Dorsey, 1940, p. 455), and it cannot contain deep subglacial streams as warmer ice does. Nevertheless, the features produced here are similar to end moraines in many parts of North America, and it is hoped that the study of moraine formation here will shed light on that of the Pleistocene (Flint, 1947, pp. 127–130).

PAST EXPANSION AND INITIAL RETREAT

The south end of Barnes Ice Cap at 69°33′ N. to 69°55′ N. on north-central Baffin Island is as large today as it has been for several centuries. This is shown strikingly by the plants. On coarse soils right up to the ice edge thick tufts of *Cassiope* (*C. tetragona*) and mosses are established. Expedition botanists and others advise that this takes at least 25 years or even a century to become established.[4] Several species of lichen grow

[1] Manuscript received June 2, 1951.

[2] General account by Baird (1950), leader of the expedition.

[3] Mean annual temperature at Clyde, 100 miles away on the seacoast, is 11°1 F.

[4] P. Dansereau was in charge of botanical studies, and M. E. Hale was lichenologist. Both men visited the Barnes Ice Cap briefly and expressed opinions verbally.

profusely on bouldery areas right up to the moraine of the active glacier. The lichenologist estimated 50 years as a minimum time for these dense large colonies to develop.

Two geological features indicate some expansion in the recent past. One is the to the low ice edge (below 2,000 feet). Yet, immediately adjacent to the southeastern margin, there are *roches moutonnées*, striae, and lunoid furrows, expressing northeastward motion parallel to the ice edge (fig. 1). They were made by ice with a very different center of motion.

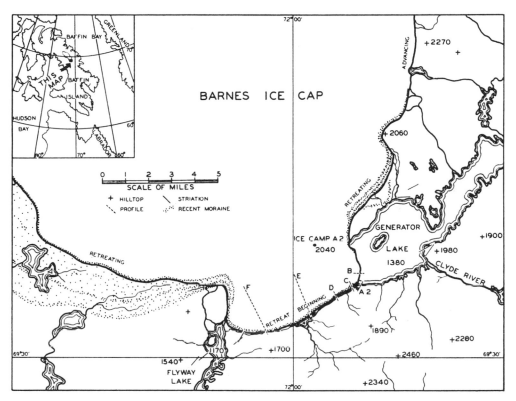

FIG. 1.—Sketchmap of the southern end of Barnes Ice Cap, showing the steep glacier margin, areas of accumulating moraine, and altitudes of hills (Paulin).

great disparity between past and present directions of ice motion at the very southeastern edge of the icecap. It is fair to assume that motion today is radially southward and southeastward from the high central dome (at 3,700 feet altitude) As centers of ice accumulation must shift slowly, these features were formed probably several centuries ago. Second, the southern third of the ice margin is immediately adjacent to fully formed, sorted polygons and stripes of boulders as

PLATE 1

A, Belts of recent retreatal moraine along the southwestern side of Barnes Ice Cap.
B, Shear planes filled with dirt, just east of profile *D* at the southern end of Barnes Ice Cap.

A

B

Retreatal moraine belts and dirt-filled shear planes, Barnes Ice Cap

PLATE 2

A, Upper edge of the black-ice slope, showing veneer of creeping dirt on the right. The uppermost dirt-laden shear crack overhangs by shove from the main glacier on the left.

B, Rock table consisting of a 4-foot boulder on a 15-inch-high pedestal of ice retaining original shear planes. Near profile *E*.

A

B

Black ice slope and rock table

well as fissure polygons. Granted that some boulders may be gathered into miscellaneous groups rapidly, recent studies in Alaska[5] and observation of contrasts on fossil and active outwash trains on Baffin Island indicate that patterned soils and even large ice wedges take one or more centuries to form. These many areas of patterned soil, identical in form to those in Alaska, could not have been disturbed by glacier movement for many decades.

In addition, there is climatological evidence of expansion along the northeastern side today. The inclination of the outer 500 feet of the icecap averages $10°–20°$, and the sharp angle between ice edge and adjacent land catches deep snowdrifts in the winter. These did not completely melt in the summer of 1950; it was estimated that half the northeastern ice edge was still covered by snow in late August. Air photographs taken in July, 1948, show some drifts for that year.[6] Thus there is some net addition of snow, which turns to ice along the very edge; therefore, Barnes Ice Cap is growing here and there by marginal accretion.

This general expansion has already turned to initial retreat on the south and southwest. On these sides the outer 500 feet of icecap generally slope only $5°–10°$. For 20 miles there are three belts of retreatal moraine parallel to active ice (pl. 1, A). These are not yet fully inhabited by local tundra plants, and annual rings in dwarf willow (*Salix arctica*) on the outer moraine show a maximum

of 25 years.[7] At the southestern end of this belt the inner two moraines join in one area 2,500 feet broad, in which the ice is heavily covered with ablation moraine. Comparison of· the photographs taken in the summer of 1948 with the ice margin as it was in 1950 suggests thinning and additional dirt cover. Probably this retreat set in locally on the southern and southwestern sides a few years ago. The net effect of a general recession on the southwest and advance on the northeast is a very slow northeastward shift of this end of the icecap.

SOURCE OF MORAINAL DEBRIS

All the debris which makes up the present superglacial moraine came up on shear planes through the thin edge of the glacier from the rock bottom. All ice up to 100 or 200 feet above the steep toe is layered with myriads of dirt-filled thin fractures, striking roughly parallel to the ice edge (pl. 1, B). The dirt filling in each shear crack is from $\frac{1}{100}$ inch to 1 inch thick; generally it is less than $\frac{1}{10}$ inch, although dirt is smeared widely at the exposure. Small pebbles and even large boulders are seen occasionally where they were in transit up these shear planes. Shear planes above and below such boulders and till clots swell around them in "augen" structures. Some of these bands crop out horizontally along the steep ice

[5] Personal communication with R. F. Black, U.S. Geological Survey, January, 1951.

[6] Much use was made of trimetrogon photographs flown by the Royal Canadian Air Force in 1948 and 1949.

[7] Other plants collected indicate brief stable conditions but no great age: *Cerastium alpinum, Luzula nivalis, Oxyria* (sp. ?), *Saxifraga oppositifolia, Stellaria longipes.*

edge for thousands of feet, but most of them die out in 10–100 feet. The clear ice layer between seams varies between $\frac{1}{8}$ and 6 inches in thickness. Multitudes of meltwater channels, 3–15 feet deep, transect these shear planes, revealing that they dip 10°–36° back under the main icecap (fig. 2).

It is apparent that the layers of ice between shear planes move very slowly over one another, like a deck of cards pushed up an incline. The 30° dip back under thicker ice corresponds to faults by

this increase of motion higher in the ice, however; so it may be that motion in any one year is dominantly along a selected few shear planes here and there, while intermediate fractures suffer no displacement.

Dirt filling in the shear planes is restricted to the very edge of the icecap. On the air photographs and in measured profiles the average width of the dirt-smeared surface ice is 450 feet, which is about 1 per cent of the whole area of active icecap. The upper limit of this dirty

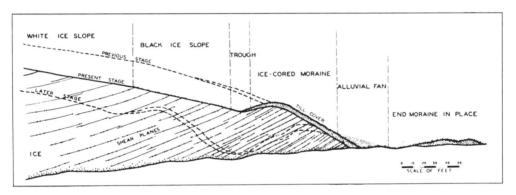

FIG. 2.—Diagram of the retreating margin of Barnes Ice Cap, based upon averages measured. Dots indicate dirt in transport up inclined shear planes, or gathering on the surface as ablation moraine, and finally deposited as end moraine (*right*).

mass thrust in rocks. On one ice cliff at the extreme southeastern end of Barnes Ice Cap (pl. 1, *B*), dowels were set in and measured, but motion in a few days was less than the errors of the method.[8] In many places visited, the upper ice layers were seen to jut out 1 or 2 inches over layers immediately below. Some shear cracks were clearly older, having been transected and displaced by younger and often steeper fractures. In some of the dirty fractures tabular pebbles and boulders showed a preferred orientation parallel to the plane of shear, suggesting differential motion. Not all layers exhibited

[8] Most dowels fell out or were broken within a week; motions were insignificant.

ice stands out sharply after the winter snow disappears (pl. 2, *A*). The ice cliffs in six lakes show dirt-filled ice at the very lake shores. Above the dirty ice rim, one may travel for tens of miles without finding a sand grain—even less a boulder. There are many clean shear planes in this upper ice, but they are farther apart and less regular in strike and dip as one gets a quarter-mile or more back from the margin. Slightly open steep fractures were encountered, to 12 feet in depth, in an excavation at Ice Camp A2 (fig. 1). The belt of black ice around the rim rises and falls from 1,300 to 2,000 feet in altitude with the marginal topography, but it rarely rises more than 200 feet vertically above the flat land around the icecap. In a few places it seems to reflect topography buried just back under the ice edge, for it arches up perceptibly where no hill is exposed. In a few shallow depressions seen in flight, single spots of dirt outcrop as much as a mile back from the ice edge, and from each one a narrow fan of dirt-covered ice reaches to the edge. This suggests a nunatak just submerged.

Because the shear planes slope $10°-36°$ back under the ice and dirt outcrops generally no more than 200 feet above the ice toe, it stands to reason that the raising of the dirt is from basal ice not very far back under the glacier—theoretically, no more than 1,600 feet from the edge (assuming that the dip of shear planes seen at the surface continues at depth). Turbulence of plastic motions may raise some dirt into the lower few feet of the ice across the whole base of the glacier, because erratics identified as having formerly traveled up to 15 miles toward the northeastern coast were found; but it seems that concentrations of dirt along shear planes do not take place underneath more than 280 feet of ice. This is the depth of ice 1,600 feet back from the toe.[9] Commonly, dirt does not crop out half so high as 200 feet, nor do shear cracks slope so gently as $10°$; so the depth of common shearing is more nearly 140 feet. No wonder that most of the rock material raised is of the same lithology as that of outcrops at the ice margin!

COLLECTION OF DIRT DOWN THE ICE SLOPE

The black-ice slope is covered with a film of dirt which is moving toward the

[9] Based upon bottom profiles worked out by C. A. Littlewood, of Dominion Observatory, Canada, from a gravimetric survey. Parts of these are shown in fig. 3, profiles C, E, and F.

ice toe (pl. 2, A). This smear of dirt does not average $\frac{1}{10}$ inch thick; an easy blow of the ice ax brings white ice. On hot days ($50°-60°$ F.) the silt and sand could be seen fairly oozing and creeping in small rills. Black-ice areas had an average incline of $15°$ from horizontal; variations from $6°$ to $20°$ were recorded (fig. 3). Fine clay moves off rapidly into regular small streams, spaced 50 feet apart or less. Stones more than 6 inches thick do not transmit much sun's heat through to the ice; hence, while surrounding ice melts down a foot or two, each larger stone stands on a shaded pedestal. Between the end of June, when snow cover melted, and July 19, rock tables 3–15 inches high developed (pl. 2, B). Sooner or later each pedestal melts thin, and each stone becomes unseated. It slides off downhill toward the ice margin. Cracks or hollows on the black ice catch the sliding dirt and accumulate a few inches of half-washed silt, sand, and stones. This also "protects" the ice beneath by reducing melting, and after a warm season it becomes a dirt cone or ridge or chain of cones left 20–120 inches high, while surrounding ice melted down. The sides of these cones melt back with uniform slopes at $30°$ to $38°$, suggesting regulation of melting by the angle of repose of the slipping and sliding sediments undermined at the top.[10] By all these motions dirt moves generally toward the ice edge.

Ablation of this black-ice slope is the most rapid of all parts of the glacier because the thin film of dirt absorbs and traps far more of the sun's radiant energy than does the white ice above. Dowels set into drilled holes along profile C on July 18, 1950, and remeasured until ablation slowed down (August 14, 1950),

[10] Similar in form and development to dirt cones recently described by Swithinbank (1950).

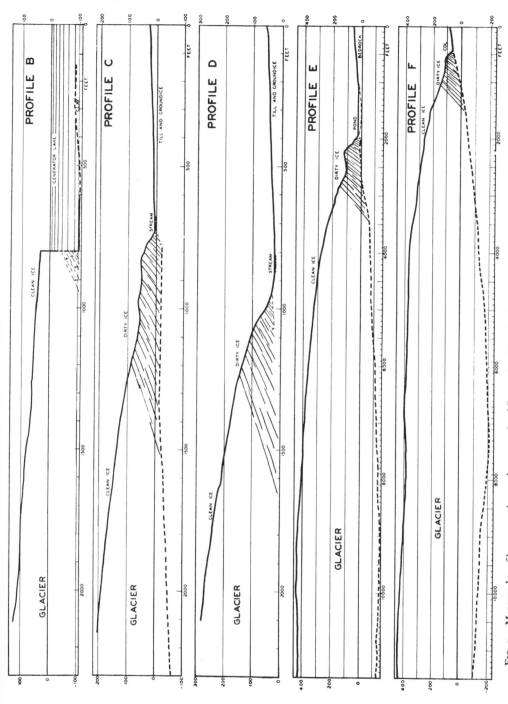

Fig. 3.—Measured profiles across the southern margin of Barnes Ice Cap. Surface determined by hand-level and tape (*B*, *C*, *D*) or by rod and transit (*E*, *F*). Base of the glacier from gravimeter survey by C. A. Littlewood along approximate lines *C*, *E*, and *F*. Base-level = 0 is Generator Lake. Vertical exaggeration $2\frac{1}{2}$ times in *E* and *F* only.

201

A

B

Stages of marginal moraine formation

melted out at the rate of 8 inches per week on clear ice and 13 inches per week on dirt-filmed ice. As the warmest season between July 1 and August 11 produced the only severe thawing and some snow had to be melted off from the ice surface in this period, the annual loss of black ice is estimated at 60–80 inches, while that of white ice just above is 40–60 inches. This difference in melting rates contributes to the steepness of the ice edge. Tangents to the curved profiles (fig. 3) on black ice were 3° steeper than those on the white-ice slope just above. This change in slope was often a visible sharp break (pl. 2, *A*).

Toward the foot of this steepened black-ice slope the creeping dirt gathers so thick that it protects the ice from annual melting. In some places, where recession may be just beginning or where ice motion just equals ablation, dirt was stacked steeply against the ice toe. Where the recession had gone on further, the ice was coated with 2–4 feet of till for at least 100 feet up onto the steep slope.

DEVELOPMENT OF A TROUGH

Along many miles of the southern and southeastern margins of Barnes Ice Cap there is a depression in the ice at the base of the black-ice slope and parallel to the steep, dirt-covered ice edge. It is frequently 30–90 feet above the flatter ground beyond the ice (fig. 2, "trough"). Between this trough and the outermost steep edge of the glacier there is a "false" ice-cored moraine (pl. 3, *A*). No significant motion was detected in

the thick ice of this core. Since the depression, 2–50 feet deep, lies between this ice-cored moraine and the active black-ice slope, it stands to reason that the depression did not exist at first and that there was formerly a continuous convex slope from the black ice to the top of the ice-cored moraine (fig. 2, "previous stage"). Only in this way could the dirt cover of the moraine collect over the lower half of the dirty ice rim for a period of years.

The ice-cored moraine is remarkable, in that the thickness of till cover on the ice rarely varies more than ± 1 foot from the average of 3 feet seen at cut after cut and dug or drilled at a dozen localities. This probably reflects the fact that the thaw zone approached this depth early in August and that this is the requisite mantle for effective insulation from summer temperatures. The steep outer face of the moraine, which slopes at 30° (fig. 3) indicates the steepest slope upon which such a protective cover can cling for a few decades. The mechanical composition of the till on profile C has a median near $\frac{1}{2}$ mm. and consists of about 14 per cent silt and clay (below $\frac{1}{16}$ mm.), 20 per cent fine sand ($\frac{1}{16}-\frac{1}{4}$ mm.), 34 per cent coarse sand ($\frac{1}{4}-1$ mm.), 15 per cent granules (1–4 mm.), and 17 per cent pebbles (over 4 mm.). Such sandy till with boulders is to be expected in an area of coarse crystalline rocks, especially after ablation seepage has washed out some fines.[11]

[11] Fine sediment brought up from the bottom of Generator Lake, 2,000 feet from the cliff of Barnes

---

PLATE 3

A, Trough parallel to the ice margin containing a shallow pond. Beyond it is the ice-cored moraine with a breach where the pond has its outlet. Profile *E*.

B, Undermining of the ice core of marginal moraine accumulation. As the dirt-smeared ice cliff melts back, a 3-foot layer of till on top avalanches down the melting slope. At the base, the end moraine is being emplaced.

Why does this trough develop in the black ice? Of the various processes observed, the contrast in ablation rate of the ice-cored moraine, which melted down hardly at all, and the black-ice slope, which melted excessively, is judged to be the major factor. Provided that the annual motion of active black ice out over the toe of dirt-covered ice is not so rapid as the annual ablation of the black ice (60–80 inches per year), the black-ice slope will recede. Thus it seems that, in order to initiate the ice-cored moraine—which is to become a belt of end moraine—there is first a rough balance of discharge of dirty ice and ablation on the black-ice slope (fig. 2, "previous stage"). The duration of this determines the volume and extent of the protective dirt cover. The accumulation of a 3-foot-thick cover over the lower half of the former black-ice slope—this is the common thickness—would take on the order of 25 years. Then it must be followed by a decreasing flow of dirty ice or increased ablation, coupled perhaps with increasing height of dirty shear cracks, to allow the black-ice slope to recede. In so doing, it leaves a shelf or trough between black ice and the ice-cored moraine (pl. 3, A).

Once initiated, this trough enlarges by several processes. For one thing, the air in the trough seemed warmer on sunny days. When katabatic winds arose, the trough afforded slight protection. To some extent at least, the quieter air in the trough is heated by radiation from

dark slopes on both sides.[12] In addition, most smaller radial streams from the vast interior ice surface flow into the trough wherever there is one. Some are ponded in enclosed hollows in the trough, and this ‘slightly warmed water melts some ice beneath. In the trough each stream flows parallel to the ice edge until it reaches a breach in the ice-cored moraine. Absorption of the sun's heat by dirt in the streams and corrasion by the moving dirt both deepen the channel. Thus the trough may deepen vertically.

The gathering of dirt in the trough from the black-ice slope counteracts the deepening processes to some extent. If recession of the black-ice slope is rapid, there are no extensive dirt accumulations in the trough, so that it widens and deepens continually; in one area the trough was enclosed on both sides to a depth of 74 feet and was about 500 feet broad. This produces a backslope on the ice-cored moraine which is inclined 13°, on the average, into the icecap (fig. 2, "present stage"). If recession of the black-ice slope is relatively slow, sufficient dirt collects in the trough to protect it, and a broad irregular shelf of dirt-protected stagnant ice is added to the ice-cored moraine. One example was 3,000 feet broad; since the dirt cover averaged about 3 feet thick, this might require several centuries to form. Wherever the cover of ablation moraine was broad, the topography was irregular, with minor ridges and hollows marking spasms of retreat of the active black ice.

In most sectors the dirt cover in the bottom of the trough is irregular. It gathers by at least three means: (1) early in July, bands of poorly sorted sand came down the black-ice slope as slush avalanches, ending on snowdrifts and later melting down into the trough; (2) better-sorted layers of silt, sand, and fine gravel

Ice Cap, was 16 per cent clay (to 0.002 mm.) and 84 per cent silt (0.002–0.063 mm.). Washing out of such clays resulted in the very sandy material (less than 3.3 per cent silt and clay) sampled from one dirt cone on a mountain ice field and from two end moraines near Clyde Inlet.

[12] Measurements taken by W. H. Ward (1951) "show that screen air temperature over debris-covered ice was sometimes as much as 2° C higher than over clean ice 70 m away."

washed spasmodically into temporary pools 10–300 feet long in the trough; and (3) till slipped and slid into the trough from melting dirty ice on both sides.

After a few seasons of melting, spots of thick dirt concentration stand higher as dirt cones. Most dirt cones observed in the trough area were larger than those far up on the black-ice slope, suggesting that they formed under a thick dirt blanket and lasted several years. Some dirt-covered ice cones are capped with a till mass, others with the bedded sand of a former pool.

EMPLACEMENT OF THE END MORAINE

When and how does this accumulated load on ice become a deposit on solid ground? The high ice-cored moraines with 30° slopes contrast markedly with the subdued ridges of till and boulders on solid ground (fig. 2, "end moraine"). It becomes evident that ice-cored moraines can lead to these puny end moraines when one realizes that the solid mantle material of the ice-cored moraine is less than one-tenth its height. Invariably, steepness was associated with a high ice core exhibiting the inclined shear planes of formerly active glacier ice. Under the subdued ridges of end moraine, 100 feet to 50 miles away from the steep ice toe, there are layers of clear crystalline ground ice, 1–5 feet thick. In fact, such ice, common to all arctic regions, was also seen under ground moraine and fine outwash generally.[13] But this ice does not contribute to the relief of end moraine.

Various processes destroy the core of glacier ice in the ice-cored moraine. Wherever the circumferential streams in the trough had ponded and broken over the ice-cored moraine a deep, sharp gash developed (pl. 3, A). These outlet openings occurred at roughly every 1,000 feet along the ice front. On warm days each one delivered a rushing cataract of water to the outer face of the ice-cored moraine and exposed ice walls with slumping till and boulders.

Elsewhere along the steep outer face of the ice-cored moraine, landslides (mudflow and rock avalanche) were effective means of destroying ice. As thawing of the dirt mantle reached 2–3 feet in depth in July, the soggy till became unstable. In a few places the till crept slowly by solifluction; in others it oozed rapidly by mudflow. The core ice exposed on the upper slope then proceeded to melt rapidly (pl. 3, B). On warm days it undermined boulders and stones of the till cover above, and these skidded and bounded down the exposed ice every minute or two. It is necessary only to remove the protective dirt cover here and there to undermine the glacier ice buried under the moraine. Heat penetration of the 3-foot till cover is not necessary, although it may occur. In two places excavation through the covering dirt of the ice-cored moraine to the glacier ice suggests that the thaw layer penetrates to the ice itself by the end of July—at least here and there. Thus the ice core is destroyed, sometimes rapidly, but in any case irregularly.

The texture of the dirt in the final moraine is understandably varied. For the most part, it is the same as the sandy till

[13] Described by many authors and summarized by Black (1950). This interpretation does not agree with Ward (1951), who refers to most buried ice excavated at some distance from the Barnes Ice Cap as "glacier ice." In a few places it was clear that glacier ice was or might be trapped beneath ablation moraines for centuries. The steep topographic rise on ablation moraines covering unequivocal glacier ice was in such contrast with the smooth topography covering widely spaced exposures of clear granular ice layers that most buried ice is believed here to be ground ice (permafrost).

of the ice-cored moraine. Samples from old moraines 40 miles from the Barnes Ice Cap were even more coarse: 8 per cent clay and silt, 6 per cent fine sand, 18 per cent coarse sand, 14 per cent granules, and 54 per cent pebbles. In addition, there are many lenses of well-sorted sand. Logically, these are derived by streams washing them into pools in the trough. Such deposits are too thin and nondescript to merit the topographic description of kame. There is an additional source of coarse gravels and sands on the outer fringes of the ice-cored moraines. The streams debouching through each gorge deposit extensive local fans (fig. 2, "alluvial cone").

Big boulders tend to become concentrated in certain irregular areas of the moraine as it "lets down." At the site of each gorge passing through the ice-cored moraine, large boulders slip into the cut which is transverse to the moraine axis. On the exposed slopes of core ice there is rude sorting, for large boulders bound farther out and fine sediments pile up at the high edge of the heap next to the ice. Thus some initial irregularities of boulder concentration in the low end moraines seem explainable.

Most of the ridges of end moraine near Barnes Ice Cap were no more than 5–15 feet high and 50–500 feet broad. They were first to appear above the snow cover as it melted in June. These ridges occur in nesting groups, but rarely could any one undulating ridge be traced more than a mile across the broad hills. It is to be noted that the distinction between these end-moraine ridges and the intervening ground moraine is only one of relief and magnitude of the local roughness. End moraines representing one simultaneous position of the ice edge stand higher merely because of greater volume. In content, certainly, the upper till of the ground moraine is the same as the till in the end moraine.

The probability that the lower till is plastered onto the rock from beneath the glacier and that it is found under both ground and end moraines is great, but it was not confirmed

Many low end-moraine ridges did not harmonize with the orientation of the present ice-cored moraine. These were covered with well-established tundra vegetation as well as century-old patterned soils, so they are considered very old. However, one area of moraines parallel to the present ice edge along the southwestern side shows three definite belts of recent moraine (pl. 1, A): (1) An outer sharp, sandy, boulder-strewn compound ridge from 10 to 50 feet above adjacent older drift and covered with lichen and many plants. No ice was discovered by digging, and there was no sign of recent slump or of steep, ice-cored slope. (2) An intermediate lower belt, dotted by a few very high humps and covered with some lichen. May contain ice locally because some slopes are steep (?). (3) An inner hummocky high belt, continuous and ice-cored, as exhibited in stream gorges. Very few small lichens were present. Dwarf willow was absent. Here perhaps is the "typical" recessional moraine series formed by accelerating and decelerating retreat (fig. 2, "later stage"). In another area on the eastern side of Barnes Ice Cap three belts of moraine, roughly 300 feet apart, all showed buried ice at places. Both by geological estimate in the field and by an independent estimate of the lichenologist, the relative ages of these moraines are twenty-five, ten, and three years, respectively. The chances are that they are double or treble this and may not be fully "let down" to final resting places for one or more centuries still.

CONCLUSIONS

Assuming that the foregoing observations have been synthesized into a proper relationship, these deductions may be made:

1. The material of the end moraines is nearly all local, being derived chiefly from bedrock only a few hundred feet back under the ice from the glacier margin. The few far-traveled erratics must have moved most of the way near the base of the ice.

2. Most of the material is raised to the ice surface along shear planes, which rise toward the glacier margin and seem to penetrate commonly no more than 250 feet below the ice surface.

3. Nearly all the material of end moraines becomes superglacial load (ablation moraine) prior to its deposition.

4. Some of the clay and silt is removed from the ablation moraine by meltwater seepage over the surface on warm days. The till must necessarily become more sandy and stony.

5. Large volumes of end moraine entail many decades or centuries of accumulation, since dirt concentration in the ice is very lean and much ablation must occur.

6. Irregular hummocky final deposition is implicit in the sloughing-off and avalanching down the sides of an enduring ice core.

7. Rude sorting into surficial boulder-rich zones is accomplished wherever the material is emplaced by avalanching down steep faces of the ice core or in meltwater notches through the ice core.

8. Lenses of sorted sand and gravel may be collected in catchment basins in the trough of ice, chiefly on the side of the moraine toward the main icecap. Fans of dipping gravels are often banked up on the opposite outer side of the moraine.

9. The end moraine and the upper till of the ground moraine differ only in the thickness of concentration. Dividing lines between the two are arbitrary and signify principally decelerating or accelerating recession of the ice edge.

10. These sequences and principles apply only to an "active" ice edge in which some motion takes place to within 100 or 200 feet of the ice toe. No wide area of stagnant ice was studied.

ACKNOWLEDGMENTS —Thanks are due to William H. Ward, of England, and John Waller, of Canada, for helpful comments, criticism, and discussion in the field. Much detailed information was gathered by these men at the ice-edge location of Camp A2. The meticulous planning and patient leadership of P. D. Baird made this work possible. Grants-in-aid of this study were made from funds granted by the Ohio State University Research Foundation to the university, for aid in fundamental research, and from the Penrose Fund of the Geological Society of America.

REFERENCES CITED

BAIRD, P. D., et al. (1950) Baffin Island Expedition, 1950: A preliminary report: Arctic, vol. 3, pp. 131–149.

BLACK, R. F. (1950) Permafrost; in: TRASK, P. D., Applied sedimentation, New York, John Wiley & Sons, Inc., pp. 247–275.

DORSEY, N. E. (1940) Properties of ordinary water-substance in all its phases: water vapor, water, and all the ices, New York, Reinhold Publishing Corp.

FLINT, R. F. (1947) Glacial geology and the Pleistocene epoch, New York, John Wiley & Sons, Inc.

SWITHINBANK, C. (1950) The origin of dirt cones on glaciers: Jour. Glaciology, vol. 1, pp. 461–465.

WARD, W. H. (1951) The physics of deglaciation in central Baffin Island, Brussels, Belgium, Internat. Union Geodesy and Geophysics.

12

Reprinted from *Jour. Glaciol.*, **3**(30), 965–978 (1961)

MECHANISM FOR THE FORMATION OF INNER MORAINES FOUND NEAR THE EDGE OF COLD ICE CAPS AND ICE SHEETS

By J. WEERTMAN*

(U.S. Army Cold Regions Research and Engineering Laboratory [formerly U.S. Army Snow, Ice and Permafrost Research Establishment], Corps of Engineers, Wilmette, Ill.)

ABSTRACT. A new mechanism is described which explains the formation of moraines in the ablation areas of cold ice sheets. The mechanism involves the freezing of water onto the bottom surface of an ice sheet. This water comes from regions of the bottom surface where the combination of the geothermal heat and the heat produced by the sliding of ice over the bed is sufficient to melt ice. A number of criticisms are made of the shear hypothesis, which has been advanced to explain moraines occurring on Baffin Island and near Thule, Greenland. It is concluded that this older hypothesis may be inadequate to account for these moraines.

Although in theory the mechanism proposed in this paper undoubtedly will lead to the formation of moraines, the existing field data are insufficient to prove conclusively that actual moraines have originated by means of this mechanism.

RÉSUMÉ. L'auteur décrit un mécanisme nouveau qui explique la formation des moraines dans les zones d'ablation des indlandsis de glace froide. Le mécanisme implique la congélation de l'eau sous la glace en contact avec le socle sous-glaciaire d'un indlandsis. Cette eau provient des zones de contact de la glace avec le socle sous-glaciaire où la chaleur géothermique et celle produite par le glissement de la glace sur son lit suffisent à fondre la glace. L'auteur formule un certain nombre de critiques de l'hypothèse du mouvement de cisaillement, hypothèse ayant été avancée pour expliquer la formation des moraines que l'on rencontre dans l'Ile de Baffin et près de Thule au Groenland. L'auteur conclut que l'ancienne hypothèse ne peut donner une bonne explication de la formation de ces moraines.

En théorie, le mécanisme proposé ici explique la formation des moraines, mais les mesures faites *in situ* sont encore trop peu nombreuses pour confirmer la formation des moraines suivant le mécanisme proposé.

ZUSAMMENFASSUNG. Zur Erklärung der Bildung von Moränen im Abschmelzgebiet kalter Eisschilde werden neue Vorgänge herangezogen. Sie setzen das Auffrieren von Wasser an der Unterseite eines Eisschildes voraus. Dieses Wasser kommt aus Gebieten der Grundfläche, in denen das Zusammenwirken von Erd- und Reibungswärme zum Schmelzen des Eises ausreicht. Eine Reihe kritischer Bemerkungen gilt der Scher-Hypothese, die zur Erklärung des Auftretens von Moränen in Baffin Island und in der Nähe von Thule auf Grönland herangezogen wurden. Sie führen zu dem Schluss, dass diese ältere Hypothese der Bildung solcher Moränen nicht gerecht werden kann.

Obwohl die in dieser Arbeit beschriebenen Vorgänge theoretisch zweifellos zur Bildung von Moränen führen, genügen die verfügbaren Beobachtungen nicht zu einen schlüssigen Beweis für die Entstehung vorhandener Moränen durch diese Vorgänge.

INTRODUCTION

In this paper a new explanation is offered for the formation of a special kind of moraine which has been studied in great detail in the Thule area of Greenland [1,2,3] and on Baffin Island.[4,5] This type of moraine also occurs in the Antarctic.[6]

The structure of the Thule-Baffin type of moraine† has been known for a long time and is discussed in detail by Chamberlin and Salisbury.[7] They proposed the shear mechanism hypothesis to explain the origin of this phenomenon. Figure 1, which is taken from Bishop's article [1] (a similar picture is given in Ward's paper [4]), serves to illustrate the mechanism as well as to describe the Thule-Baffin moraine. Figure 1 shows a cross-section of the edge of a cold ice sheet which contains a number of Thule-Baffin moraines. It is assumed that "active" ice from the interior of the ice sheet moves outward to the edge, where it is blocked by a zone of "dead" ice. The active ice rides over the dead ice by slippage over active shear planes. It is further assumed that these active shear planes extend to the bottom of the ice sheet and that debris from the bed can be scraped up, carried into the ice to form debris layers, and ultimately be transported to the upper surface along these planes. Of course, the edge of the ice sheet must

---

* Permanent address: Department of Materials Science, Technological Institute, Northwestern University, Evanston, Illinois.

† Since this type of moraine has been investigated so thoroughly on Baffin Island and near Thule, Greenland, I prefer the name "Thule-Baffin moraine" to the commonly-used designation, "shear moraine." The latter name implies that the mechanism of formation is well established, which is not the case.

be in an ablation zone if the debris is to be exposed at the upper surface. In time, a layer of dirt, rock, sand, etc. will accumulate at the upper surface. This layer then protects the ice underneath from further extensive ablation. A debris layer whose thickness is only of the order of 0·5–1·0 m. will furnish such protection and can lead to the formation of ice hills approximately 15 m. in height. These debris-veneered ice hills are the Thule-Baffin moraines which are considered in this paper. They are often called shear moraines, because it has been taken for granted that the mechanism advanced in Figure 1 for their formation is the correct explanation.

It is highly plausible to use the shear mechanism to explain the Thule-Baffin moraine. Yet there are serious difficulties associated with the theory which have not been pointed out in the literature and which will be discussed in the next section. Because of these difficulties, it is believed that another mechanism must be considered in order to give a satisfactory explanation of the phenomenon. In this paper it is proposed that the debris-carrying layers of ice are not formed by a scraping action of cold ice over the bed of an ice sheet, but rather are frozen into the ice by the freezing of water onto the bottom of the ice sheet.* The source of this water is ice which has been melted in the interior of the ice sheet (at the bottom surface) by geothermal heat and heat produced by the sliding of ice over the bed of the ice sheet. This water is forced by the pressure head to flow to the edges of the ice sheet, where it is refrozen to the bottom.

A third explanation for the formation of the Thule-Baffin moraine can be ruled out as highly unlikely. This hypothesis proposes that the debris-carrying ice layers originally were formed farther inland at the upper surface of the ice cap from rock, dirt, etc., which was derived from a protruding nunatak. This debris was buried later by snowfalls and eventually reappeared at the surface because of the ablation at the edge of the ice sheet. If this explanation were correct, one would not expect to find the Thule lobe of the Greenland Ice Sheet so completely fringed with Thule-Baffin type moraines. Instead one would see only an occasional moraine, each traceable to some local nunatak. It appears extremely probable that the debris which causes the Thule-Baffin moraines comes from the bottom of the ice sheet and not from the top surface. The fundamental question to be answered is: How does this debris become incorporated at the bottom surface into the ice mass?

COMMENTS ON THE SHEAR HYPOTHESIS

Appearance of the debris layers

There are a number of criticisms that can be made of the shear hypothesis. Perhaps the most serious of these is the appearance of the debris layers themselves. In the Thule area, tunnels have been dug into the ice sheet, through these layers. As a consequence, the layers are easily examined. Their appearance varies greatly. They can occur as solid layers of stone, sand, etc. up to about 0·5 m. in thickness. At the other extreme they may consist of a layer of slightly dirty ice containing a very fine dispersion of sand or dirt particles. These slightly dirty layers, which may be up to 1–2 m. in thickness, are quite common in occurrence.

Now it may be plausible that ice can scrape layers of solid debris into a shear plane. But it is highly unlikely that a meter-thick layer of very slightly dirty ice would be formed by such

---

* Mr. J. Hollin (private communication) has informed me that he also has been considering the possibility that ice may be formed at the bottom surface of a polar ice sheet.

After writing this paper, it came to my attention that E. von Drygalski [22] has mentioned the possibility of debris layers being frozen onto the bottom surface of an ice sheet. His idea (he bases it on some work of A. Blümcke and S. Finsterwalder [23] on glacial erosion) is that the pressure at the bottom of an ice sheet could cause melting and the water from this melting will flow into the sand or dirt underneath the ice and refreeze. It is not clear how this mechanism would lead to the formation of alternate layers of clear and debris-carrying ice. A criticism that can be made of it is that, although pressure will depress the freezing point of ice, it is still necessary to supply heat, if appreciable melting is to take place. Pressure by itself will not cause the melting envisaged in Drygalski's mechanism.

a scraping action. What would be the mechanism for dispersing the particles picked off the bed through such a thickness of ice? Yet, if the shear hypothesis is to be accepted, it must account for layers of ice containing a fine dispersion of particles. On the other hand, if water were being frozen to the bottom surface of an ice sheet, one would not be surprised to find that thick layers of this refrozen ice were sometimes slightly dirty.

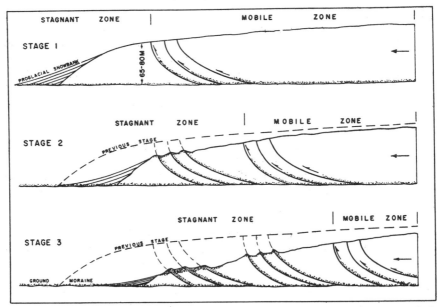

Fig. 1. Diagram showing formation of successive moraines by the shear hypothesis. Figure is taken from Bishop's paper [1]

The possibility of cold ice scraping up debris

A question that the shear hypothesis must answer satisfactorily is: Can debris be scraped from the bed? The present answer to this question is that it cannot if the bottom of the ice sheet is below the freezing point. Direct observation has shown that cold ice of an ice sheet does not slide over a boulder-strewn bed. Goldthwait [2] has made this observation from the bottom of a vertical shaft dug from a horizontal tunnel which was excavated into the edge of the ice sheet in the Thule area. Although he found appreciable differential flow in the cold ice just above the ice sheet bed, there was no sliding over the bed. Even if an immeasurably small amount of sliding does occur, it is very ineffective in picking up debris. If sliding does occur, it is unable to remove even the moss covering the boulders at the bed. This moss was carbon dated as about 200 yr. old.[2]

If, in the interior of the ice sheet, the bottom is at the melting point, it could be possible for debris to be picked up. Direct observation by McCall [8] has shown that sliding does occur at the bed of a temperate glacier. McCall's study was made at the end of a tunnel dug to bed rock in a small Norwegian glacier. However, if one assumes that the debris is being scraped up in the interior of an ice sheet by temperate rather than cold ice, then one also must consider the mechanism that will be proposed for the formation of the debris layers.* This theory i

* Dr. H. Roethlisberger (private communication) has pointed out that debris might be brought into the ic mass in the boundary region at the bottom surface between temperate and cold ice. The thickness of dirt, sand etc. which is frozen to the bottom, could be expected to be small in the region where the freezing point isotherm just begins to descend into the ground. If the shear strength of the interface between frozen and unfrozen dirt small, it might be possible, if the shear hypothesis mechanism can operate, to have this thin layer of dirt carried into the ice mass along a shear plane. If a shear plane did become active in this boundary region, it could als carry debris picked up by sliding temperate ice and transported to this region.

based on the assumption that temperate ice does exist somewhere at the bottom of ice sheets which contain Thule-Baffin moraines. It might be mentioned that, from theoretical considerations [9] as well as field observations, one expects that cold ice will not slide over a rock interface, but that ice at its melting point will.

Since direct observation has shown that cold ice does not slide over a rock when subjected to a stress which is sufficient to produce appreciable differential flow in the ice itself, it would appear highly unlikely that the shear hypothesis can work in an ice sheet which is cold everywhere. Against the field observation just mentioned, one might argue that cold ice *will* slide over a rock–ice interface if the shear stress in a particular area is sufficiently high. However, experiments of Raraty and Tabor [10] have shown that the shear strength of an ice–solid interface is usually of the order of 10 bars if the solid is stronger than cold ice and if the solid is wetable. This shear strength of the interface is an order of magnitude larger than the shear stresses which occur at the bottoms of glaciers and ice sheets ($0 \cdot 4$ to $1 \cdot 0$ bar).

The shear across a debris layer

It is a requirement of the shear hypothesis (although not sufficient proof of the validity of this mechanism) that appreciable discrete shear displacements occur across the debris layers (or at least that a very large amount of differential flow occurs in a layer of finite thickness). The evidence for the existence of these shear displacements is conflicting. Butkovich and Landauer [11] observed no discrete shearing motion across a dirt layer exposed in a tunnel in the Thule area, although there was differential flow in the ice. Hilty [12] made a similar observation in another tunnel in the Thule area. These measurements are evidence against the shear hypothesis. On the other hand, Ward [4] has published a photograph of the surface of the Barnes Ice Cap which shows a discrete shear across a dirt band. It can be argued, however, that Ward's observation was made in the summer time and was only a surface observation. During the Summer the surface layer is at the melting point. According to the sliding mechanism here proposed,[9] a discrete shear displacement of ice across a solid debris layer can occur when the ice is at the melting point. The shear seen by Ward may have occurred, therefore, only to those depths for which the ice is at the melting point.*

Even if further observations do show that appreciable shear does occur across debris layers, this result in itself would not be sufficient to prove the shear hypothesis. A condition necessary to the proof of the shear hypothesis would indeed have been met, but not a sufficient condition. Suppose that the debris layers were formed by a mechanism other than scraping. Then, let a layer be subjected to a stress which has a shear component such as is always found within a glacier or ice sheet. If the debris layer has the property that discrete shear takes place within it when such a stress is applied, this shear will occur regardless of how the debris layer was formed originally. A measurement of the shear across a debris layer thus yields information on a property of debris layers, but reveals nothing of their formation.

The geometry of the debris layers

Another criticism of the shear hypothesis is that it suffers from an "embarrassment of riches." If one accepts the shear hypothesis, one can account for a few debris layers. But can one really explain the very large number of debris layers that actually are observed? For example, in the ice tunnel constructed in 1959 next to Camp TUTO in the Thule area, one can see countless numbers of distinct dirt and debris layers within a distance of about 20 m. These layers often are separated from each other by distances of the order of millimeters. How could such a fine spacing be achieved by the shear model of Figure 1? One way might be for one "active" shear plane to start operation and carry debris to the surface. It then becomes "inactive" and another plane very close to it becomes active and carries a load of debris to the

---

* McCall, Nye and Grove [24] have shown that differential ablation can lead to an apparent thrust plane such as was observed by Ward.

surface. In turn, the second plane stops and another nearby plane becomes active. An explanation along these lines would be difficult to accept since one must explain why one shear plane is active and another inactive when the planes are separated by distances of the order of only 0·01 to 1·00 m.

On the other hand, one might propose that many closely spaced shear planes commence at about the same time to carry debris to the surface. But again one is in difficulties. The sequence that would be followed in this case is shown in Figure 2. Since it must be assumed that the spacing of the planes is close, one shear band will catch up with another, as shown in Figure 2. Thus, the slope of the shear planes will increase the farther one goes back into an ice sheet. In the tunnel dug near Camp TUTO the slope of the "shear" planes actually decreases the farther one goes into the tunnel, the opposite of the behavior predicted by this explanation.

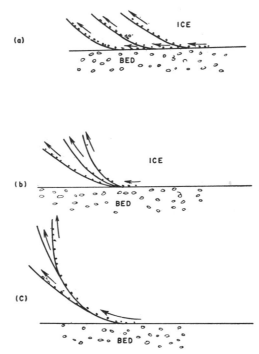

Fig. 2. *Sequence of events after the start of several closely spaced shear planes : (a) Planes just starting to bring up debris from the bottom; (b) Intermediate stage, the shear layers have just caught up to each other by shearing over the bed; and (c) A late stage, in which the layers have joined each other and only a single layer can exist at the bottom*

Although the shear hypothesis might account for debris planes which are separated by distances of the order of the thickness of an ice sheet (near the edge of the Thule ice lobe the separations would be of the order of 50 to 100 m.), it appears to be difficult to explain* with this mechanism the small scale separations which are actually observed. There really are too many "shear" planes present to be accounted for by the shear hypothesis.

* Dr. H. Roethlisberger has suggested to me that the small separations might be explained if a single debris layer can become highly folded over a period of time. A multilayered debris layer then could be produced from a single debris layer. Of course, such a possibility cannot be ruled out. If this is the explanation of the field observa tions, the shear hypothesis must be a much more complex process than it has been pictured hitherto. It also would lose that simplicity of conception which is such a strong point in its favor. It is difficult to imagine the processe that would lead to a high degree of folding and at the same time cause the folded layers to assume the overal orientation pictured in Figure 1. Since moraines are general features of an ice sheet, this difficulty cannot be explained by local peculiarities in the shape of the ice sheet bed. A general rather than special cause will have to be invoked to explain the multilayered debris bands.

The actual upward-curving form of an individual debris layer, which is approximately that shown in Figure 1, can be explained without the assumption that discrete shear occurs across a debris layer. It has been known for a long time from arguments based on the principle of conservation of mass that the flow lines of ice in an ablation area must go upwards. (The shape of flow lines in glaciers and ice sheets has been discussed recently by Nielsen and Stockton.[13]) Thus, even without the shear hypothesis, it is possible to understand the general shape of the debris layers.

FREEZING MODEL

As an alternative to the shear hypothesis the mechanism illustrated in Figure 3 is proposed to explain the formation of Thule-Baffin moraines. Figure 3 shows a cross-section of an ice sheet whose edge is frozen to its bed, the actual situation existing at the Thule ice lobe and at Baffin Island. It is assumed that farther inland from the edge the bottom is at the melting point. This second region can be divided into two parts. In the furthermost inland part, the combination of geothermal heat and heat produced from sliding is greater than can be conducted down a temperature gradient in the ice. As a consequence ice is melted to water. This water is forced by the pressure gradient outward to the edge of the ice sheet. As it moves outward, it enters into a region where the temperature gradient in the ice can conduct away more heat than is produced by any sliding or comes from the geothermal heat. In this region the water refreezes to ice and rejoins the ice sheet. The bottom of the ice sheet will still be at the melting point in this region; the extra heat required to keep it at this temperature comes from the latent heat of freezing which is given up as the water freezes.

The scheme presented in Figure 3 should always occur in any ice sheet whose edge is frozen to the bed and whose bottom surface at some positions in the interior is at the melting point. Thus, if direct measurements show that in the interior of a cold ice sheet the bottom surface is at the melting point, and that no water is escaping underneath the ice sheet, one can be sure that ice is being formed at certain places at the bottom surface.

Under steady-state conditions, the model of Figure 3 probably would not lead to the inclusion of solid dirt layers into the ice sheet, since ice would merely form at the ice surface. However, no glacier or ice sheet is ever in a completely steady-state condition. The thickness, among other things, varies as a function of time. A change in thickness changes the stress acting on the bed and, hence, affects the speed of sliding and, thus, the amount of heat available to melt ice or retard the freezing of water. This change will cause a shift in the position of the border between the region where the ice sheet is frozen to the bottom and where it is at the melting point. Figure 4 shows how a cyclic shift in the amount of heat* produced by sliding will lead to a freezing-in of debris. Since the border being considered is simply the point at which the $0°$ C. isotherm ceases to coincide with the bottom of the ice sheet, a shift of this border inland simply causes the zero point isotherm to descend into the material upon which the ice is resting. Debris can thus be frozen onto the bottom of the ice sheet. If the border shifts towards the edge of an ice sheet, the region in which water is being frozen to the bottom of the ice sheet is extended. If debris has already been frozen to the bottom, the new ice will have to start forming underneath this frozen-on debris, for it is here that the temperature is at the melting point. The debris will therefore be incorporated into the ice sheet with ice surrounding it, as shown in Figure 4. Through numerous repetitions of the cycle, a larger number of debris layers can be incorporated into the ice of an ice sheet lying on unconsolidated material. The thickness of these layers depends on the time frequency of the cyclic change and can have any value. The length of the debris layers depends on the magnitude of the

---

* Temperature changes at the upper surface also affect the amount of water being frozen or melted at the bottom surface. However, because of the extremely great length of time involved for temperature changes at the top surface to penetrate to the bottom surface, this variation would be unimportant compared to that produced by changes in the thickness or slope of an ice sheet.

change in slope or thickness of the ice sheet, as this change controls the distance the border in Figure 4 will shift.

Once horizontal layers of debris are incorporated into the ice, they flow with the ice. If the edge of an ice sheet is an ablation zone, as is the case in the Thule area and on Baffin Island, the flow lines of ice have to come up to the surface in this region. Hence the debris layers will become exposed in the ablation area and can form Thule-Baffin moraines.

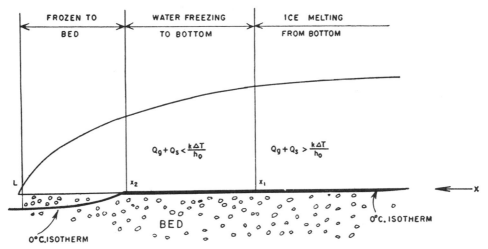

Fig. 3. The freezing model. A cross-section of an ice sheet is shown. The edge of the ice sheet is frozen to the bed and the 0° C. isotherm reaches the bottom of the ice sheet inland from the edge

THEORY

In this section are set down some formulae which will indicate the conditions under which the freezing mechanism will operate. Of course, it would be desirable to be able to take known field data and prove conclusively that this mechanism has led to the formation of existing moraines. That hope, however, cannot be satisfied with the measurements at present available.

Consider an ice sheet whose thickness across a particular cross-section is given by $h(x)$, where x is horizontal distance. Let $a(x)$ be the accumulation or ablation at the position x; this function has a positive value for accumulation and a negative value when there is ablation.

Under steady-state conditions, the heat flowing down the temperature gradient at the bottom surface is exactly equal to the sum of the geothermal heat, Q_g, the heat produced by any sliding, Q_s, and any heat given off or absorbed by the freezing of water or melting of ice at the bottom surface. Values of the geothermal heat vary between different places on the Earth.[14] An average value listed by Bullard [14] is 39 cal. cm.$^{-2}$ yr.$^{-1}$. The heat of sliding is given by the equation

$$Q_s = \frac{V\tau}{J},\qquad(1)$$

where J is the mechanical equivalent of heat ($J = 4\cdot185 \times 10^7$ erg cal.$^{-1}$), V is the velocity of sliding, and τ is the shear stress acting at the bottom surface which produces the sliding. To a good approximation, this shear stress is equal to $\rho g h \alpha$, where ρ is the average density of ice and α is the slope of the upper surface of the ice sheet. The velocity, V, itself also depends on shear stress. A theoretical expression [9] for the velocity is

$$V = B\tau^m,\qquad(2)$$

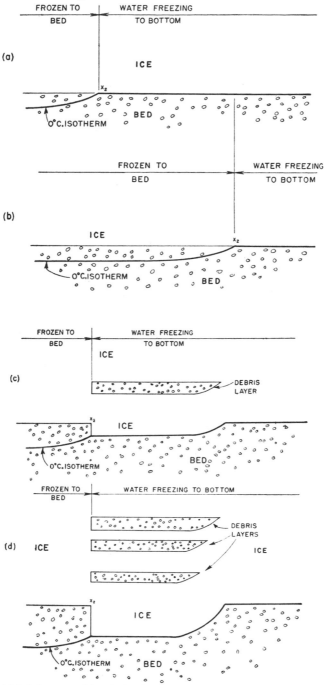

Fig. 4. Illustrating how debris is incorporated into the ice sheet under non-steady state conditions: (a) The starting point. To the left of point x_2, the 0° C. isotherm descends into the debris which makes up the bed. To the right of point x_2, the bottom is at the melting point and water is being frozen to the bottom surface. Water flowing from the right permits the bottom to be kept at 0° C., but the water supply is exhausted at point x_2; (b) Less water is flowing from the right and the point of descent of the 0° C. isotherm has shifted to the right; (c) A greater supply of water now flows from the right. Water now freezes to the 0° C. isotherm position given in (b) and pushes up into the ice the debris frozen on the bottom shown in (b). Point x_2, which marks the limit of the flow of water, has moved to the left; and (d) The cycle repeated after many times with a number of debris layers fixed into the ice

215

where m and B are constants. Reasonable values of these constants [15] are $m = 2$ and $B = 81$ m. yr.$^{-1}$ bar^{-2}. Since typical values [16, 17] of the shear stress on the bed of glaciers and ice sheets are in the range of $0 \cdot 4$ to $1 \cdot 0$ bar, equation (2) would give velocities of the order of 13 to 81 m. yr.$^{-1}$. From equation (1), therefore, one would expect the heats of sliding to be of the order of 13 to 190 cal. cm.$^{-2}$ yr.$^{-1}$. These values are about the same order of magnitude as the geothermal heat.

Under steady-state conditions, the heat flowing through the bottom interface of an ice sheet must balance the heat from the geothermal heat flow, the heat produced by sliding, and the heat given up or absorbed by any melting or freezing at the bottom. If $A(x)$ represents the thickness of ice being frozen to the bottom surface (it would be negative if the ice is melting) and $\gamma(x)$ represents the vertical temperature gradient at the bottom surface, the condition for steady-state heat flow at the bottom interface is given by the following equation:

$$k\gamma = Q_g + Q_s + SA, \tag{3}$$

where S is the heat of fusion ($S = 72$ cal. cm.$^{-3}$ ice) and k is the coefficient of heat conductivity of ice ($k = 1 \cdot 7 \times 10^5$ cal. cm.$^{-1}$ yr.$^{-1}$ ° C.$^{-1}$).

The temperature gradient at the bottom surface can be calculated if the temperatures at the top and bottom surfaces are known. This calculation has been done by Robin [18] for the case where heat flow in the horizontal direction (the x direction) can be neglected and where the longitudinal strain rate of the creep flow of ice is not a function of vertical distance. Since this longitudinal strain rate is approximately a/h, Robin was able to obtain the following equation for γ:

$$\gamma = \frac{\Delta T}{h^*}, \tag{4}$$

where ΔT is the temperature difference between the top and bottom surfaces and is considered to be positive if the bottom is warmer than the top, and h^* is a "compensated" thickness given by†

$$h^* = \int_0^h \exp\left(-\frac{cay^2}{2kh}\right) dy, \tag{5}$$

where c is the specific heat of ice ($c = 0 \cdot 45$ cal. cm.$^{-3}$). When the longitudinal strain rate is zero (and there is no melting or freezing at the bottom surface) this "compensated" thickness is equal to the actual thickness of the ice sheet.

If equation (4) is substituted into equation (3) one obtains

$$\frac{k\Delta T}{h^*} = Q_g + Q_s + SA. \tag{6}$$

If the bottom surface is below the freezing point, so that $A = Q_s = 0$, this equation will determine the bottom temperature, if the values of the temperature at the upper surface and the geothermal heat are known. On the other hand, if the bottom surface is at the melting point, so that $A \neq 0$, equation (6) will determine the amount of melting or freezing once the temperature of the top surface and the amount of sliding are fixed.

---

† This equation of Robin neglects any melting of ice or freezing of water at the bottom surface. If this factor is taken into account, the term inside the integral becomes

$$\exp\left\{-\frac{c}{k}\left(\frac{d}{2h}y^2 - Ay\right)\right\}.$$

The additional term in the exponent usually can be neglected if $|a|$ is much greater than $|A|$.

From equations (5) and (6) the conditions can be determined under which the bottom of an ice sheet can be at the melting point. As a very simple example, suppose that the accumulation rate is zero. It can be seen from equation (6) that the bottom surface will be at the melting point when the ice thickness is equal to h_0, where

$$h_0 = k \frac{\Delta T}{Q_g + Q_s} \qquad (7)$$

and ΔT is taken to be the difference in temperature between the melting point of ice and the upper surface temperature. When equation (7) is valid, the heat conducted through the ice is exactly equal to the geothermal heat plus the heat of sliding. The ice at the bottom is at the melting point, but neither ice is melting nor is water freezing there. If the ice thickness is greater than h_0, less heat can be conducted through the ice than is being supplied by the geothermal heat and the sliding heat. In this situation, ice melts off the bottom surface to make up the heat balance. If the ice thickness is less than h_0, more heat is conducted away than is being supplied. The bottom can remain at the melting point only if water is being frozen onto the bottom surface. However, if no water is available to give up its latent heat by freezing, the bottom surface has to be at a temperature lower than the melting point of ice and the ice sheet is frozen onto its bed. Thus, when the accumulation is zero, the thickness h_0 separates the condition of freezing from that of melting.

Now consider the more general case when the accumulation or ablation is not zero. Corresponding to any particular set of values for ΔT, Q_g, and Q_s, there will be a curve on a plot of a versus h which divides freezing conditions from melting conditions. This curve is determined by setting A equal to zero in equation (6). One obtains the equation

$$h_0 = \int_0^h e^{\beta y^2} dy, \qquad (8)$$

where $\beta = -ca/2kh$ and h_0 is given again by equation (7). This equation can be written as

$$h_0 = h e^{\beta h^2} \left\{ 1 - \frac{(2\beta h^2)}{3} + \frac{(2\beta h^2)^2}{5 \cdot 3} - \frac{(2\beta h^2)^3}{7 \cdot 5 \cdot 3} + \frac{(2\beta h^2)^4}{9 \cdot 7 \cdot 5 \cdot 3} \cdots \right\}. \qquad (9)$$

For large values of h and a, equation (8) reduces to

$$a = \frac{h \pi k}{2 c h_0^2}. \qquad (10)$$

Once h_0 has been found from equation (7), equations (9) and (10) determine a curve of a versus h. Figure 5 shows such curves for various values of h_0. Also indicated in Figure 5 are the temperature differences, ΔT, which correspond to these values of h_0 when Q_g is the average geothermal heat of 39 cal. cm.$^{-2}$ yr.$^{-1}$ and Q_s is taken to be equal to Q_g. This value of the sliding heat corresponds to a shear stress of 0·58 bar and a sliding velocity of 27 m.yr.$^{-1}$ (or any other combination of stress and sliding velocity whose product is the same).

If the values of the ice thickness and accumulation or ablation in a region of an ice sheet are such that they correspond to a point lying to the *right* of the appropriate curve in Figure 5, the bottom of the ice sheet is *melting* away. If the thickness and accumulation are such that the point lies to the *left* the bottom is either frozen to the bed or water is *freezing* to the bottom. Hence, if the temperature at the upper surface and the geothermal heat and heat of sliding are known, it is possible to tell from the measurements of ice thickness and rate of accumulation or ablation whether or not it is possible for ice at the bottom to be melting or freezing.

The bottom surface of an ice sheet which is located in a region corresponding to a point in the freezing zone can remain at the melting point only if there is a water supply available. The melting zone can supply this water if a hydrostatic pressure head exists to push water from one region into the other. From Figure 5 it can be seen that, other things being equal, the thicker parts of an ice sheet are more likely to be in a melting zone, if a melting zone does exist. The greater weight of ice in the melting zone will supply the necessary pressure head to move water into the freezing zone.

The principle of the conservation of mass enables one to calculate the extent of the region in the freezing zone where the bottom of an ice sheet is at the melting point of ice. That is, all the water created in the melting zone must be turned back to ice in the freezing zone if no

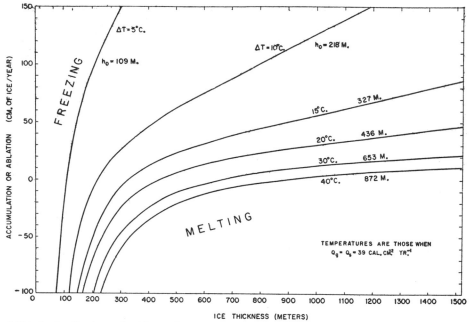

Fig. 5. *A plot showing freezing and melting conditions at the bottom of ice sheets. Each curve is calculated for a different value of* $h_0 = k\,\Delta T/(Q^{\circ}+Q_s)$, *where* k *is the thermal conductivity of ice,* ΔT *the difference between the melting point of ice and the upper surface temperature of an ice sheet, and* Q_g *and* Q_s *are the geothermal heat and the heat produced by the sliding of ice over the ice sheet bed. The region of the figure to the right of a particular curve corresponds to conditions where melting occurs, and the region to the left where freezing can take place*

water is to escape from underneath a cold ice sheet. The amount of ice, $A(x)$, which is either being melted from, or frozen to, the ice sheet in the regions where the ice is at the melting point is given by equation (6):

$$A(x) = \frac{1}{S}\left\{\frac{k\Delta T}{h^*} - (Q_g+Q_s)\right\}. \tag{11}$$

Consider how equation (11) can be applied to an actual problem. Suppose a two-dimensional ice sheet (center at $x = 0$ and edges at $x = +L$) has a melting zone extending over the distance from x_0 to x_1, where $x_0 < x_1 < L$. What would be the extent of the freezing zone in which the ice at the bottom of the sheet is at the melting point? In order to satisfy the principle of conservation of mass, when no water escapes from underneath the ice sheet, the following equation must be obeyed:

$$\int_{x_0}^{x_1} A(x)dx = -\int_{x_1}^{x_2} A(x)dx, \tag{12}$$

where the first integral covers the melting zone and the second the freezing zone. This equation fixes the value of x_2 ($x_2 < L$). Since ΔT (the difference between the melting point of ice

218

and the surface temperature) and h^* (determined from the thickness of ice and the rate of accumulation) can be determined as functions of distance, once the heat of sliding and the geothermal heat are known, equations (11) and (12) determine uniquely the extent of the freezing zone at the melting point. At distances greater than x_2 the bottom of an ice sheet is below the freezing point of ice and the sheet is frozen to its bed. In this region, there is no water available to freeze onto the bottom and hence raise the ice temperature to the melting point.

DISCUSSION

In the previous section, it has been shown that under certain conditions a part of the bottom surface of a cold ice sheet may be at the melting temperature of ice and that, in this part, ice can be melting away from the bottom in some parts and water can be freezing onto the bottom in the other parts. Of course, one would like to know if the ice sheets which contain Thule-Baffin moraines meet the special conditions that lead to the ice at the bottom surface being melted in one place and refrozen in another. Since the edges of the Barnes Ice Cap on Baffin Island and of the Thule lobe of the Greenland Ice Sheet are known to be frozen to their beds, it is only necessary to show that in some inland regions the conditions are such that the bottom is at the melting point (or more exactly, *was* at the melting point during the time the debris which forms the moraines on the edges of these ice sheets was brought into the ice). Unfortunately, from the field evidence at hand, it is not possible to prove conclusively whether or not there exist regions on either of these ice sheets where the bottom surface is at the melting point of ice.

Consider first the Barnes Ice Cap on Baffin Island. The upper surface of the Barnes Ice Cap appears to be at $-10 \cdot 7°$ C. from one measurement made by Ward;[4] its thickness goes from 0 to about 450 m.;[16] its accumulation ranges from -170 to $+20$ cm. of ice yr.$^{-1}$, the value depending on the elevation of the upper surface. From these data, Figure 5 would predict that the bottom surface in the interior of the ice cap definitely is at the melting point. In order to arrive at this conclusion, consider the curve of Figure 5 marked $\Delta T = 10°$ C. This curve represents a temperature close to the actual temperature of $-10 \cdot 7°$ C. Any thickness greater than 280 m. corresponds to a point in the melting region, even for the maximum rate of accumulation of 20 cm. of ice yr.$^{-1}$. One must remember, however, that the curves in Figure 5 were calculated on the assumption that the heat of sliding is equal to the geothermal heat, and it was assumed further that the geothermal heat is equal to 39 cal.cm.$^{-2}$yr.$^{-1}$. Suppose the estimate of the heat passing through the bottom is too great by a factor of 2. Then the curve to be considered is not the one marked 10° C., but rather the curve marked 20° C. In this case, it is open to question whether any part of the Barnes Ice Cap lies in the melting region and it may well be that the entire ice cap is frozen to the bottom. The only way one could really be sure is to sink bore holes to the bottom of the ice cap and actually measure the temperature.

The same uncertain situation occurs in the Thule ice lobe. The thickness of this ice has been only partially measured. Roethlisberger[19] found a thickness of 260 m. at a distance of 5 km. inland from the edge. (The ice sheet is about 33 km. wide and a somewhat greater thickness can be expected farther inland.) Accumulation and ablation rates[20, 21] depend on position and range from about -100 cm. of ice yr.$^{-1}$ close to the edge to 70 cm.yr.$^{-1}$ in the central region of the ice sheet. The temperature in the ablation zone[20] near the edge of the ice sheet is about $-12°$ C. Temperatures have been measured[20] in the inland region at two points down to a depth of 9 m. The temperatures at 9 m. seem to be of the order of $-3°$ to $-6°$ C.* If these inland temperatures are representative values and if the curves of Figure 5

---

* The inland ice appears to be warmer than the ice at the edge because in the ablation region the melt water produced in the Summer simply runs off the upper surface and does not soak into the solid ice surface. On the other hand, in the inland region, the melt water produced in the Summer does not run off the surface, but trickles into the firn layers below, where it refreezes and gives up its latent heat. Thus, in the Summer, the accumulation area is able to warm up more than the ablation area.

are valid, there is no question but that the bottom surface in the interior of the Thule ice lobe is at the melting point. On the other hand, if these temperatures are not reasonable and if other values of the heat of sliding and the geothermal heat are used, it is quite possible that the ice sheet is frozen everywhere to its bed.

Another problem connected with the determination of the temperature at the lower surface of a cold ice sheet concerns the question as to whether or not the ice is actually sliding over the bottom bed. If no sliding occurs, no heat will be produced by this mechanism. It is thus conceivable to have an ice sheet which would be frozen to its bottom surface, if the heat of sliding were not available, but whose lower surface would be at the melting point if this heat were present. Once the ice sheet started to slide, the heat produced would keep the bottom surface at the melting point and permit the sliding to continue. If the sliding were stopped, the ice sheet would refreeze to its bed and no further sliding would occur. In the absence of direct measurements, a knowledge of the past history of such an ice sheet would be necessary in order to determine whether or not the bottom is at the melting point.

Conclusion

It is concluded that there are a number of objections that can be raised against the shear hypothesis for the formation of a type of moraine found on Baffin Island and in Greenland. It is further concluded that in certain situations it is possible for ice to be formed at the bottom of an ice sheet and that this accretion of new ice can lead to a freezing-in of loose debris lying on the bed of the ice sheet. In turn, this frozen-in debris can result in the formation of moraines in the ablation areas of an ice sheet.

Although one can demonstrate the possibility that the moraines found on the Thule ice lobe in Greenland and on the Barnes Ice Cap of Baffin Island are formed by the mechanism proposed here, it is not possible from the field data at hand to prove conclusively that these moraines have been so formed.

Acknowledgements

I wish to thank Mr. James Bender for the opportunity of seeing at first hand the moraines near Thule, Greenland. I am indebted to him and to Dr. Hans Roethlisberger and Dr. George Swinzow for informative discussions and arguments on the formation of these moraines.

MS. received 20 March 1961

REFERENCES

1. Bishop, B. C. Shear moraines in the Thule area, northwest Greenland. *U.S. Snow, Ice and Permafrost Research Establishment. Research Report* 17, 1957.
2. Goldthwait, R. P. Formation of ice cliffs. (*In* Study of ice cliff in Nunatarssuaq, Greenland. *U.S. Snow, Ice and Permafrost Research Establishment. Technical Report* 39, 1956, p. 139–50.)
3. Nobles, L. Investigations of structures and movement of the steep ice ramp near Red Rock Lake, Nunatarssuaq, Greenland. *U.S. Snow, Ice and Permafrost Research Establishment. Report on Contract No. DA-11-190-ENG-12*, 1960.
4. Ward, W. H. The glaciological studies of the Baffin Island Expedition, 1950. Part II: The physics of deglaciation in central Baffin Island. *Journal of Glaciology*, Vol. 2, No. 11, 1952, p. 9–23.
5. Goldthwait, R. P. Development of end moraines in east-central Baffin Island. *Journal of Geology*, Vol. 59, No. 6, 1951, p. 567–77.
6. Hollin, J. T., *and* Cameron, R. L. I.G.Y. glaciological work at Wilkes Station, Antarctica. *Journal of Glaciology*, Vol. 3, No. 29, 1961, p. 833–42.
7. Chamberlin, T. C., *and* Salisbury, R. D. *Geology. Vol. I. Second edition, revised.* New York, Henry Holt and Co., 1904, ch. 5.
8. McCall, J. G. The internal structure of a cirque glacier. *Journal of Glaciology*, Vol. 2, No. 12, 1952, p. 122–30.
9. Weertman, J. On the sliding of glaciers. *Journal of Glaciology*, Vol. 3, No. 21, 1957, p. 33–38.
10. Raraty, L. E., *and* Tabor, D. The adhesion and strength properties of ice. *Proceedings of the Royal Society*, Ser. A, Vol. 245, No. 1241, 1958, p. 184–201.

11. Butkovich, T. R., *and* Landauer, J. K. A grid technique for measuring ice tunnel deformation. *Journal of Glaciology*, Vol. 3, No. 26, 1959, p. 508–11.
12. Hilty, R. E. Measurements of ice tunnel deformation, Camp Red Rock, Greenland. *U.S. Snow, Ice and Permafrost Research Establishment. Special Report* 28, 1959.
13. Nielsen, L. E., *and* Stockton, F. D. Flow patterns in glacier ice. *Journal of Applied Physics*, Vol. 27, No. 5, 1956, p. 448–53.
14. Bullard, E. The interior of the earth. (*In* Kuiper, G. P., *ed. The Earth as a planet*. Chicago, University of Chicago Press, 1954, p. 57–137.)
15. Weertman, J. Equilibrium profile of ice caps. *Journal of Glaciology*, Vol. 3, No. 30, 1961, p. 953–64.
16. Orvig, S. The glaciological studies of the Baffin Island Expedition, 1950. Part V: On the variation of the shear stress on the bed of an ice cap. *Journal of Glaciology*, Vol. 2, No. 14, 1953, p. 242–47.
17. Ward, W. H. Studies in glacier physics on the Penny Ice Cap, Baffin Island, 1953. Part IV: The flow of Highway Glacier. *Journal of Glaciology*, Vol. 2, No. 18, 1955, p. 592–98.
18. Robin, G. de Q. Ice movement and temperature distribution in glaciers and ice sheets. *Journal of Glaciology*, Vol. 2, No. 18, 1955, p. 523–32.
19. Roethlisberger, H. Seismic survey 1957, Thule area, Greenland. *U.S. Snow, Ice and Permafrost Research Establishment. Technical Report* 64, 1959.
20. Schytt, V. Glaciological investigations in the Thule Ramp area. *U.S. Snow. Ice and Permafrost Research Establishment. Report* 28, 1955.
21. Griffiths, T. M. Glaciological investigations in the TUTO area of Greenland. *U.S. Snow, Ice and Permafrost Research Establishment. Technical Report* 47, 1960.
22. Drygalski, E. von. *Grönland-Expedition der Gesellschaft für Erdkunde zu Berlin, 1891–1893*. Bd. 1. Grönlands Eis und sein Vorland. Berlin, W. H. Kuhl, 1897, p. 109.
23. Blümcke, A., *and* Finsterwalder, S. Zur Frage der Gletschererosion. *Sitzungsberichte der Kgl. Bayerischen Akademie der Wissenschaften zu München*, Math.-phys. Klasse, Bd. 20, 1890, p. 435–44.
24. Lewis, W. V., *ed. Investigations on Norwegian cirque glaciers*. London, Royal Geographical Society, 1960, p. 49. (R.G.S. Research Series, No. 4.)

13

Reprinted from *Geog. Ann.*, **47A**(2), 73–82 (1965)

MAPPINGS AND GEOCHRONOLOGICAL INVESTIGATIONS IN SOME MORAINE AREAS OF SOUTH-CENTRAL SWEDEN

BY BO STRÖMBERG

Department of Physical Geography, University of Stockholm

INTRODUCTION

Moraines of the type variously known as *annual moraines, small end moraines* or *De Geer moraines* are common in parts of south-central Sweden where deglaciation took place in a subaquatic environment. In discussions concerning the origin of such ridges their geochronological value has played an important rôle in Sweden (cf. *Hoppe*, 1957, p. 1). The assumption that the ridges, or some of them, were formed one each year has limited investigators to interpretations that postulate an annual rhythm of the retreating ice front and the formation of such ridges at (in front of) the ice edge. Neither the rhythm of recession nor the formation of ridges in front of the ice edge, however, have been satisfactorily proved—good evidence for such assumptions should of course be present if the annual nature of the ridges is to be convincingly demonstrated.

In a few cases careful mapping of the moraines was combined with special geochronological investigations, but the results have not been in agreement. (*1 st*) After a detailed investigation in the Stockholm region *De Geer* (1912, p. 250) at first considered that each little end moraine corresponded to the proximal limit of a specific annual varve in the glacial clay. Later it was shown that in addition to the "annual", so-called *winter moraines*, in places short parallel ridges appeared forming *twin moraines*. These probably formed during the same winter as the adjacent main ridges. In the higher parts of the terrain smaller so-called *interannual moraines* are present —those in lower areas are presumably buried beneath clay—and they are believed to have formed in connection with calving during the summer (*De Geer*, 1932, pp. 14–15; 1940, p. 129). (*2nd*) *Anrick's* (1915) geochronological study in an end moraine area near Odensala, Uppland, resulted in the conclusion that the moraines, in places only about 80 m apart on an average,

were "annual" (cf. p. 79 in the present paper). (*3rd*) North of Stockholm, in the Ullna district, *Möller* (1962, Fig. 1, p. 126) has shown by careful mapping and detailed geochronological investigations that a series of eight De Geer moraines were formed one each year. In other parts of the area smaller and lower ridges appear here and there between the larger ones. Möller suggests that the larger ridges mark winter pauses in the retreat of the ice front, whereas the smaller ridges should be regarded as interannual.

Studies in Canada have contributed interesting material to the discussion of subaquatically formed moraines. *Mawdsley* (1936) described patterns of moraines similar to those in some central Swedish areas and probably formed during recession in glacial lakes. He says (p. 11): "The regular pattern of the forms is like a system of cracks in relief, and may have originated through some type of regular fracturing suffered by the ice sheet prior to its disappearance." This interpretation implies that the ridges were formed *inside* the ice border, and that the resulting pattern and the distances between the ridges were controlled by the pattern of crevasses in the borderzone of the ice. Subaquatically formed moraines, probably genetically related to De Geer moraines, have been studied in detail by Andrews. He proposes various hypotheses to explain the formation of the ridges ("cross-valley moraines"), some of which suggest a frontal formation, one a formation in crevasses (*Andrews*, 1963a, pp. 74–76). Examination of ridge pattern and a study of till fabric seems to favour the belief that they are formed in crevasses (*Andrews*, 1963b, p. 126). In Sweden the possibility for De Geer moraine formation in crevasses has been emphasized by *Hoppe* (1957, p. 5; 1959, p. 197).

In addition, moraines interpreted as having developed in basal crevasses have been described also from areas where deglaciation took place in a supra-aquatic environment (e.g., *Hoppe*, 1952; *Gravenor and Kupsch*, 1959).

A problem in studies and discussions of moraines, presumably owing to the paucity of information we have about such features, is the lack of a definite classification. For the present the term *moraine* will be used in general discussions of ridges whether they are oriented normal to the direction of ice movement or in other directions. The available field material does not make a definite grouping of the moraines possible.

MAPPING AND VARVE STUDIES

In connection with field work in south-central Sweden between 1959 and 1964 I have studied *the patterns of moraines* and *the geochronology* in areas where these patterns occur. All areas examined (Fig. 1) are situated below the highest shoreline—the deglaciation took place subaquatically, and the water depth was probably between 110 and 150 m (*Lundqvist*, 1961, Fig. 35, p. 101).

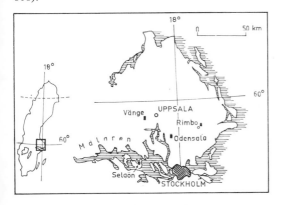

Fig. 1. Sketch map of the investigated region.

Mapping of the rather low ridges—generally only 2–5 m high—had to be made from the ground as many of them, especially when they occur in dense woodland, are not visible on air photographs. Normally the moraines were drawn on economic maps at a scale of 1:10,000 or on enlargements of them by means of pace and compass. In some forest districts, difficult of access, auxiliary lines were layed out by theodolite and rod to make more accurate determinations of position possible. The accuracy of the mapping varies. In two of the districts investigated, Rimbo and Selaö, the account must be regarded as more

schematic than in the Vänge and Odensala areas. Correctness has been sought, but small errors may of course occur. They are, however, not likely to be of fundamental importance.

It has been my intention to include on the maps *all distinct moraines* in the various areas, irrespective of orientation or eventual relation to bedrock. However, distal accumulations of the type known as "lee-moraines" in relation to bedrock outcrops (*Munthe*, 1905, p. 82) are not shown if they do not form parts of distinct moraine patterns.

The geochronological datings are based upon *varve measurements* in glacial clay according to De Geer's classic method (e. g., see *De Geer* 1940 pp. 19–21). In this part of Sweden the varves can normally be regarded as annual, which for instance is confirmed by radiocarbon dating (cf., *Fromm*, 1963, p. 46). The majority of my varve measurements have been made in excavated pits; thus it has been possible to determine the absolute bottom varve with confidence as well as to notice any visible disturbances in the stratification. A problem in almost all pits was water which had to be removed by a motor-operated pump. In one locality in the Vänge area and in all Odensala localities a simple sampling device, which takes up undisturbed, 50 mm-wide cores from a depth of up to 4 m, has been used.

The sampler is adapted from a Swedish standard sounding rod and post hole digger (older type, 19 mm solid extension rods). The sampling is carried out in the following manner: After sounding, a 90 mm hole is dug through the clay down to the substratum (usually till). The sampler, consisting of a sheet-iron box fixed to a U-formed profile of light metal, is lowered into the hole and is pressed into the clay by a wedge mounted on the extension rods. The wedge is exchanged for a knife which cuts the box free from the wall, and the sample is then taken up.

Sounding to determine the depth preceded sampling, and two or more cores were taken up at most localities. The risk of disturbed stratification is always present in connection with such sampling. Normally, however, any disturbances are discovered when the varve series from different localities in the area are connected. If the substratum is not a big stone or boulder the sample frequently contains some till in the bottom. As soundings are made in the same hole as the sample is taken I have considered it possible to mark till as substratum in the graphs.

74

223

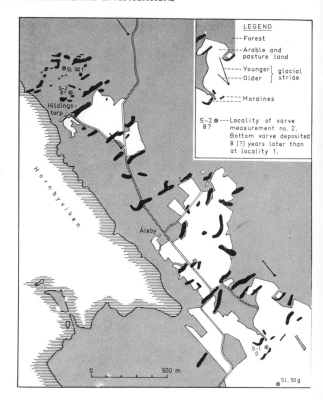

Fig. 2. Moraines and varve measurement localities in the Äleby district, Selaön. Sl. 50 f and g according to De Geer 1940.

1. The Selaö area

A map showing moraines on Selaön was published by *De Geer* in 1940 (Fig. 25, p. 101). In an area around Äleby, east of Hornbyviken, a new map of the moraines was made in 1962 by *G.*

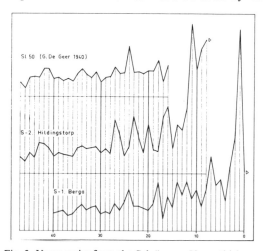

Fig. 3. Varve series from the Selaö area. Varve thickness at a scale of 1:4. Triangle = till substratum. Local time scale.

Andersson and *W. Åhlin* as a seminar project under my direction (Fig. 2). This map agrees with De Geer's. The ridges are all rather short and in the field no clear difference in size seems to be visible so as to permit a separation into "annual" and "interannual" ridges. As the sediment cover between the ridges has proved to be rather thick in many places (e.g., 4 m deep only 30 m from a ridge in the southern part of the area), a number of soundings are needed in order to find all ridges.

Unfortunately the geochronological investigations failed almost entirely. Varve measurements in five pits in different parts of the area showed that everywhere the proximal varves (lower varves) were more or less disturbed. Only two complete varve series, which appeared to be correct in the field, were obtained. In attempts to connect them only about the upper 15 varves could be correlated, whereas the lower disagree (Fig. 3). According to this attempt at connection the ice receded at a rate of about 260 m a year. The datings given by *De Geer* (1940; varve measurings by H. Eriksson in 1915) show 10 years of recession (about 250 m a year) between Berga and Äleby

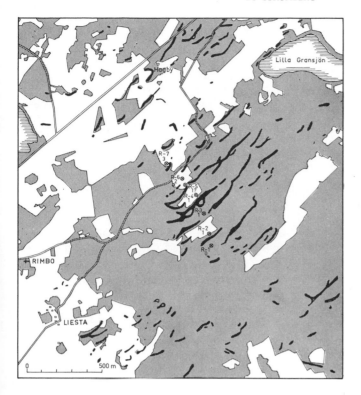

Fig. 4. Moraines and varve measurement localities in the Rimbo area. For explanation see Fig. 2.

according to the map (Fig. 25, p. 101), 13 years of recession, according to the main time scale (p. 255). For comparison the proximal part of De Geer's (1940, Pl. 78) graph Sl. 50 from Selaön is included in Fig. 3. Unfortunately this graph is not complete; the proximal varves are missing, perhaps due to disturbances in the pits.

The result of the field studies on Selaön is that *nothing in the material permits the conclusion that the ridges should have originated one each year.*

2. The Rimbo area

The map in Fig. 4 shows part of an area of moraines, about 4 × 7 km in size, situated in the district east of Rö-Rimbo. This area was mapped schematically in 1959–60 (*B. Strömberg* 1962, pp. 43–45). About one km NE. of Rimbo church the moraines lie only 60–100 m apart in a series. The latest ice movement according to observations of striae immediately east and west of the area had a direction from N. 18–35° W, that is generally normal to most of the ridges.

The varve measurements were made at seven localities 50–150 m apart in a line. The proximal varves were normally very thick, and they characteristically enclosed sandy-silty layers up to a few centimeters in thickness. The varves decrease rapidly in thickness upwards. The varve limits were difficult to distinguish; several measurements under different light conditions and varying degrees of moisture of the walls had to be made. Series longer than 14 to 22 varves could nowhere be obtained in pits up to 2.4 m deep. As a result of the short series and of the difficulties in delimiting the varves, the connections, or attempts at connections, must be regarded as quite uncertain. The most probable chronology is presented in Fig. 5. Identical bottom varves seem to be present at localities R–2 and R–3, and perhaps at R–4. At R–4, however, the bottom varve is incomplete and consists only of a dark winter layer resting directly on till. According to these data the ice receded at a rate of 200–300 m a year in the area investigated.

Thus in this area, also, it is difficult to get definite conclusions about the geochronological value of the moraines. *The field material does not support an assumption of such a value.*

76

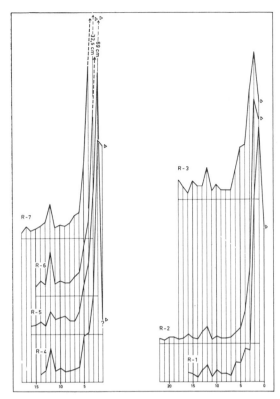

Fig. 5. Varve series from the Rimbo area. Varve thickness at a scale of 1:4. Numbers indicate real thickness in cm of the thickest varves. Triangel = till substratum. Local time scale.

3. The Vänge area

This area, situated about 15 km west of Uppsala, has been the subject of earlier discussions on the geochronological value of end moraines. On the geological map, sheet Uppsala (S.G.U., Ser. Aa 199), which appeared in 1956, a general account is given of some of the moraines in the district. From a geochronological investigation of ice recession in the Uppsala area *Järnefors* (1956, p. 310; 1958, p. 9–10) concluded, that the end moraines in the south-western part of the area were to a large extent annual moraines. *Hoppe* (1957, p. 7) has disagreed with this conclusion.

Detailed mapping of the moraines, including all the ridges in the area irrespective of their orientation, was carried out in 1961 to 1963. The map (Fig. 6) demonstrates the complicated patterns in the area; branching or net-like ridges are not uncommon. In fact, there seems to be no possi-

bility of separating the ridges into different categories in the field as their morphology is frequently uniform. If we ignore the thick sediment cover in places (soundings have indicated sediment depths of more than 5 m close to moraines), variations in size exist between the visible parts of the ridges— some are short and low, others long, wide and high.

Successful varve measurements were made at 12 localities (Fig. 7). V–1 to V–11 were made in excavated pits, V–12 by clay sampler (four independent samplings were necessary in the latter place because of disturbances in the proximal layers). The glacial clay is very distinctly varved in this area, and the varve limits are easy to determine in most cases. The agreement in variations of varve thickness from one locality to another is very good; thus the connection can be regarded as reliable.

The deglaciation from locality V–1 in the south to V–11 in the north took 19 years according to the varve measurements; that is, the ice front receded at a rate of about 180 m per year. Attempts to count "end moraines" between these localities did not give the same result; the number of ridges one after another, oriented normal to the direction of ice movement is more than 25, even if the small insignificant ridges are omitted. Only in the southern part of the area there may be agreement between the probable yearly positions of the retreating ice border and 3 to 4 of the larger and longer moraines. In the area shown in the north-western corner of the map, where the ridges are mainly distinguished as "end moraines" (Fig. 6, V–9 to V–12), such agreement does not seem to be present. In this section, where the ice recession lasted for three years, there are two rather distinct moraines and one intervening smaller ridge. In the northern part of the section four shorter ridges occur in addition, most of them are, however, well-developed. It does not seem possible to regard the whole group of proximal, smaller ridges as "interannual".

Good agreement has been obtained between the varve series presented here and those from the same district published by *Järnefors* (1956, Pl. 5; no. 15 Myrtorp and no. 14 Långtibble). The agreement with no. 15 Myrtorp would be practically perfect if an alteration in some of the graphs is made with regard to a varve of the year 13 according to the local time scale in the present paper (Fig. 8). In order to connect to series 14 Långtibble however, this has to be displaced in

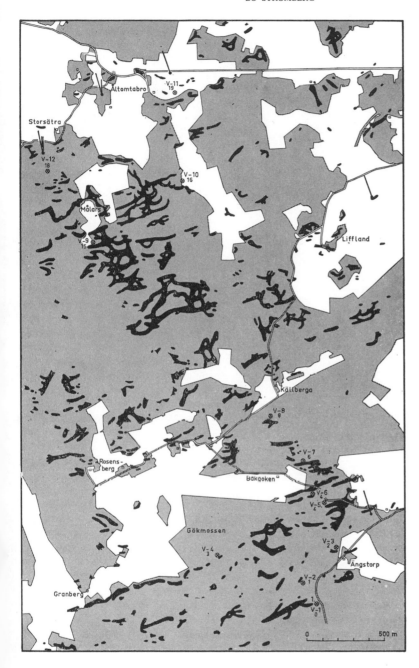

Fig. 6. Moraines and varve measurement localities in the Vänge area. For explanation see Fig. 9.

relation to the Myrtorp-series, resulting in a prolongation of the time of recession between the localities from 16 to 24 years.

Thus *little possibility exists in the Vänge area of fitting moraines in a consistent manner into yearly* positions of the retreating ice border. The ridges can only exceptionally be regarded as "end moraines" — rather, the complicated patterns they form in most parts of the area indicate an entirely different origin.

78

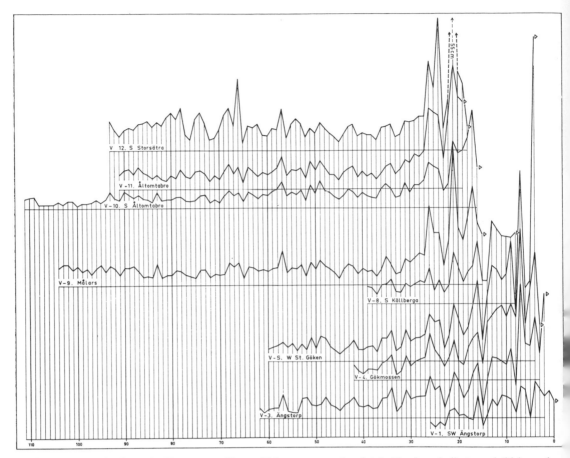

Fig. 7. Varve series from the Vänge area. Varve thickness at a scale of 1:4. Numbers indicate real thickness in cm of the thickest varves. Triangle = till substratum. Local time scale.

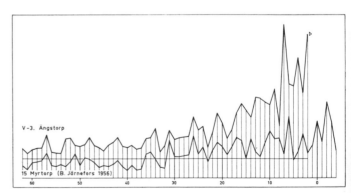

Fig. 8. The varve series V–3 from the Vänge area connected to the proximal part of series 15 Myrtorp (*Järnefors* 1956). Varve thickness at a scale of 1:4.

4. The Odensala area

The moraines in this area, situated about 3 km NE. of Odensala church, were mapped in 1964 by E. Hådell and P.-E. Tonell as a seminar project under my direction. In 1915 Anrick mapped the moraines and investigated the clay varves in the same area. The new map (Fig. 9) includes a somewhat larger area to the south and east where many ridges were found. *Anrick* (1915) concluded that the moraines were "annual", although the

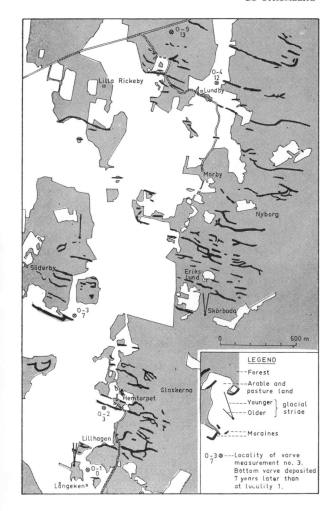

Fig. 9. Moraines and varve measurement localities in the Odensala area.

distance between successive moraines was only about 80 m. Thus the ice must have receded very slowly, but the proposed chronology is extremely doubtful. The varve series are very short and the connections are not convincing. Furthermore *De Geer* (1940, pp. 148–149; Pl. 78) obtained an annual ice recession of about 350 m at the same ice border position within the part K1—K5 of the main Swedish time scale. These localities are situated 4–5 km west of the investigated moraine area.

In the present investigation varve measurements were made at five localities (Fig. 10). Since all measurements were made on clay taken by clay sampler more than one sample was studied at some of the localities. One core was taken at Långeken and one at Söderby, two cores at both

NW. Lundby and Hemtorpet and four at Lundby (one core of which is devoid of the seven lowest varves). The series with accurate bottom varves — or if such are present in more than one series from each locality, the longest varve series—are presented in the graphs.

The agreement in variation of varve thicknesses from one locality to another is very good according to the proposed connections. Remarkably no positive correlation could be successfully made to any of *Anrick's* varve series.

The chronology strongly favours a deglaciation period from Hemtorpet to Lundby (O–2 to O–4) of nine years (about 250 m recession per year as a mean value). It can be seen, however, that the ice receded at a much slower rate in the southern (O–1 to O–3, only about 130 m per year)

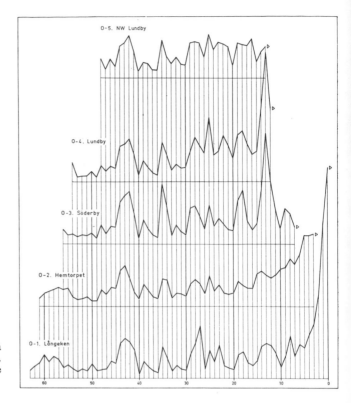

Fig. 10. Varve series from the Odensala area. Varve thickness at a scale of 1:4. Triangle = till substratum. Local time scale.

than in the northern part of the area. Within the section Hemtorpet to Lundby *at least 19* well developed moraines are found one after the other normal to the direction of ice movement and in addition a number of intervening smaller ridges occur. *Thus the moraines in this area cannot be separated into "annual" and "interannual" ridges.*

CONCLUSIONS

In the areas investigated no evidence has been obtained for an interpretation of moraines lying at right angles to the last direction of ice movement as "annual moraines". Nor can a division be made into "annual" and "interannual" ridges.

The moraines, particularly in the Vänge and Odensala districts, are *arranged in patterns* such that *formation of all ridges at the ice front seems most improbable.* Morphologically all the ridges appear similar. Two possible origins of the moraines in these areas seem worthy of consideration: 1) Some ridges are frontal formations whereas others are formed in basal cracks; or 2) all ridges originated in basal cracks. In my opinion the second hypothesis seems worthy of consideration even in areas where the moraines are oriented predominantly normal to the direction of ice movement. No material gathered in the field so far has suggested that moraines cannot form in this way. The resemblance between the patterns of moraines in some of the areas investigated and patterns of cracks on the surface of present-day calving glaciers is striking (e.g., cf. *Hoppe* 1957, Pl. 1). There are reasons to believe that the basal pattern of cracks is similar.

ACKNOWLEDGMENTS

The field works were financially supported by the Swedish Society of Anthropology and Geography. Support was also received from Prof. *Gunnar Hoppe*, who made valuable critical comments to the manuscript and from Dr *Weston Blake Jr*, who corrected an early edition of the English text. The maps and diagrams were drawn by Mrs *Birgit Hansson*.

REFERENCES

Andersson, G. and *Åhlin, W.*, 1962: Studier av isrecessionen på Selaön i Mälaren. *Department of Physical Geography*, Stockholm. 11 pp. (typewritten).

Andrews, J. T., 1963a: Cross-valley moraines of the Rimrock and Isortoq river valleys, Baffin Island, N. W. T. A descriptive analysis. *Geogr. Bull.*, no. 19, 49–77.

— 1963b: The cross-valley moraines of north-central Baffin Island: a quantitative analysis. *Geogr. Bull.*, no. 20, 82–127.

Anrick, C. J., 1915: Morän- och isrecessionsstudier i Odensala socken, Uppland, *Geol. För. Förh.* 37, 688–694.

De Geer. G., 1912: A geochronology of the last 12 000 years. *Compte Rendu du XI:e Congr. Géol. Int.*, Stockholm 1910, 241–253.

— 1932: Stockholmstraktens kvartärgeologi. *Sver. Geol. Unders.*, Ser. Ba. 12, 89 pp.

— 1940: Geochronologia Suecica Principles. *K. Sv. Vetenskapsakad. Handl.* III Ser., 18:6, 367 pp.

Fromm, E., 1963: Absolute chronology of the late Quaternary Baltic. Review of Sw. investigations, *Baltica* I, 46–59.

Gravenor, C. P. and *Kupsch, W. O.*, 1959: Ice-disintegration features in western Canada. *Journ. of Geol.*, 67, 48–64.

Hoppe, G., 1952: Hummocky moraine regions with special reference to the interior of Norrbotten. Geogr. Annaler 34, 1–72.

— 1957: Problems of glacial morphology and the ice age. 1–3. *Geogr. Annaler* 39, 1–17.

— 1959: Glacial morphology and ice recession in northern Sweden. *Geogr. Annaler* 41, 193–212.

Hådell, E. and *Tonell, P.-E.*, 1964: Studier av ett De Geer-moränområde i Odensala socken. *Department of Physical Geography*, Stockholm, 22 pp. (typewritten).

Järnefors, B., 1956: Isrecessionen inom Uppsalaområdet *Geol. För. Förh.* 78, 301–315.

— 1958: Beskrivning till jordartskarta över Uppsalatrakten. *Sver. Geol. Unders.*, Ser. Ba 15, 46pp.

Lundqvist, G., 1961: Beskrivning till karta över landisens avsmältning och högsta kustlinjen i Sverige. *Sver. Geol. Unders.*, Ser. Ba 18, 148 pp.

Mawdsley, J. B., 1936: The wash-board moraines of the Opawica-Chibougamau area, Quebec. *Roy. Soc. Can. Trans.*, Ser. 3, 30:4, 9–12.

Munthe, H., 1905: Beskrifning till kartbladet Sköfde. *Sver. Geol. Unders.*, Ser. Aa 121, 158 pp.

Möller, H., 1962: Annuella och interannuella ändmoräner *Geol. För. Förh.* 84, 134–143.

S.G.U., 1956: Geologiska kartbladet *Uppsala*. *Sver. Geol. Unders.*, Ser. Aa 199.

Strömberg, B., 1962: Studier av isrecessionen i nordöstra Uppland och trakterna kring Ålands hav. *Department of Physical Geography*, Stockholm, 103 pp. (typewritten).

ESSELTE AB, STHLM 65

14

Reprinted from *Jour. Geol.*, **70**(5), 582–594 (1962)

ICE-THRUST RIDGES IN WESTERN CANADA[1]

W. O. KUPSCH

Department of Geological Sciences, University of Saskatchewan, Saskatoon, Saskatchewan, Canada

ABSTRACT

Numerous parallel, generally sharply crested ridges constitute a distinctive element of several portions of the Missouri Coteau, which is a prominent upland in Saskatchewan and Alberta, western Canada. The map pattern of the ridges, which in places are over 200 feet high, is linear or curvilinear with the concave side to the east, northeast, or north.

Exposures reveal that the ridges are composed mainly of complexly folded and faulted unconsolidated sediments of Cretaceous age. The deformations resulted from ice push by the Wisconsinan glacier where it moved upslope against the Missouri Coteau, especially where it entered preglacial valleys and pressed against the sides of these pre-existing, transverse sags of the upland. The ridges are thus regarded as *Stauchmoränen* or ice-thrust ridges, which represent slices of frozen ground deformed and dislocated by the glacier but left more or less intact during the final melting of glacier and ground ice. The ice-thrust ridges, which trend parallel to the glacier front, formed in the zone of glacier thrusting by compressive flow near the terminus under a light static pressure exerted by a rather thin mass of ice.

INTRODUCTION

The occurrence of ice-thrust ridges in western Canada has been mentioned before in the literature, but, as far as the writer is aware, the areal distribution of these ridges has not yet been shown on a regional map. The present study deals with the distribution of ice-thrust ridges in western Canada, with their morphology and its dependence on the structural geology, and with the conditions of origin and the mechanics of deformation. Field observations of known and suspected ice-thrust ridges were made and in addition, air photos were examined, and topographic maps studied.

DEFINITION

Deformations in drift and non-glacial sediments caused by thrust of an overriding active glacier are termed ice-thrust features (Flint, 1957, p. 88). They are evident geologically in exposures of the deformed deposits and, in some places, topographically as distinctive ridges which are the surface expression of the deformations produced by the glacier.

Ridges resulting from ice-thrust have been termed variously *Stauchendmoränen, Stauendmoränen, Stauchmoränen, Staumorä-*

[1] Manuscript received June 15, 1961.

nen, *Stauchrücken* in German, *stuuwmorenen, stuuwwallen* in Dutch, and pseudo-moraines, push moraines, thrust moraines, ice-pushed ridges, push ridges, or ice-thrust ridges in English. The ridges are parallel to the front of the glacier that formed them. They resemble end moraines, but they are distinguished by being structural rather than constructional or depositional in origin. Gradations to ridges with substantial amounts of incorporated morainic material are known, and thus a distinction between push ridges with no morainic material and push moraines composed mainly of drift could be made. Although such a distinction is made in Europe, the writer uses the term ice-thrust ridges for the less well-known ridges in western Canada which, as far as can be determined from outcrops, represent deformed bedrock, with only occasional incorporated basal till but in places covered by ablation till.

PREVIOUS AND PRESENT WORK

Ice-thrust ridges have been described from several European localities (Woldstedt, 1954, p. 101–110) and a voluminous literature is available on these occurrences, but those in western Canada have received only scant attention. Hopkins (1923) appears to have been the first serious student of ice-thrust features in western Canada. Slater (1927) made a detailed structural study of ice-disturbed beds in some parts of Alberta.

The complex surface structures of the Dirt Hills, Saskatchewan, have been known since 1873 when they were first observed by Bell (1874, p. 76). Subsequently, attention was directed to this area by Ries and Keele (1912, p. 84–92), who showed some photographs of the structures, and by Fraser *et al.* (1935, p. 62), who presented theories regarding their origin. Their formation was generally not understood, and, like those of other countries, the structures were long regarded as of tectonic or gravitational (slump) origin. Detailed structural mapping by Byers (1960) showed convincingly that only ice thrust could have caused the deformations.

In previous studies topographic expression was generally not considered, mainly because neither detailed topographic maps nor vertical air photographs and photo mosaics were available for the whole of western Canada until about 1955.

Studies by the present writer during the last five years revealed the presence of well-defined ridges in the known outcrop areas of ice-disturbed bedrock in Saskatchewan and Alberta. Similar ridges were noted away from the outcrop areas, but, because exposures are not present, it is not everywhere possible to verify the contention that the ridges are the surface expression of bedrock structures caused by overriding ice.

DISTRIBUTION

The ice-thrust ridges of Saskatchewan and Alberta are conspicuously well developed on the Missouri Coteau Upland and its northwestern extension from southern Saskatchewan to east central Alberta (fig. 1). The Missouri Coteau is a prominent, northwesterly trending upland of unconsolidated Upper Cretaceous shales, silts, sands, and clays. In places the upland has a distinctive northeasterly facing escarpment. South of Moose Jaw, Saskatchewan, this escarpment rises as much as 800 feet (224 m.)[2] above the lowlands to the northeast. The continental ice in its encounter with this pre-existing upland deformed the soft sediments into complex structures. South and east of Moose Jaw, Saskatchewan, sublobes of a large southeasterly moving lobe (the Weyburn lobe of Christiansen, 1956, p. 27) pressed deeply into the bedrock upland along preglacial valleys to form two sets of horseshoe-shaped ridges outlining the sublobes (pl. 1). Preglacial valleys, now filled with drift, in the Regina area (which comprises parts of the Missouri Coteau) are well shown on the bedrock topography map presented by Christiansen (1961, pl. 2).

The regional distribution of the ice-thrust ridges shows that they extend intermittently for at least 350 miles (560 km.) along the Missouri Coteau and its northwesterly extension (fig. 1). In this region they occur in fields which range from a few ridges, whose track can be followed for only 2–3 miles (3.2–4.8 km.), to hundreds of well-defined ridges traceable for over 70 miles (112 km.).

Local areas visited by the writer in the field are in the following paragraphs referred to by name and number which corresponds to the number shown in figure 1. The ridges suggesting an ice-thrust origin located in the Radville area (1), Saskatchewan, are distinctive on the ground as well as from the air. They are sharp-crested and form a pattern which trends east–west and is concave to the north where it is obliterated by abla-

[2] Because most literature on ice-thrust ridges originated in countries employing the metric system, all measurements from western Canada will be presented in both the English and metric system.

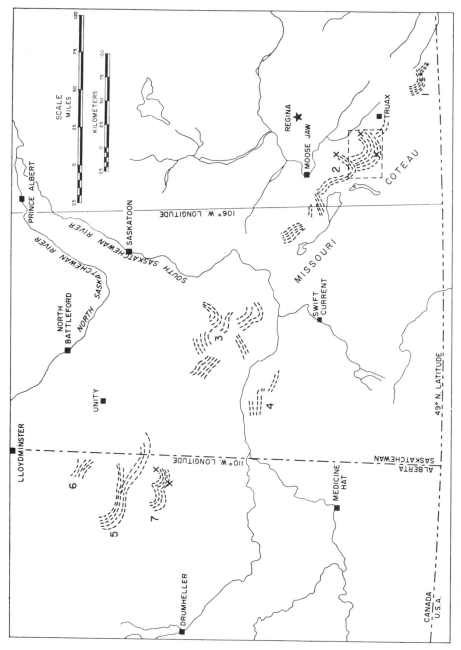

Fig. 1.—Distribution of known and strongly suspected, well-developed ice-thrust ridges in Western Canada. Exposures are indicated by ×. The numbers refer to localities mentioned in the text: (1) Radville area, (2) Dirt and Cactus Hills, (3) Clearwater Lake area, (4) Lancer area, (5) Neutral Hills, (6) Killarney Lake area, (7) Monitor area. The area covered by pl. 1 is indicated by a broken line.

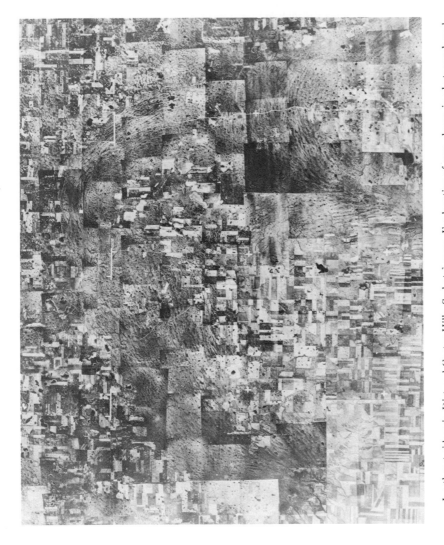

Ice-thrust ridges in Dirt and Cactus Hills, Saskatchewan. For location of area shown in photograph and for position of outcrops see fig. 1 and text. Mosaic compiled by Imperial Oil Co. Ltd., from air photos taken by the Royal Canadian Air Force.

tion moraine. The ridges, of which the internal structure is nowhere exposed, show on their flanks some erosion consisting of short, parallel gullies from the crest down to the troughs.

The Dirt and Cactus Hills (2), Saskatchewan, comprise the most extensive, continuous known area of ice-thrust ridges in western Canada. They extend from near Truax to northwest of Moose Jaw. The ridges are distinct and can easily be mapped from the air-photo mosaics (pl. 1). They are only thinly covered by ablation till which in most places does not obliterate the underlying ridge pattern of two distinct lobes. The bedrock interior of the ridges is well exposed in the northeastern extremity of the Dirt Hills near Claybank (sec. 28, T. 12, R. 24, W. 2; pl. 1) as well as in the Cactus Hills (T. 13, R. 26, W. 2; pl. 1) and in sec. 28, T. 10, R. 26, W. 2 (pl. 1). Surface observations indicate that the strike of the exposed, deformed strata coincides with the trend of the ice-thrust ridges as determined from air photographs (Byers, 1960, p. 7).

The Clearwater Lake (3), Saskatchewan, ice-thrust ridges lie north of the South Saskatchewan River and east of Clearwater Lake. They are very well developed in this area and are clearly evident on air-photo mosaics. No bedrock exposures are known to substantiate the contention that these ridges are the result of ice thrust in bedrock and that only minor amounts of till are incorporated. That till may be actually quite thick in places is suggested by a test hole drilled 42 feet (12.8 m.) into a ridge which encountered mainly till and minor amounts of gravel and sand (Christiansen, 1959, p. 12). However, because the ridges are morphologically similar to proved ice-thrust ridges in bedrock with some till in the deformations they are here regarded as such, and the thick till section reported is believed to be either an ablation cover over disturbed bedrock or local infolded basal till. The arcuate pattern, which consists of a rhythmic sequence of subparallel ridges rising 10–100 feet (3–30 m.) above the intervening troughs, can be traced for over 15 miles (24 km.).

The Lancer area (4), Saskatchewan, shows possibly the best developed sharp-crested ridges in western Canada in a narrow band about 2 miles (3.2 km.) wide and 15 miles (24 km.) long, which trends northwesterly and is slightly concave to the northeast. This band comprises about five ridges in 1 mile (1.6 km.), which means that the ridges are spaced about 1,050 feet (320 m.) apart. Postglacial gullying is most severe at the southeastern extremity of the field. There are no bedrock exposures in the Lancer area known to the writer, but from evidence gathered from nearby boreholes it is likely that Upper Cretaceous sands form the core of the ridges.

The ice-thrust ridges of the Neutral Hills (5), Alberta, were described by Gravenor and Bayrock (1955, p. 10–11) and later figured by Gravenor (1956, p. 17–19), who mentions that the northwest-trending ridge pattern seen on air photographs is the result of faulting and folding in bedrock by ice pressure as proved by shallow drilling carried out to determine the geology of the ridges.

South of Killarney Lake (6), Alberta, the bedrock is strongly ridged, but there are no exposures to demonstrate the internal ice-thrust structures.

South of Monitor (7), Alberta, in the Mud and Tit Hills, several disturbed bedrock outcrops can be noticed in the field. Although the ice-thrust origin of these structures has long been recognized (Hopkins, 1923; Slater, 1927), the topography of the hills received little attention. An examination of the air photographs reveals distinctive ridges with an over-all northwest trend and divided into two lobes, concave to the northeast. A blanket of ablation till, which shows a dead-ice disintegration pattern, covers the extreme parts of the ridges.

It appears that Rutten (1960, p. 296) must have had some of the occurrences of ice-thrust ridges shown in figure 1 in mind when he mentioned the regional development of ice-pushed ridges in Canada and referred to Slater's (1927) work on the Monitor area (7). It should be pointed out, however, that these ridges can hardly be considered to lie in "northwestern" Canada as Rutten stated.

It should be emphasized here that there are many areas of ice-thrust structures known to the writer in Saskatchewan and Alberta which are not shown in figure 1, because they have no or only insignificant topographic expression. It is intended to show only ice-thrust features forming ridges in figure 1 and to show only ridges resultant from known or strongly suspected ice-thrusting in bedrock with only occasional incorporated till.

CHARACTERISTICS

The characteristics of ice-thrust ridges in western Canada can best be studied by examining both air photographs and, where available, the most detailed topographic maps which are on a scale of 1:50,000 with a contour interval of 25 feet (7.6 m.), in addition to field studies. Where little or no drift lies on the ridges near exposures of deformed bedrock, the ridges are characteristically well defined and sharp crested. The sharp crests become rounded and the ridge less well defined where more drift obliterates the underlying ridge. Finally, the trend of a ridge may become lost under a cover of ablation till. The arrangement of typical disintegration features in this ablation till may, however, still show some control by the underlying ice-thrust ridges (Gravenor and Kupsch, 1959, p. 53).

Topographic maps (fig. 2) show that in places the ridges may rise as much as 200 feet (61 m.) above the intervening troughs which are commonly occupied by shallow lakes, swamps, or just more luxuriant, treed vegetation. From this maximum local relief the ridges, which are generally grass covered and treeless on top, grade down to hardly noticeable swells that are part of the local, short-grassed prairie upland. The ridges are commonly broad at the base which is on an average as wide as 600 feet (183 m.) but may be over 1,000 feet (305 m.) measured

from trough to trough. In map view the ridges are linear or, when a larger area is viewed, more characteristically curvilinear with the concave side toward the east, northeast, or north, facing the direction of local ice flow.

ORIGIN

At various times the structures now considered to be the result of ice thrust were regarded as due to landsliding, to gravitational sliding, or to deep-seated forces such as diastrophism. That only ice thrust can afford a reasonable explanation of the structures, and therefore of the ridges, is evident from the following observations by Byers (1960, p. 9) and the writer:

1. The concave side of the curvilinear, or lobate pattern of the ridge fields faces in a general northerly direction (fig. 1).
2. The presence of ridges well back from the main escarpment of the Missouri Coteau (pl. 1) argues against the interpretation of the ridges as slump block ridges.
3. The seismic records, logs of boreholes, and data from outcrops show that the deformations are confined to the near surface and surface and therefore rule out deep-seated forces as causes of the deformations.
4. The occurrence, in places, of compact basal till involved in folded strata rules out penecontemporaneous deformation of the Upper Cretaceous beds (fig. 3).
5. The structural characteristics of the deformations (fig. 4) show overturned folds, with absent inverted limbs, associated with imbricate thrusts dipping at about 40°–45° and steepening at the rear of the folds; the thrusts presumably flatten out and become essentially bedding plane faults in the shales underneath the deformed sediments; this type of structure is characteristic of superficial folds (de Sitter, 1956, p. 405).
6. The ice-thrust structures in western Canada are quantitatively similar to those elsewhere (table 1).

To these arguments could be added the regional distribution of the ice-thrust structures and the ridges. Byers (1960, p. 9) mentions that the maximum development of the structures occurs along regional slopes facing in a general northward direction, such as the northeast side of the Missouri Coteau. On these regional slopes the structures appear to be especially well developed along the borders of preglacial valleys such as the sag between the Dirt and Cactus Hills (fig. 1; pl. 1). In these areas of well-developed structures the ridges are also very prominent, but where a minimum development of ice-thrust structures has been observed, as over the relatively flat surfaces such as the lowlands to the northeast in front of the Missouri Coteau (fig. 1), the ridges are absent. It should be pointed out, however, that this regional distribution may be more apparent than real because ridges may now be absent where they formerly existed owing to glacial and possibly postglacial erosion after the ridges were formed. Also well-developed ice-thrust features and ridges may now be covered by

FIG. 2.—Topography of ice-thrust ridges. For location of area refer to grid system on pl. 1. After map sheet 72-H-13 (east half), Mossbank, Saskatchewan, produced by the Canada Department of Mines and Technical Surveys, Ottawa, on a scale of 1:50,000, contour interval 25 feet, 1957.

later deposits, particularly the front of the Missouri Coteau where glacial lake sediments obliterate the underlying bedrock and till. The existence of ice-thrust features, which show the same intense deformation and which possess the characteristic imbricate thrusts of the structures on the Missouri Coteau, can be noticed in front of the Coteau near Halbrite to the east of the Radville area (1). Prominent ridges are absent in the Halbrite area. However, careful examination of air photographs reveals some poorly defined low swells in a lobate pattern concave to the east, which, since it is so inconspicuous, is not shown in figure 1. It is believed that the absence of prominent ridges in this area is mainly due to later erosion which has erased a once distinctive local relief.

On the Missouri Coteau the ridges are,

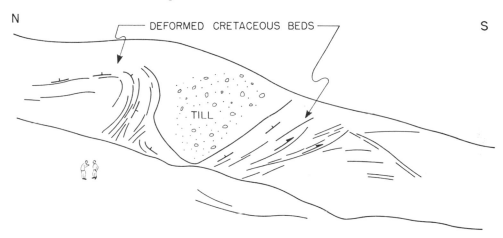

FIG. 3.—Basal till in syncline of deformed bedrock. Location: sec. 28, T. 10, R. 26, W. 2d mer., Saskatchewan. Sketched after photograph by the writer, 1960.

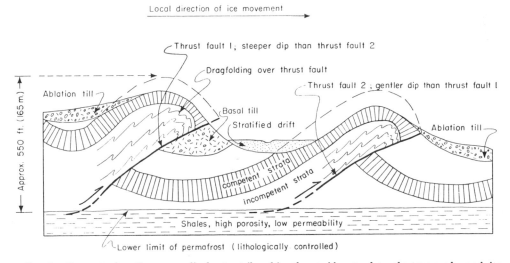

FIG. 4.—Conceptual section perpendicular to strike of ice-thrust ridges to show the topography and its relation to typical structures, stratigraphy, and permafrost. Diagrammatic after structure sections by Slater (1927) and by Byers (1960).

as far as known from outcrop studies, a direct or first-order morphologic expression of the folding and thrust-faulting produced by ice push. The ridges appear to correspond to the major structures, the imbricate thrust blocks. The surface trace of the thrust fault commonly lies on the flank of the ridge which is downstream with respect to the local direction of ice movement. This generalized relationship is shown in figure 4 from which also the stratigraphic-lithologic control of the ridges is apparent. The ridges seem to form where relatively competent beds are thrust over incompetent strata. In

are shown as missing. Such beveling of tops of folds may be due to flowing ice which carried away some of the deformed beds as Brinkmann (1953, p. 241) assumed for the structures on the Island of Rügen. In western Canada, where no corroborative evidence for a glacial readvance has been found, it may have resulted from meltwater erosion, either underneath the still active ice or more probably during the time when continued down-wasting had rendered the ice immobile. During this dead-ice phase, when the glacier started to break up or disintegrate into small blocks, ablation till was

TABLE 1

MEASUREMENTS OF ICE-THRUST STRUCTURES

MEASUREMENT	DIRT HILLS (2), SASKATCHEWAN (BYERS, 1960)		ISLAND OF RÜGEN (BRINKMANN, 1953) (METERS)
	Feet	Meters	
Minimum stratigraphic thickness involved.....	208 Cretaceous strata	63	≤100 Cretaceous strata 40 Deformed Pleistocene sediments
Maximum visible height of deformation in outcrops..................................	356	108	100
Projected maximum vertical extent of folds....	500–600	152–183	150
Dip-slip component of movement on larger thrust faults...........................	150–200	46– 61	No quantitative statement, but "nearly autochthonous"
Dip of thrust faults.......................	40°–60°		20°–90° Most common: 40°–70°

western Canada the competent beds of the deformed Cretaceous sediments are slightly indurated siltstones, sandstones, or concretionary layers, whereas the incompetent strata are generally clay or coal. Drag folds are common in these incompetent beds and the thrust planes are almost everywhere confined to them.

Figure 4 also shows that the steeper side of the ridges is generally in the downstream direction of glacier movement and that the upstream side is essentially a dipslope. This appears to be opposite from what has been observed on ice-thrust ridges in the Netherlands (Maarleveld, 1960, p. 299). In part, this could be explained by the various modifications which the ridges may undergo after they were formed. It should be noted that in figure 4 the upper parts of the structures

let down in places, thus further modifying the topography of the ridges. Some of the finer material, which was carried away from the ablation till by trickles and small streams of meltwater, was redeposited as stratified drift, mainly in the troughs between the ridges. The ablation till differs from the basal till in physical characteristics as well as structural position. It is stony whereas the basal till has fewer stones in a dense, fine-grained matrix. The ablation till lies disconformably on the beveled bedrock structures, but the basal till constitutes a conformable part of the deformations.

Still later changes, mainly after the melting of both glacier and ground ice was completed, were probably slight under the semiarid conditions which became gradually established and which are now prevalent in this region. However, some erosion and deposition took place, including the accumulation of plant material in the ponds between the ridges, and some gully erosion on the flanks of the ridges.

TIME AND CONDITIONS

Although ice-thrust origin of the ridges appears to be well substantiated, much less definite answers can be given to the questions of time of origin and conditions under which the ridges were formed.

The observed till core of a synclinal structure in the Dirt Hills (fig. 3) shows that deformation took place in Pleistocene time or later. Because so far no tills definitely older than Wisconsinan are recognized by geologists working in this area, this time range can be narrowed down to Wisconsinan or younger. Still unpublished radiocarbon dates (University of Saskatchewan, Saskatchewan Research Council) of postglacial marl and charcoal in this part of the Missouri Coteau require that the deformation be older than about 10,000 years. It appears therefore that deformation took place during the Wisconsinan but present information does not allow this time range to be narrowed further.

Viete (1960), in his comprehensive and excellent study of ice-thrust features in central Germany, enumerates and attempts to evaluate the complex factors which conditioned the occurrence and type of deformations caused by glacier push. To these belong the static and dynamic pressure of the ice, which in turn are functions of ice thickness, load, mode and velocity of flow, and pre-existing topography. Physical characteristics of the floor, such as the stratigraphic sequence and thickness of rock units, their lithology, water content and its pressure, and presence and kind of ground ice (which may depend again on lithology and hydrological factors) also influence the formation of ice-thrust features. The quantitative evaluation of most of these various factors is usually not possible and only general descriptions of prevailing conditions can be made. For some, not even qualitative statements are warranted. Conditions about which some observations can be made in the area of ice-thrust ridges in western Canada should, however, be considered.

Rutten (1960, p. 296) points out that ice-pushed ridges were not formed in Europe during the Würm (= Wisconsinan) glaciation but only during the Riss (= Illinoian) glaciation. He explains this by the difference in drainage pattern existing during these times, which was away from the ice during the Würm but toward the ice during the Riss glaciation. He follows this by the observation that in "northwestern" Canada (apparently meaning Alberta and Saskatchewan in western Canada) the drainage was toward the ice during the Wisconsinan glaciation. In his opinion, the direction of drainage, toward the ice, is a controlling factor in the formation of ice-thrust ridges, because the water-logged area in front of the glacier favored the development of permafrost.

It should be noted that the term "permafrost" as defined by Black (1954, p. 839) encompasses any "perennially frozen mantle or bedrock . . . whether the material is actually consolidated by ice or not." From Rutten's (1960) discussion it is evident, however, that he uses the term permafrost in synonymy with ground ice, and the present writer also adopted this more popular, though admittedly less exact, definition.

That the presence of ground ice holding the mineral grains in the sediment together is an essential condition of the formation of ice-thrust ridges follows from the lithological characteristics of the deformed sediments. In their present unfrozen state it would not be possible to deform the friable and uncemented sands and silts that comprise an appreciable part of the disturbed strata, and to maintain the continuity of bedding which they display in the ice-thrust structures of the exposures in western Canada.

The structural geology of the deformations is typical of hard rock deformation with clearly defined rock boundaries. Gradational boundaries which are to be expected if the disturbances took place in uncemented, unconsolidated sediments are absent (Byers, 1961, personal communication). Also it should be noted, as shown in figure 4, that all main failures (the thrusts) originated from one horizon and that similar planes of principal failure did not initiate above this horizon, but only continued above it mainly in the incompetent clays. This can be explained by the assumption that the material above the horizon of principal failure, with the probable exception of the clays, was held together by ground ice and that the lower boundary of permafrost determined the position of the plane of *décollement* of the thrusts.

From these observations it appears that a sheet of ground ice underwent deformation by push, which caused folding and ultimately the characteristic imbricate thrust-faulting, and that the structures were preserved on final melting of the ice. It seems that the permafrost layer in essence constituted the lower part of the glacier, the bottom of which was not the top of the bedrock but the lower limit of the ground ice sheet. The structures are then to be regarded as having originated in the basal part of such a glacier. The same conclusion was reached by several other geologists for similar structures elsewhere, mainly in Europe.

Because the drainage was toward the ice in western Canada during the Wisconsinan, when the ice-thrust ridges were apparently formed, the hydrological conditions and the resulting development of a thick layer of permafrost should receive further consideration. The permafrost was probably discontinuous and the local development and character of the ground ice may have depended on physical differences of the surface sediments related to their water-bearing capacities. This is suggested by the seemingly better development of ice-thrust ridges in sandy and silty sediments than in less permeable shales which constitute the surface sediments in some parts of the Missouri Coteau. Where deformed strata overlie these shales the latter constituted the lower limit of the permafrost layer and caused a *décollement* of the overlying deformed stra-

ta. It should be understood that the concept of figure 4 explains the absence of ice-thrust features in the shales at the bottom of the section as the result of the absence of permafrost in the shales. It is suggested that the absence of permafrost in the shales, but its presence in the sands and silts, is a function of the water-bearing characteristics of the sediments, principally the porosity and permeability. Where the shales constitute the bedrock surface in western Canada, ice-thrust ridges are not observed.

A zone of discontinuously distributed permafrost still exists today in the northernmost parts of Saskatchewan and Alberta (Black, 1954, p. 841). The southern boundary of this zone roughly follows the 0° C. isotherm of October, which in Alberta and Saskatchewan is near the sixtieth parallel North latitude. The occurrence of ice-thrust ridges near the forty-ninth parallel (fig. 1), which is only about 60 miles (96 km.) north of the southernmost limit of glaciation in the adjacent part of North Dakota during Wisconsinan time, suggests a southward shift of the isotherms of at least 750 miles (1,200 km.), from the sixtieth parallel to the forty-ninth parallel during the time when the ice-thrust ridges in southern Saskatchewan were formed. It is interesting to note that this shift is of the same order as the 10° latitude displacement of the isolation-variation curve at the time of the last glacial maximum (Emiliani, 1955, p. 568).

The approximate maximum thickness of the glacier ice where ice-thrust ridges were formed can be estimated for the Missouri Coteau in the Dirt Hills area (2), if it is assumed that the general region of the Coteau in southern Saskatchewan was not affected by postglacial differential movements. The Wood Mountain area just west of the Missouri Coteau along the forty-ninth parallel and between the 106° and 107° meridian W. longitude (fig. 1) apparently escaped glaciation above approximately 2,800 feet (855 m.) elevation (Wickenden, 1931, p. 45). Because, according to Byers (1960, fig. 3), ice-disturbed bedrock in the Dirt Hills occurs as high as at least 2,350 feet

(715 m.) elevation a thickness of the glacier above the bedrock surface of about 450 feet (137 m.) is arrived at, if the amount of surface slope of the glacier between the Dirt Hills and Wood Mountain is regarded as negligible. At the base of the Missouri Coteau, where the bedrock surface is at 1,900 feet (578 m.) elevation, the ice above the bedrock may have been as thick as 900 feet (275 m.).

The static pressure exerted by a glacier 200 m. thick varies between 18 kg/sq cm for clean ice to as high as 36 kg/sq cm for ice choked in debris (Viete, 1960, p. 54). This pressure alone is too small to have caused the structural disturbances of the magnitude observed in ice-thrust ridges, as was pointed out by Huizinga (1944, p. 269) in his remarks on the soil mechanics of ice-thrust ridges in the Netherlands. He regards a minimal dynamic pressure or lateral push of approximately 300 kg/sq cm sufficient to form the ridges. A pressure-melting curve for ice shows that such a pressure can be attained by ice at a temperature of $-3°$ C. The postulated value of dynamic pressure therefore does not require abnormally low temperatures in the glacier ice or the permafrost.

Granite and limestone boulders fractured by ice pressure in drumlinized kames described from near Dollard in southwestern Saskatchewan (Kupsch, 1955) present corroborative evidence for the existence of permafrost in parts of western Canada during Wisconsinan time. They could also provide some measure of the dynamic pressure exerted by the ice except for the lack of experimental data on rock failure under conditions prevailing underneath the Wisconsinan glacier. Moreover, these conditions, such as the confining pressure and particularly the length of time during which the stresses were operative, are only inadequately known and in part non-reproducible in experiments.

In summary, it can be stated that in western Canada ice-thrust ridges apparently developed parallel to the ice front in the marginal zone of a relatively thin but actively flowing glacier where the ice moved up-

slope, particularly into pre-existing valleys, where it pushed against the valley walls. It also seems that, possibly because of ponded drainage, a thick layer of permanent ground ice developed where physical conditions in the surface materials were suitable, and that slices of frozen ground which were part of the glacier were deformed and dislocated but left more or less intact during the final melting of glacier and ground ice.

MECHANICS OF DEFORMATION

Even fewer definite statements can be made regarding the actual mechanics of deformation of the ice-thrust ridges than about the conditions under which deformation took place.

Scale models have been devised in which structures, similar in general characteristics to those of ice-thrust ridges, can be produced by pushing a wooden glacier model against a horizontal sequence of gypsum layers in a matrix of paraffin and sand (Köster, 1957). In these experiments imbricate thrust blocks formed in front of the glacier model and drag structures underneath. Folds are produced first in the gypsum. On further advance of the glacier model the folds become recumbent and finally transform into imbricate thrusts. The thrust fault farthest away from the glacier is the youngest. The thrusts have shovel-shaped fault planes.

Although the structures produced in experiments are similar to those observed in the ice-thrust ridges of western Canada it should not be concluded that the latter were necessarily formed in front of the advancing ice. That the Dirt Hills (2) structures originated, at least in part, underneath the glacier is shown by the syncline in which basal till constitutes the core (fig. 3).

Brinkmann (1953, p. 238) showed that on the Island of Rügen several deformational zones, characterized by different types of structures, can be related to the distance back from the ice front. Closest to the terminus, and partly in front of it, rather steeply dipping imbricate thrust blocks form, such as those in western Canada, but farther back flat-lying folds originate by frictional drag underneath the thick ice mass. Slater

(1927, p. 727) mentions that within the zone of imbricate thrusting the angles of dip of the thrust planes "progressively increase from south to north" and that this is "a feature common to all glacially disturbed deposits." The same may hold true for the Dirt Hills (2), although here this tendency is not very pronounced as an examination of the cross sections by Byers (1960, p. 5) will show. The experiments by Köster (1956, p. 567, fig. 1*a*) also show a progressive steepening of fault planes in the deformed "proglacial" layers toward the older thrust blocks at the edge of the glacier model.

It has been observed in existing glaciers that thrusting in the ice is best developed near its margin, where the rigid upper surface of the glacier, which farther upstream overlies plastic ice, extends to the base and is thrust upward over a sheet of dead ice in front of the glacier. These thrusts also show a steepening dip away from the terminus (Flint, 1957, p. 73). The same seems to hold true for areas of obstructed or compressive flow, where the ice is thrust upward over a topographic obstruction. Ice-thrust ridges are believed to have formed in such an area of thrusting in the marginal zone of a glacier in which the upper part of the bedrock down to the lower limit of permafrost was incorporated in the glacier, the base of which was therefore not the top of the bedrock but the lower limit of permafrost or the top of the water-bearing, unfrozen shales beneath the ground ice.

The frictional resistance to sliding of the upper shales underneath the deformed layers, as postulated in figure 4, may have been reduced by anomalously high pressures of interstitial ground water. Such pressures in the shales can be considered abnormally high because of the gravitational compaction exerted on the strata by the weight of the overlying permafrost layer and glacier ice, which also prevented the escape of the water from the shales. Reduction by pore water pressure of the critical value of the shear strength required to produce sliding is discussed by Hubbert and Rubey (1959). The principle of a reduction of shear strength by increased pore pressure was applied to subglacial thrusting by Mathews and MacKay (1960), who point out that (1) the shear strength of frozen sediments is commonly considerably greater than that of unfrozen material, (2) the lowest critical shear strength lies at the base of the permafrost, (3) high pore pressures may develop below the permafrost, (4) shearing may take place in clays in the permafrost because they may remain plastic at subzero temperatures even in spite of included ice.

In summary, it appears that where exposures reveal the internal structure of ice-thrust ridges in western Canada the deformations, which all belong to the zone of thrust blocks with gradual steepening of fault planes upstream, suggest an origin in the zone of glacier thrusting near the terminus under thin ice exerting a light static pressure, and also possibly in front of the ice. Fold structures, without imbricate thrust-faulting, which could have formed at the same time farther back from the glacier front, have not yet been recognized in western Canada, probably because (1) they lack the distinctive topographic expression of the imbricate structures, (2) no good exposures are present, (3) not enough detailed field work has as yet been done. At the present it is therefore impossible to distinguish distinctive zones of structural deformation as was done by Brinkmann (1953) on the Island of Rügen.

ACKNOWLEDGMENTS.—The writer gratefully acknowledges the stimulating company at various times of Dr. A. R. Byers, University of Saskatchewan, of Dr. E. A. Christiansen, Saskatchewan Research Council, and of Mr. R. R. Parizek, Pennsylvania State University, while examining disturbed outcrops in Saskatchewan. Dr. C. P. Gravenor and Dr. L. A. Bayrock, Research Council of Alberta, joined the author on his visits to the known localities in Alberta. A grant received from the Saskatchewan Research Council enabled the writer to make the field trips and to undertake the air-photo study in which he was ably assisted by K. N. Gulstene. The staff of the drafting section of the Saskatchewan Department of Mineral Resources and D. Delorme and O. Miedema assisted with the illustrations.

REFERENCES CITED

BELL, ROBERT, 1874, Report on the country between Red River and the South Saskatchewan, with notes on the geology of the region between Lake Superior and Red River: Geol. Survey Canada Rept. Progress 1873–1874, p. 66–93.

BLACK, R. F., 1954, Permafrost—a review: Geol. Soc. America Bull., v. 65, p. 839–856.

BRINKMANN, R., 1953, Über die diluvialen Störungen auf Rügen: Geol. Rundschau, v. 41, Sonderband, p. 231–241.

BYERS, A. R., 1960, Deformation of the Whitemud and Eastend formations near Claybank, Saskatchewan: Royal Soc. Canada Trans., v. 53, ser. 3, sec. 4, p. 1–11.

CHRISTIANSEN, E. A., 1956, Glacial geology of the Moose Mountain area, Saskatchewan: Saskatchewan Dept. Mineral Resources Rept. 21.

————1959, Glacial geology of the Swift Current area, Saskatchewan: Saskatchewan Dept. Mineral Resources Rept. 32.

————1961, Geology and ground-water resources of the Regina area, Saskatchewan: Saskatchewan Research Council Rept. 2.

DE SITTER, L. U., 1956, Structural geology: New York, McGraw-Hill Book Co.

EMILIANI, CESARE, 1955, Pleistocene temperatures: Jour. Geology, v. 63, p. 538–578.

FLINT, R. F., 1957, Glacial and Pleistocene geology: New York, John Wiley & Sons.

FRASER, F. J., et al., 1935, Geology of Southern Saskatchewan: Geol. Survey Canada Mem. 176.

GRAVENOR, C. P., 1956, Air photographs of the plains region of Alberta: Research Council of Alberta, Prelim. Rept. 56-5.

———— and BAYROCK, L. A., 1955, Glacial geology of Coronation district: Research Council of Alberta, Prelim. Rept. 55-1.

———— and KUPSCH, W. O., 1959, Ice-disintegration features in western Canada: Jour. Geology, v. 67, p. 48–64.

HOPKINS, O. B., 1923, Some structural features of the plains area of Alberta caused by Pleistocene glaciation: Geol. Soc. America Bull., v. 34, p. 419–430.

HUBBERT, M. K., and RUBEY, W. W., 1959, Mechanics of fluid-filled porous solids and its application to overthrust faulting: Geol. Soc. America Bull., v. 70, p. 115–166.

HUIZINGA, T. K., 1944, Geologie en grondmechanica: Geol. Mijnbouwk. Genootschap Nederland, en Koloniën, Geol. Ser., v. 14, p. 259–275.

KÖSTER, ROLF, 1957, Schuppung und Faltung im glazialtektonischen Experiment: Geol. Rundschau, v. 46, no. 2, p. 564–571.

KUPSCH, W. O., 1955, Drumlins with jointed boulders near Dollard, Saskatchewan: Geol. Soc. America Bull., v. 66, p. 327–338.

MAARLEVELD, G. C., 1960, Glacial and peri-glacial landscape forms in the central and northern Netherlands: Tydschrift Koninkl. Nederl. Aardrijksk. Genootschap, v. 77, no. 3, p. 298–304.

MATHEWS, W. H., and MACKAY, J. R., 1960, Deformation of soils by glacier ice and the influence of pore pressures and permafrost: Royal Soc. Canada Trans., v. 54, ser. 3, sec. 4, p. 27–36.

RIES, HEINRICH, and KEELE, JOSEPH, 1912, Preliminary report on the clay and shale deposits of the western provinces: Geol. Survey Canada Mem. 24-E.

RUTTEN, M. G., 1960, Ice-pushed ridges, permafrost, and drainage: Am. Jour. Sci., v. 258, p. 293–297.

SLATER, GEORGE, 1927, Structure of the Mud Buttes and Tit Hills in Alberta: Geol. Soc. America Bull., v. 38, p. 721–730.

VIETE, GÜNTHER, 1960, Zur Entstehung der glazigenen Lagerungsstörungen unter besonderer Berücksichtigung der Flözdeformationen im mitteldeutschen Raum: Freiberger Forschungshefte C78, Berlin, Akademie Verlag.

WICKENDEN, R. T. D., 1931, An area of little or no drift in southern Saskatchewan: Royal Soc. Canada Trans., v. 25., ser. 3, sec. 4, p. 45–47.

WOLDSTEDT, PAUL, 1954, Das Eiszeitalter (Erster Band): Stuttgart, Ferdinand Enke.

IV
Ice Contact: Eskers and Kames

Editor's Comments on Papers 15 Through 19

15   **De Geer:** Excerpts from *Geochronologia Suecica: Principles*

16   **Price:** *Eskers near the Casement Glacier, Alaska*

17   **Hellaakoski:** Excerpts from *On the Transportation of Materials in the Esker of Laitila*

18   **McKenzie:** *Observations on a Collapsing Kame Terrace in Glacier Bay National Monument, South-Eastern Alaska*

19   **Jewtuchowicz:** *Kame Structure West of Zieleniew*

The striking long ridges of sandy gravel called eskers (or oses in Europe) have inspired many regional studies during the last 100 years and yielded some deduction about circumstances of deposition (Upham, 1876; Stone, 1881; Fairchild, 1896; Giles, 1918; Chadwick, 1928; Howarth, 1966; Jewtuchowicz, 1967). In addition, it was popular in the middle part of that century to write on the origin of eskers in general (Crosby, 1902; Trowbridge, 1914; Flint, 1928; Tanner, 1932). None of these were based directly upon studies of the glacier ablation zone itself, but they do agree on glaciofluvial high-energy deposition between ice walls. I picked the most influential early study (not fully published until 1940) which De Geer developed in conjunction with varved-clay studies (Paper 15). The eskers that he studied were built under the sea in hydrostatic tunnel conditions. Unfortunately, De Geer's English leaves a little to be desired, but it is faithfully reproduced. Paragraphs omitted here involve some sea-level evidence, irrelevent early theories, and several repetitive examples.

Gravity flow on or under ice is the dominant characteristic of most of the situations described by others, so I chose a more recent study by Price which treats eskers both on and under the ice of an Alaskan glacier (Paper 16). Price has also done several studies in America and Iceland (see Price, 1966, and Part V). Others contributing to our knowledge at actual glacier sites would be Lewis (1949), Stokes (1958), and Mickelson (1971). As Flint (1928) and others point out, eskers must be polygenetic.

Eskers are economically valuable, which is a chief reason for many studies past and present. Older eskers left by Pleistocene ice are a chief source of coarse aggregate in most glaciated areas. Roads have utilized the crests of some as ready-made well-drained sites. Two recent studies (Grano, 1958; Lee, 1965) utilize eskers for finding ores, and many more are unpublished or lie ahead.

The ice movement plus meltwater transportation means that pebbles and cobbles travel further in eskers than in adjacent ground-moraine till. Hellaakoski in Finland was perhaps the first to work this out carefully, so most of his original article in English is reprinted here (Paper 17). Sections on methods of mechanical analysis, tables of countings, and a few figures had to be omitted. A similar and shorter article (Trefethen and Trefethen, 1944) started the same treatment in America. Hydrological lab studies and modeling of flow have only just begun (Hanson, 1943).

Kames are usually associated with eskers to the point that they evinced little separate literature (e.g., Upham, 1876; Gregory, 1912; Jewtuchowicz, 1967). The original idea of glaciofluvial accumulation on or against irregular ice (Close, 1867) has persisted to the present. Holmes (1947) may be closest to a modern definition. In any case, all students deduce from kame groups or "kame fields" that the ice was locally stagnant or thin and partially buried ice masses existed. During the vigorous arguments of the late 1920s and 1930s about how an ice sheet melts away, the width of that stagnant ice melt, i.e., the simultaneous kame field, became an issue (see Part V).

Finally, from these deductive studies grew a few studies along actual decaying ice edges. McKenzie's study of thermal regime and melting of a kame terrace in southeast Alaska (Paper 18) was one of the first. Others on Martin River Glacier, Alaska (Clayton, 1964), are most pertinent to disintegration (Part V).

In the older literature Cook (1946) anticipated some of the details of water–sediment–ice relations in his "perforation hypothesis." The concept of honeycombed marginal ice, almost completely stagnant, had evolved slowly out of early New England–New York studies (Stone, 1881; Shaler, 1888; Fairchild, 1896). The interpretation of kames reached a climax in a series of papers in Poland in 1969 (e.g., Jahn, 1969). Jewtuchowicz's article (Paper 19) was translated rather freely for this purpose. Possibly this is the first really detailed interpretation of complex kame structure.

References Cited

Chadwick, G. H. (1928) Adirondak eskers: Geol. Soc. America Bull., v. 39, p. 923–929.

Clayton, L. (1964) Karst topography on stagnant glaciers: Jour. Glaciol., v. 5, p. 107–112.

Close, M. H. (1867) Notes on the general glaciation of Ireland: Roy. Geol. Soc. Ireland Jour., v. 1, p. 207–242.

Cook, J. H. (1946) Kame complexes and perforation deposits: Amer. Jour. Sci., v. 244, p. 573–583.

Crosby, W. O. (1902) Origin of eskers: Amer. Geologist, v. 30, p. 1–38.

Fairchild, H. L. (1896) Kame areas in western New York, south of Irondequit and Sodus Bay: Jour. Geol., v. 4, p. 129–159.

Flint, R. F. (1928) Eskers and crevasse fillings: Amer. Jour. Sci., v. 240, p. 410–416.

Giles, A. W. (1918) Eskers in the vicinity of Rochester, N.Y.: Rochester Acad. Sci. Proc., v. 5, p. 161–240.

Grano, O. (1958) The Vasso esker of south Finland and its economic importance: Fennia, v. 82, p. 3–33.

Gregory, J. W. (1912) The relation of eskers and kames: Geograph. Jour., v. 40, p. 169–175.

Hanson, G. F. (1943) A contribution to experimental geology: the origin of eskers: Amer. Jour. Sci., v. 241, p. 447–452.

Holmes, C. D. (1947) Kames: Amer. Jour. Sci., v. 245, p. 240–249.

Howarth, P. J. (1966) An esker, Breiomerkurjökull, Iceland: British Geomorphology Research Group, Report of Symposium at St. Andrews, p. 6–9.

Jahn, A. (1969) Kame terraces in the Sudetes: Folia Quaternaria, v. 30, p. 17–21.

Jewtuchowicz, S. (1967) Description of eskers and kames in Gashamnöyra and on Bungebreen south of Hornsund, Vestspitsbergen: Jour. Glaciol., v. 5, no. 41, p. 719–725.

Lee, H. A. (1965) Investigation of eskers for mineral exploration: Geol. Survey Canada Paper 65–14, p. 1–17.

Lewis, W. V. (1949) An esker in the process of formation, Böverbreen, Jotunheim, 1947: Jour. Glaciol., v. 1, p. 314–319.

Mickelson, D. M. (1971) Glacial geology of the Burroughs Glacier area: Ohio State University, Institute of Polar Studies, Rept. 40, 149 p.

Price, R. J. (1966) Landforms produced by the wastage of the Casement Glacier, southeast Alaska: Ohio State University, Institute of Polar Studies, Rept. 9, 24 p.

Shaler, N. S. (1888) On the origin of kames: Boston Soc. Nat., Hist. Proc., v. 23, p. 36–44.

Stokes, J. C., (1958) An esker-like ridge in the process of formation, Flötisen, Norway: Jour. Glaciol., v. 3, p. 286–290.

Stone, G. H. (1881) The kames of Maine: Boston Soc. Nat. Hist. Proc., v. 20, p. 430–469.

Tanner, V. (1932) The problem of eskers: Fennia, v. 55, p. 1–13.

Trefethen, J. M., and H. B. Trefethen (1944) Lithology of the Kennebec Valley esker: Amer. Jour. Sci., v. 242, p. 521–527.

Trowbridge, A. C. (1914) The formation of eskers: Iowa Acad. Sci., v. 21, p. 211–218.

Upham, W. (1876) On the origin of kames and eskers in New Hampshire: Amer. Assoc. Adv. Sci., Proc. v. 25, p. 216–225.

15

Reprinted from *Kungl. Sv. Vet. Akademiens Handlingar,* **18**(6), 45–47, 51–53, 54–55, 58, 65–69, 74, 75 (1940)

Geochronologia Suecica: Principles

GERARD DE GEER

Oses.

Introduction.

Perhaps the most striking and fascinating Quaternary feature in the Swedish landscape is meeting the eye by the marked, long, winding, forest-covered ridges of gravel and sand which, as a rule, are following the bottom of valleys, or also passing out across the plains, lakes and rivers or even across meeting hills (Pls. 2—14, 54—58, 68—71).

Already from prehistoric times those smooth and dry ridges often were used as natural roads, just as the Indians in America used the deserted beaches from their glacial ice-dammed lakes.

From their importance for any easy communication through regions often rather stony or swampy, the oses often determined the foundation of new settlements and very early must have attracted the attention of the aborigines.

But from a practical appreciation to a real understanding of the ose phenomenon it was a long way.

A good expression of the difficulty of finding a natural explanation of the ose-

Editor's Note: A row of asterisks indicates material deleted owing to lack of space.

phenomenon is given in the following tale in the style of folk-lore from the plains of the province of Nerike.

»It was in the olden time of great dragons. The warm southern seas crawled of enormous sea-serpents. Once they were struck by a terrifying plague. Those who could escape, fled, struck by awe, towards the north and cooling waters. But so terrified they were, that many of them rushed up on land, before dying. There they became petrified. Their scales and bones were all changed into pebbles, and that was the origin of the long, winding oses, which really remind of some exotic, strange monsters, which have crawled up, invading our nordic landscape.»

The great pace-maker of modern natural science in Sweden, CARL LINNAEUS, already in 1747 said: »When shall our Swedes become as attentive as to describe all the oses of Sweden and thereby to lay a foundation for scientists with respect to the first emergence and origin of Sweden.» (CARL LINNAEUS, Wästgöta-resa. Reprint, 1928. Göteborg, 8:o. P. 14.)

Evidently LINNAEUS suspected that the pebbly deposits of the oses had some connection with the land-emergence, and this also was the assumption of several other scientists, who speculated upon the origin of the oses.

But during the childhood of geology, when careful observations in Nature still were very insufficient, it was a natural consequence that every attempt at an explanation of such complicated phenomena as oses must be rather hypothetical and that every hypothesis, having explained only some part of the phenomenon, must cede its place to some other explanation, until the series of tentative explanations could be substituted by direct, detailed observations, showing up all essential parts of the normal ose-anatomy and of its real origin.

By following, in the vicinity of Stockholm and Uppsala, the successive cutting of several great ose-sections by detailed levellings and measurings, I had the opportunity of gradually getting a large collection of facts, fixing as well the inner structure of the original ose-deltas as their actual, secondary, exterior shape.

What made it possible in this part of the glaciated area to discuss and illustrate more fully as well the formation as also the partial deformation of the oses, was that here they were deposited below the sea-level, whereby not only the coarser, real ose-sediment, but also its finer constituents of sand and clay were deposited close by. Here, consequently, the whole of the imposing annual phenomenon of late-glacial ice-melting could be studied in its totality, with its different parts completing each other, and here, furthermore, the later postglacial marine deposits clearly certified the secondary, partial degradation and ose-masking, which is necessary to bear in mind in order really to understand the great phenomenon of ice-melting which put an end to the Ice Age.

It appeared that the original ose-deposits were delta-like and flat-topped with current-bedding, having its general dip towards the distal, or here about to the southern side (Fig. 6).

Along both sides of the delta, and also around so-called *åsgropar* or sink-holes within the deposits there often occur dislocations, sometimes with a considerable throw. Thus along the west side of the ose at Uppsala I could fix in detail two faults, having together a throw of more than 9 m, indicating that along this

Fig. 6. Uppsala ose. Section at Asknäs, Ekerö, W of Stockholm. Primary distal ose lamination.
On top: redisposited wave-rolled gravel. Photo J. Dylik, 1934.

side of the ose a subbasement of at least that thickness had disappeared, being
evidently the ice-basis of the glacier-vault within which the ose-delta was
deposited.

Not seldom such basal remnants of the marginal ice-vaults by a covering of
ose-sediment were prevented from flowing away and could be covered by several
tenths of annual varves, before they melted and caused a local sinking down and
a considerable dip of the ose-varves. This is the reason why the varves in the
exposures along the sides of the oses as a rule are so dislocated that their thicknesses are quite falsified. As the temperature of the water at the sea-bottom in
those masses of melt-water must have been near the freezing-point of fresh water,
it is easily understood that the melting of such imbedded bodies of ice must have
required many years, as shown by the number of dislocated varves, sometimes
amounting to more than seventy.

Besides the dislocations by melt-sinking also other ones, caused by slidings
out from the original steep delta-sides are quite common. They commenced no
doubt very often as soon as the delta shores of ice withdrew, and, as long as the sea-
water was too deep to allow any wave movement at its bottom, the slope of the
slidings may have been relatively steep, but, judging from the actual minimum
depth in the most different parts of the water, at which fine sediment is left by
natural selection without being transported farther off, it seems to follow that
— the sea-bottom may be rising or sinking — everywhere in open situations no
clay sediment will be left in peace at smaller depths than 50 m. Thus it must be
assumed that bottom-waves at very strong winds must be able to carry away
the finest sediment from more shallow parts of the sea.

*　*　*　*　*　*　*

Ose ridge with steep slopes. Typical pine forest. Crest: ancient riding path. Right: olden road and coach. Supramarine deposition, spared from later wave-degradation, the slopes only adjusted by down-sliding. Loos, Hälsingland. Photo H. Hesselman, 1903.

The explanation of the author.

Still, if this had been the case at all the deviations of the ice-rivers from the direction of the ice-movement, the morainic material following these latter ought to have been deposited below the moving ice. The definitive objection against the continuous deposition within long tunnels at the bottom of the ice derived from the new, detailed investigations especially at Stockholm and Uppsala, had

Ose ridge crossing river Dalälven. Hedesunda, Gästrikland.
Photo Aeromateriel A/B, G. 183.

fixed by maps and sections that the oses in question had a rhythmic composition with alternating centers of big cobbles, grading over to finer and finer gravel and sand and sometimes even to intervals without any sediment at all. Those rhythmical deposits I called centres and emphasised that they could not have been deposited simultaneously in one and the same river-bed as their rapid variation in coarseness required equally rapid changes in the transporting power of the rivers.

This was the main reason why in 1897 I published not a new theory but the results obtained by a long series of measurements and other observations. By these it had been possible in great sections directly to follow and to fix the continuous tectonics of quite a number of ose centers with their immediate extraglacial continuation into the adjacent sea-water of local low stone-free sandridges, an earlier overlooked feature in the ose-tectonics, which I called current-ridges, and further on into less and less sandy clay, radiating further out into half-circular fans or aprons, gradually fading out and passing over into less and less sandy, ordinary clay-varves.

By these investigations it became directly proved that the ose-centres with their pertaining finer sediment must have been deposited successively at the very mouth of the bottom rivers, like their submarginal deltas at the same rate as the ice-border retired. This explained the rhythmical structure of the ose-material and even of such hitherto unexplained ose-hills as occur at the side of the general trend.

By the preceding theory it had been assumed that the intervals on the contrary of being secondary and eroded at the very mouth of the ice-river, which was assumed to have poured out violently, while it could have had a relatively quiet course below the ice, thus just the opposite to the results derived from my investigations. The rivers in question, namely, when opening into the still-standing sea-water, lost their cutting and transporting power and, on the contrary, must have deposited the heaviest part with which they had been overburdened as long as under the ice they had been forced forward under head from the melt-water in the fissure-system of the land-ice.

At numerous good sections it had been possible to fix by measurements, as a normal feature along the sides of the oses, marked dislocations registering the original situation of the steep ice-walls, once forming the very shores of the ose-delta and the sides of its glacier-vault. At some instances this latter may have been broken up by fracture and swimmed away, thus forming a local ice-bay, sometimes indicated by terminal moraines.

Very soon, off the ice-border, the current-ridges are characterised by a total lack of stones, even the smallest, in a striking way illustrating how suddenly the great power of the gigantic melt-rivers was subdued at the outlet into the sea. Still, as to the finest part of the sedimentation, or the varved clay, it has been found, quite recently, that its very finest particles have been distributed unexpectedly far into the open sea, at least more than 170 kilometres, still showing by sufficient magnification, a quite recognisable variation in thickness, permitting an exact dating. These microdistal varves no doubt form the very summit of this magnificent process of sedimentation, or so to say, »la crème de la crème», as being deposited far out from the coast and the different sources of error which must be taken into account concerning varves from more shallow waters.

In this way it has been possible by direct observations to trace and follow up the whole of the process of sedimentation from giant cobbles over one meter in diameter unto the utterly fine microdistal clay, and it must be emphasised that this is a relation of long continued direct observations made in a region where all the parts of this sedimentation can be studied in their mutual relations. It is therefore a misconception when certain authors are believing that my explanation is but a theory, the foundation of which can be discussed by confrontation with incomplete, locally developed or even more or less dislocated ose-deposits in other regions.

Subaquatic and supraaquatic oses.
Radial or recessional, annual oses.

* * * * * * *

Ose centres (submarginal).

The dating of special parts of certain oses makes it possible to compare these with synchronous parts of other oses in different regions and thereby to get new aspects on the study of the ose problem and on the distribution of the proximal melt-sediments.

Especially in the late-glacial subaquatic regions where, by means of the occurrence also of the finer varved sediment, exact datings are possible, comparative studies of synchronous oses are very instructive as indicating the successive directions followed by the melt-rivers in the neighbourhood of their outlets. But even in supra-aquatic regions marginal and sometimes small, regular creeks of erosion may be helpful for estimating the local, annual ice-recession.

As also observed by C. MANNERFELT on the higher parts of Mt Städjan in Dalarna and by myself in Jämtland near Medstugan, the ose-ridges are sometimes rather regularly divided up into distinctly separated hillocks, which might represent annual centra and thus give some hints as to the rate of ice-recession.

Here it may also be remarked that in the more distal parts of the glaciated region of northern Europe, the oses occur in a somewhat different shape. They are often more or less covered by till, and sometimes rather dislocated. Several authors have assumed that they were deposited by melt-rivers in glacier fissures. But the sections published are generally schematic sketches only, and accurately measured, detailed maps and sections on a large scale of a representative such ose are much wanted, before this kind of oses can be satisfactorily explained. They may be distinguished as tilly oses.

By means of the varve-dated ice-border lines it has been possible to make out how within submarine or other subaquatic regions on the whole the structure of the glacifluvial oses coincides with the rate of ice-recession as expressed by their immediate continuation in the more distal varve sediment.

A greater degree of regularity cannot be expected, when it is the question of sediments deposited at the very mouth of a melt-river with varying and sometimes very intense strength of the current and further on not seldom with sudden changes of the whole river-mouth and of its ice-walls, reduced by the ice-melting or the opening of new fissures.

As a rule the deposits of such an annual ose delta or ose centre begins with finer gravel, overlaid by a zone of maximum coarseness, often consisting of bigger, well-rounded cobbles, sometimes in a most obvious way bearing witness of the grand mechanic analysis, every year executed by the melt-rivers on the raw till-material produced by the grinding and transporting action of the glaciers.

* * * * * * *

Complex structure of the oses.

The oses themselves sometimes as exactly in the Stockholm region very obviously show that they were not at all continuous in the outset, but divided up into separate hills or centres, which sometimes are totally separated as well topographically as also with respect to their inner structure. The sections show clearly that this latter depends of their original deposition and is not due to later erosion. On the opposite, when several ose-centres have been deposited in a row, the one close after the other, they are often so well tied together and smoothened by the waves during the land-emergence that generally they were considered as quite continuous ridges. The inner structure of typical oses, however, when examined in good sections, always is found to be discontinuous and characterised by an inner cyclic structure (Pl. 9 b, 10) as to every individual centre, corresponding to the sediment in a certain clay varve, both these facies of deposition being corresponding parts of one and the same reassortment of morainic matter during one and the same year, both at the same time deposited in its delta. When STRANDMARK put forth his explanation of the oses, their inner periodic structure was not known, why he still believed that they were deposited continuously along the whole of their length, and that their interruptions were secondary and caused by erosion, while, on the opposite, it has now been shown that they are originally separated and explainable only by a successive, retrograde deposition at the very mouth of the ose-rivers in such cases where these had been captured, probably by side-fissures, and so melted out new glacier vaults.

What is said above especially concerns those extended regions which at the formation of the oses were covered by water and where thus the whole complex of melt-sediment can be studied in its mutual relations.

Supra-aquatic oses which generally are more or less separated from their finer sediment, flowing away with the running melt-water, need a very detailed investigation if the minute history of their origin is to be entangled.

Probably, in lack of standing water outside the supra-aquatic river-mouths, there will not, in this case, be so marked a marginal deposition as in subaquatic regions. In such cases it seems possible that gravelly material may be deposited also at some distance inside the ice-border and that much of the sandy components of the sediment will be spread out during flood time as ordinary river-sand, while silt and clay may follow the currents unto quiet water in lakes or in the sea.

* * * * * * *

Storkullen ose hill at Haga Norra.

The next ose-hill towards the north, here called *Storkullen* or the Big Hill, rises to the greatest height in the environs, or to 51 m a. s. (Fig. 10 and pp. 48—49).

As affording good examples of many typical ose-phenomena, this hill has been especially studied since a long time and hypsometrically mapped out by Bennet Co on the scale of 1:1000, here reproduced by 1:3000 (Pl. 55).

As to the rocky subbasement of this hill, it may be mentioned that a short way from its southwest side the granitic bed-rocks are exposed close to the northern

Fig. 9. Stockholm N, Haga Norra. Total view of the ose summit with the main section, towards
NW. Nearly half of its height is due to the redeposited material, eroded and wave-transported from
other high parts, now destroyed. Photo E. H. D. G., 1936.

entrance of the royal park of *Haga* and the tramway station *Haga Norra* at a
height of about 25 m a. s.

At both sides of the proximal steeper slope of the ose-hill bed-rocks are seen
at a somewhat higher level or a little over 30 m a. s., probably being parts of a
rock ledge running about W—E below the proximal part of the ose-hill, though
probably here somewhat lower, thus affording the best passage upwards for the
subglacial ose-river. Judging from the height of the moraine it may be a fair assumption that the subbasement below the main part of the hill may be something
about 20 m a. s. Thus the remaining thickness of this ose-hill amounts to about thirty
meters. But it is evident that such a deposit of loose sandy gravel in an open
situation towards all directions necessarily during the land emergence must have
been essentially lowered by wave erosion. This has also been amply verified by a
really grand section which during the last thirty years has been carried on along
the southeast axis of the ose almost unto its remaining highest part.

The successive situation of these sections are marked on the map with dotted
lines and respective years. Thus I have had the opportunity to follow these sections
during about half a century by the help of photos and measuring. Thereby it
has turned out that a considerable part of the gravel-covering, which very easily
could have been mistaken for a discordant upper ose-delta, in reality was a marine
sediment consisting of redeposited ose-gravel with about twenty degrees dip in a
centrifugal direction against the peripheral parts of the hill (Fig. 10).

Near the basis of the hill slope, between ose-deposits *in situ* and the superjacent
down-wash remnants of downslidden dislocated proximal clay-varves were found.

Fig. 10. Stockholm ose at Haga Norra, Sth. N. in 1919, NW wall. Sharply marked discordance between fine-grained ose-centre, below, and coarser material, above; on the summit real shore gravel, redeposited by the waves. Section at south end of the ose ridge. Photo O. Halldin, 1919.

Farther up and northwards this clay-horizon could be traced practically all along the section. At a certain division of its higher part only two small series of undisturbed microdistal varves were found, as it seemed resting concordantly upon the underlying ose-bed, and superposed by fine marine gravel, upwards more and more pebbly, this latter even at this height amounting to a thickness of somewhat over 10 m. This was at about a height of 35 m near the northwestern part of the digging in the year 1937 (Fig. 18) and observed at the SW side already in 1910 (Fig. 11).

Farther to the south, about where the surface of the hill approximately may have been some 35 m, I had found somewhat weathered but well determinable remnants of *Mytilus edulis*, *Tellina baltica* and *Cardium edule* in the postglacial downwash. As those shells were only weathering residuals, their original extension could not here be traced, but at other places in the region they have been observed to a height of somewhat over 30 m, this being still a minimum figure. Yet the shell-bearing postglacial deposits continue at those places as well as here in one continuous succession at least up to about 50 m a. s. Thereby it is certified that in the Stockholm region the postglacial land-emergence must have been at least of that amount, but at the same time no proofs have been found hitherto that it had been any higher.

No doubt there are indications in this region at about the height named, or some 50 m, of a special shore activity, possibly marking a small transgression

Fig. 11. Sthm. Haga Norra SW wall. Dark beds of micro-varved clay, covering the discordance between primary ose material, below, and secondary, marine above, dipping SW also containing *Mytilus edulis*. Photo G. De Geer, 1910.

of the sea or a stop in the land-emergence, but this problem is not yet conclusively solved.

The very height of the ose-hill in question and the considerable masses of down-wash which must have been cut down by the waves until the hill was reduced to that height, may be explained by such a stop in the land-emergence.

The great sections here mentioned, namely, show that along almost the whole of the inner part, being cut down to an almost horizontal bottom at about 24 m a. s., there is a magnificent cut all through up to a little more than 10 m thick such downwash, testifying a very considerable degradation of the original ose-hill.

The remnants of the original ose-deposits are rising to a maximum towards the highest part of the actual hill, here covered only by a thinner down-wash, as mostly accumulated along the more protected southern and western slopes. But the secondary degradation of the hill has been so very considerable that even a schematic reconstruction of the original delta-form would require quite a series of great diggings like those already performed. Within this latter the postglacial downwashed material may amount to about 11 m. Being assumed that this amount represents at least a fifth part of the corresponding down-wash along the other slopes which is probably a much too low minimum figure, especially when considering the finer sand-sediments carried farther away, it will be clearly understood to what extent the ose-hills of to-day have been lowered and remodelled by postglacial wave action. This must be emphasised as important to remember when studying subaquatic oses (Pl. 10).

In this connection it may be mentioned that in a great ose-section immediately to the south of Uppsala I had the opportunity of fixing by detailed measurements a series of big, marked faults with a throw of together not less than 9 m, evidently

PLATE 10

W

a. Haga Norra in 1936. Section prolonged still more westwards into the ose.
a: Ose centre, primary deposition. — Discordance. — b: Glacial, varved clay. — Discordance. — c: Ose-gravel, in secondary position, downslidden. — Discordance. —
d: Horizon of micro-varves, occurring more distinctly on southern side, and followed into the cirque of the back-ground. — Discordance. — g: Wave-transported, redeposited ose-gravel with *Litorina* and *Mytilus*.
From outer, left part of horizon d, SW, are derived the specimens of Figs 18—20, dated by micro-varves to the years c. –800 to –300 b.Z.   Photo E. H. D. G., 1936.

b. Haga Norra: inner, western wall of big ose section in 1937. The microdistal discordant zone d, from Fig. 10 a. above fine ose-sand of distal type, probably upper part of the lower ose-centre, a. Thereabove, middle, a pocket of current-bedded, coarser sand, probably of next annual centre e, cut off and planed by discordance f. Wave-transported dip, g, towards NE. Discordance h, and shore-gravel, i.
Photo E. H. D. G., 1937.

263

caused by a final melting away of imbedded remnants from the original base-ramparts of ice at the sides of the glacier-vault. Finally the wave-action has smoothened the ose-surface by at least the amount named but probably considerably more, or some 10 m of later sediment.

The relation between these faults and the actual surface of the hill, being degraded more than nine meters by the postglacial waves, is shown on Pl. 62, where the sandy, distal part of an ose centre is cut through by the faults. In the left, proximal part of the section I counted 500 cobbles mainly of local rocks with a diameter of more than 0.3 m.

The influence of the waves at Storkullen during the postglacial upheaval of the land must have successively remodelled the shore-material upon the surface of the down-wash already mentioned.

Upon the flat upper surface of the hill the action of the breakers has been marked by a series of pebbly beaches, the axes of which often exhibit a moderate slope from the eastern portion of the hill towards the SW, as being the less exposed side. The figures on the map, Pl. 55, indicate the height determined by levelling above sea-level of the beach summits above the isohypse of 40 m a. s.

On the steeper slopes of the hill no beaches could be deposited and no special lines of wave action as a rule are registered.

* * * * * * *

Ose centres and sediment isopachytes.

By these investigations it is definitively certified that quite naturally there is an uninterrupted connection between the coarser and finer parts of the melt-sediment from 1. every individual ose-centre or submarginal delta, 2. into its distal continuation in form of an extramarginal current-ridge of stone-free sand, in its turn 3. gradually passing over into an ordinary clay-varve, proximal and distal (Pls. 13, 14) which 4. ultimately has a very considerable microdistal extension before totally evanescing.

As the whole of such an assortment has a very great extension, it cannot be well represented graphically on the map with all its parts together.

Thus ose-hills and current-ridges can be mapped out, while the finer sediments or the varves as to their succession or the annual transgression of their proximal northern borders cannot be shown on the same map. Their relative thickness is to be given by diagrams and tables as shown in the atlas.

The thinning out within every single varve can only partially be reproduced by mapping and isopachytes or thickness curves. Thus in Fig. 14 (= 18 in Stockholmstraktens Kvartärgeologi — The Quaternary geology of the Stockholm region) they are given somewhat schematically on the maps, showing the thinning out of three annual varves as to their proximal parts.

Though the observations at that time were rather few, the general direction of the thickness curves seems to indicate a dominant southeastern flow of the bottom sediment.

More detailed mapping of that kind is given in Pl. 57, representing the proximal parts of two representative varves, showing their distribution of thickness as determined by the ice-border by inequalities of the sea-bottom and probably also more general currents in the sea (years −1028 and 1029 b. z.).

The ice-borders of these three years were extended between the great ose-hill Storkullen and the northern part of Lake Brunnsviken. Influenced by the main depression of this lake, the land-ice border here had projected by a lobe about half a kilometer south of its main extension in this region, transgradiating over a couple of the iceborder-lines from the preceding years (Pl. 56).

This projecting ice-lobe together with the opposed, rather dominating ose-hill *Storkullen* render the sediment distribution of these years especially instructive.

Thus Storkullen acted like a rather effective divisor of the bottom currents in question. On the other hand a protruding ice-lobe was followed in such an adhering way by the eastern wing of the clay transport that it seems to indicate an east-running current of the bottom-water, though this indication perhaps is not sufficiently supported by the south-directed western current-wing.

*   *   *   *   *   *   *

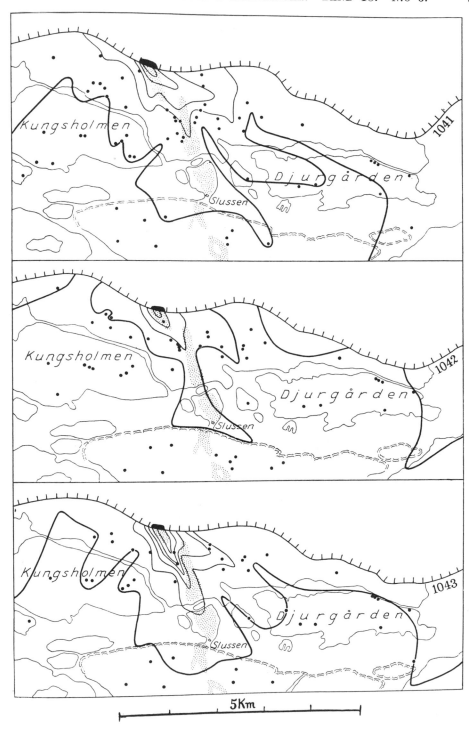

Fig. 14.   Three annual varves: distribution of proximal sediment. On each map: northernmost: landice border, dated; black: river mouth; dotted: ose deltas of previous years; curves: in dm of the varve in question; single dots: varve measurings. The maps represent the years −1041 — −1043 before Zero. Confer pp. 74, 105—106.

16

Reprinted from *Geog. Ann.*, **48A**(3), 111–125 (1966)

ESKERS NEAR THE CASEMENT GLACIER, ALASKA

BY ROBERT J. PRICE

Department of Geography, University of Glasgow

ABSTRACT. A series of eskers and esker systems in front of the Casement Glacier, Alaska, were examined in the field in 1962. The same eskers were mapped by photogrammetric methods from aerial photography taken in 1948 and 1963. Comparisons between the 1948 and 1963 maps of the eskers indicate that some of the eskers originated as the deposits of englacial or supraglacial streams and were subsequently let down on to the subglacial floor.

Introduction

During the past 50 years the Casement Glacier has retreated eastwards from the shore of Muir Inlet in the Glacier Bay National Monument in southeast Alaska (Fig. 1). The wastage of the glacier has been well documented and the development of land forms and meltwater drainage systems during the past 50 years has been discussed elsewhere (Price 1964, 1965). However, the presence of esker-like features with an approximate total crest length of 11 miles (17 km) that have been photographed with an aerial camera both in 1948 and 1963 and also examined on the ground in 1962 permits an asessment of some of the events which have led to the present form and character of the eskers.

During three months of fieldwork in 1962 all the eskers in the area across which the Casement Glacier has retreated were mapped and examined. Air photographs (1:10,000) were used for morphological mapping in the field and large scale (1:2,000) plane table maps were constructed of two esker systems. When new aerial photography, taken under the direction of Austin Post, became available in 1963 and the author had access to photogrammetric plotting machines at the University of Glasgow it was decided to produce large scale photogrammetric maps (1:20,000 and 1:8,000) of the snout of the Casement Glacier and its outwash area. The methods used in the photogrammetric work, the accuracy of the maps, the results of the measurements of ice wastage and a discussion of the general morphological changes between 1948 and 1963 have already been published, (Petrie and Price 1966).

The glacial chronology since the retreat of the Wisconsin Age glaciers in Glacier Bay is well known (Goldthwait 1963). The features to be discussed in this paper are related to the retreat of the Little Ice Age glaciers which reached their maximum extent between 200 and 300 years ago and since that time there has been rapid and well-documented retreat.

Wastage of the Casement Glacier has taken place across an area composed almost entirely of drift deposits. These deposits consist of the basal Wisconsin till, up to 300 feet (90 m) of Hypsithermal gravels, the Little Ice Age till, (up to 30 feet (9 m) thick) and the very recent fluvioglacial deposits (up to 100 feet (30 m) thick). The form of these deposits has had considerable bearing on

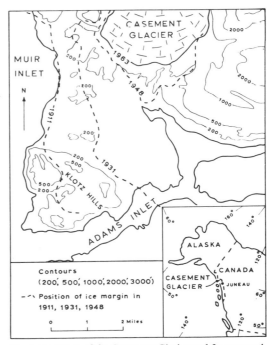

Fig. 1. Location of the Casement Glacier and former positions of its terminus.

267

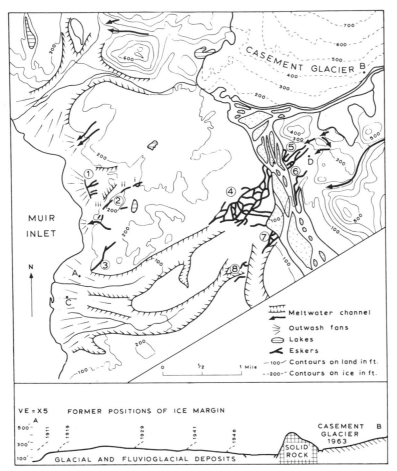

Meltwater channel
Outwash fans
Lakes
Eskers
—100— Contours on land in ft.
--200-- Contours on ice in ft.

VE = X5   FORMER POSITIONS OF ICE MARGIN

CASEMENT   B
GLACIER
1963
SOLID
ROCK

GLACIAL AND FLUVIOGLACIAL DEPOSITS

Fig. 2. Land forms near the Casement Glacier.

the development of fluvioglacial erosion and deposition during the past 50 years (Fig. 2). A thick gravel deposit topped by the Little Ice Age till forms a steep scarp behind a series of gravel fans along the east shore of Muir Inlet. This steep scarp which is generally 50 to 100 feet (15 to 30 m) high is backed by a more gentle slope rising to a north-south trending ridge generally between 200 and 300 feet (60 and 90 m) in altitude. To the east of this ridge the topography is irregular, falling to 130 feet (39 m) at the position of the ice margin in 1963.

The total of 11 miles (17 km) of esker-like features is made up of sharp-crested ridges 10 to 60 feet (3 to 18 m) high consisting mainly of coarse gravel and cobbles. Both single ridges and complex systems occur. The longest single continuous ridge is 1 mile (1.6 km) long. There were

no cross-sections in these ridges which revealed sedimentary structures and the constituents were so coarse that attempts to dig out sections resulted in the collapse of any structures which may have existed. With the exception of a few short segments in the major complex system (Fig. 2; 4) all the ridges had sharp crests and an undulating crest-line. The angles of the slopes constituting the sides of the ridges were between 25° and 30°. All ridges have a general north-east to south-west trend which is parallel to the direction of former movement of the Casement Glacier across the area. A detailed analysis of each of the esker systems will now be made. They are divided into three groups.

1. Single eskers.
2. Eskers leading to an outwash fan.
3. The complex esker systems.

112

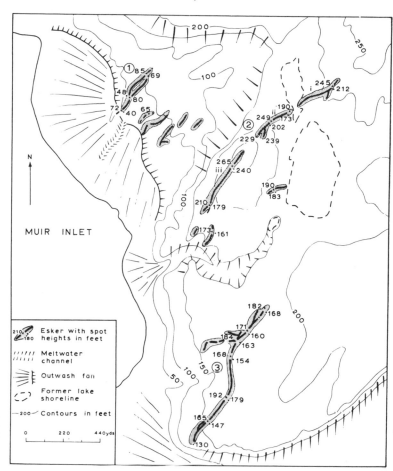

Fig. 3. Esker systems 1, 2 and 3. Spot heights obtained photogrammetrically from 1963 photography.

1. Single eskers

The total length of single eskers amounts to about 2 miles (3.2 km). They are all between 10 and 30 feet (3 and 9 m) high and consist of well-rounded gravel and cobbles with occasional lenses of sand. The only difference between each of these ridges is their position. The three segments (i, ii, iii) of esker 2 (Fig. 3) illustrates the variations in positions shown by any of the single ridges (Fig. 4). At its eastern end, esker 2 descends a minor west facing slope, crosses a dried-out lake floor and ascends an east facing slope. It dies out near the ridge crest and then continues on the west-facing slope. Most of the other single eskers occur on west facing slopes but there are a few short segments on flat areas. In explaining the origin of these single eskers two major characteristics must be considered. Firstly, the sections

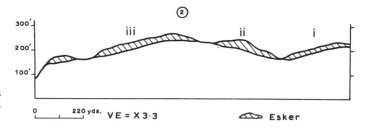

Fig. 4. Profile of esker 2. Spot heights obtained photogrammetrically from 1963 photography.

Fig. 5. An esker leading to an outwash fan.

of the eskers on the east-facing slopes (i.e. on slopes opposite to the general direction of ice movement and meltwater movement) must be accounted for and secondly, the irregular crest line of the eskers must be explained.

2. Eskers leading to an outwash fan

At the proximal side of the outwash fan (Fig. 3; 1) two eskers are connected at right angles to the ice contact face which bounds the fan on the proximal side. The eskers are sharp crested and sinuous and are 10 to 30 feet (3 to 9 m) high (Fig. 5). Several angular blocks occur on the crests and sides of the eskers. The origin of these eskers must be related to the formation of the outwash fan to which some of them are connected and the development of the meltwater channel on the floor of which they are located. The eskers must have been formed at approximately the same time as the fan, but must post-date the formation of the meltwater channel.

3. The complex esker system

There are five complex esker systems in front of the Casement Glacier (Fig. 2; 4.5.6.7.8). It can be

shown by examination of the 1948 aerial photographs that systems 4 and 5 are actually one system, part of which was visible on the glacier surface in 1948 and part of which must have been englacial at that time. Between 1948 and 1963 a major meltwater stream flowing from the Casement Glacier south to Adams Inlet cut through the esker system so producing systems 4 and 5 (Price 1965).

(I) Esker System 4

A complex system of anastomosing ridges (Fig. 2; 4 and Fig. 6), formed of gravel and cobbles occurs at the eastern end of one of the major deserted meltwater channels of the area. The system is spread over an area of 1 mile (1.6 km) from northeast to south-west and ½ mile (0.8 km) from north to south. Most of the ridges (Fig. 7) are between 20 and 40 feet (6 and 12 m) high with a few ridges attaining 40 to 50 feet (12 to 15 m). It is possible to distinguish three main types of ridges in the complex system: linear ridges, circular ridges and flat-topped ridges.

Linear ridges

There are several ridges in the system which can be traced as continuous features for almost the

114

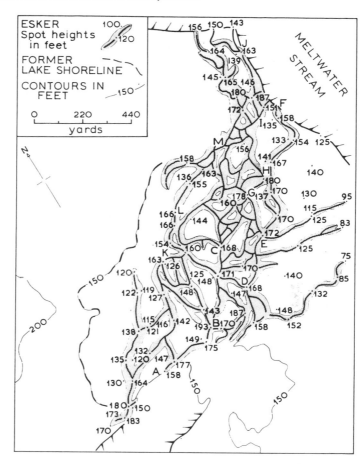

Fig. 6. Esker system 4. Spot heights obtained photogrammetrically from 1963 photography.

whole length of the system. (Fig. 6, A-B-C; D-E-F; G-H-I-J; K-L-M-). These dominant ridges are generally 20 to 50 feet (6 to 15 m) high, are sinuous and have uneven crest lines. The maximum altitudes attained by the crests of these main ridges occur at the south-western end of the system even though the general slope of the ground beneath the system is north-eastward. Several minor linear ridges join the major linear ridges, the angle of junction being usually between 30° and 70°. In general these minor linear ridges are shorter, smaller in cross-sectional area and lower in their middle sections than the major linear ridges.

Circular ridges

In the central and northern parts of this system is a series of circular ridges. The ridge crests are from 30 to 50 feet (9 to 15 m) above the floors of

the depressions they surround. Although the depressions tend to have a rounded appearance the junction of the ridges which form the depressions tend to be angular.

Flat-topped ridges

Near the south central part of the ridge system there are some well-developed, flat-topped ridges which have a common altitude of approximately 170 feet (51 m). Flat-topped ridges only form a very small percentage of the ridge system.

Associated with this ridge system is a great deal of evidence of the former existence of a lake which surrounded and partly submerged the ridges. A well-marked shore-line was observed during the construction of the plane-table maps at approximately 156 feet (47 m) on the north-west side of the system, and the presence of small beaches on the sides of the gravel ridges in the south-western

Fig. 7. Esker system 4.

half of the system indicates that the ridges were formed prior to the drainage of the lake. It appears that as the general lake level was lowered the various ridges emerged above its surface. Aerial photographs taken in 1948 (Fig. 8) show clearly the existence of this lake and at that time all ridges below 173 feet (53 m) were below the surface of the lake. The lake was dammed by the margin of the Casement Glacier.

In 1962, 12 feet (4 m) of buried ice was seen exposed in one of the gravel ridges at the extreme north-eastern end of the ridge system. After close examination of the whole system for evidence of other buried ice masses none was found.

In suggesting a mode of origin for the ridge system the following factors must be considered:

a) The presence of linear ridges with a dominant north-east to south-west trend but connected by other minor ridges and 'circular' ridges.

b) The presence of sharp-crested and flat-topped ridges in the same system.

c) The irregular crest line of most of the ridges.

d) The general increase in altitude of the ridges in the system towards the south-west, remem-bering that both the ice and meltwater moved towards the south-west even though the general slope of the sub-ice surface was towards the north-east.

e) The existence of a lake after the ridge system was developed.

f) The presence of ice beneath one ridge.

(II) Esker System 5

The eskers in this system (Fig. 2) are very similar to those of system 4. The ridges are 10 to 40 feet (3 to 12 m) high but all of them have been truncated by the main southward flowing meltwater stream (Fig. 9). In three localities at the truncated ends of this system, ice could be seen forming the cores of the esker ridges.

(III) Esker System 6

This system consists of one major sinuous ridge 30 to 90 feet (9 to 27 m) high (the largest esker in the area) and a few tributaries (Fig. 2). Most of the south-western part of this esker system has been undercut by the present meltwater stream and an ice cliff some 20 to 40 feet (6 to 12 m) high

116

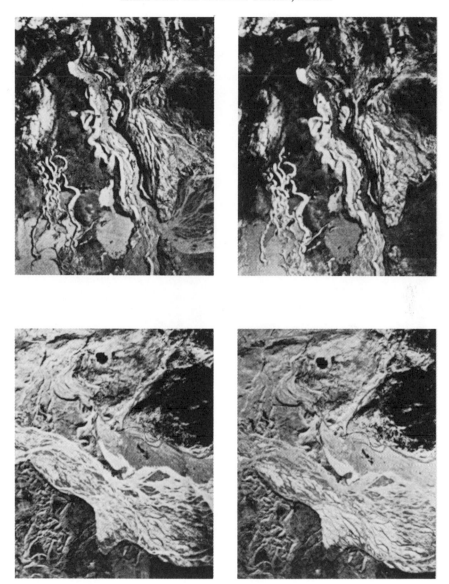

Fig. 8. Stereo-pairs of vertical aerial photographs of the same area taken in 1948 and 1963. X marks common point on each pair of photographs.

and covered by fluvioglacial deposits can be seen (Fig. 10). In suggesting the mode of origin for this system the considerable thickness of ice beneath the esker system and the marked sinuosity in its crest line must be taken into consideration. It is not possible to determine how much of these eskers has been removed by the establishment of the present meltwater stream which is flowing transverse to the eskers.

(IV) Esker System 7

The eskers in this system are 10 to 40 feet (3 to 12 m) high (Fig. 2). The longest continuous esker in this system is just over ¼ mile (0.4 km) in

117

273

Fig. 9. Esker system 5 (centre) and 6 (middle left) viewed from the North-east.

Fig. 10. Esker 6 seen from the west across the main meltwater stream. The dark areas represent the ice core which is covered by coarse gravel deposits.

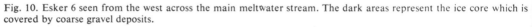

118

length. The altitude of its crest ranges from 130 to 150 feet (39 to 45 m). Where the present meltwater stream is undercutting the eastern end of the esker system, buried ice was observed in the core of one esker in 1962. The location of this esker system is most remarkable in that it is situated on the floor of a major meltwater channel. These eskers are generally 30 feet (9 m) below the eskers in system 8.

In providing an explanation of the origin of this esker system the presence of the ice core in one of the eskers as well as the location of the system on the floor of a meltwater channel must be considered.

(V) *Esker System 8*

This system could be subdivided into two parts but both parts are very similar and are separated by a meltwater channel (Fig. 2). These eskers are 10 to 30 feet (3 to 9 m) high, consist of well-rounded gravels and cobbles, and have sinuous and irregular crest-lines. Scattered over the surface of these gravel ridges there are numerous angular blocks, some of them 6 feet (1.8 m) across. Between the major ridges are depressions with irregular hummocky floors with scattered patches of fine sediment. Even within the main gravel ridges there are a few depressions 3 to 10 feet (0.9 to 3 m) deep which suggest the former presence of buried ice blocks.

In explaining the origin of this system the presence of large angular blocks on the surface of eskers and the occurrence of kettle holes in and between the esker ridges must be taken into account.

The existence of aerial photography of the Casement Glacier snout and outwash area, taken in 1948 by the U.S. Geological Survey, has permitted a reconstruction of the eskers at that time. Photogrammetric maps have been made of the eskers from the 1963 and 1948 aerial photographs and the highting accuracy permitted by this photography and the use of a Wild B8 plotting machine is of the order of ±3 feet (0.9 m). (Petrie and Price 1966). Although the ground control available for photogrammetric plotting was very limited and only consisted of natural points with no height information it was possible to obtain planimetric control from these points. The height datum was achieved by levelling the model, in the plotting machine, on the water surfaces in Muir and Adams Inlets. Owing to the lack of

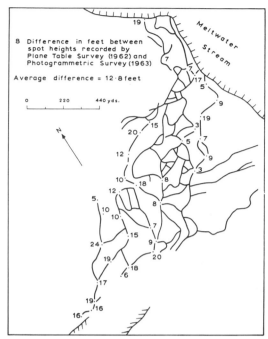

Fig. 11. The difference in values of spot heights obtained by a Kern self-reducing alidade in 1962 and by photogrammetric plotting of 1963 photography.

detailed information about the local tides it is possible that the absolute datum for the two photogrammetric maps may be in error to the order of ±5 feet (1.5 m). However, since both photogrammetric maps are related to the same datum the relative changes in altitude of land forms between 1948 and 1963 are not affected by the possible datum error, only by the error in spot heighting (±3 feet (0.9 m) previously referred to.

The agreement between the contours of the 1948 and 1963 plots, which were executed quite independently, is very good in the areas other than the glacier snout and those recently deglacierized, where comparisons can be made. The differences are very small, less than 5 feet (1.5 m) and almost all are those to be expected from two sets of independently executed contours.

A further check on the accuracy of the 1964 photogrammetric map is possible. In 1962 a plane table map, on the scale 1:2,000 and based on an unreliable datum, was made of esker system 4. Figure 11 indicates the differences between spot heights recorded on the plane table map and the attempt made to height the same locations in the

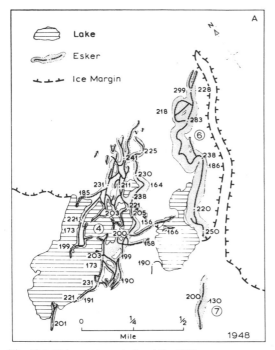

Fig. 12 a. Esker systems 4, 6 and 7 – 1948

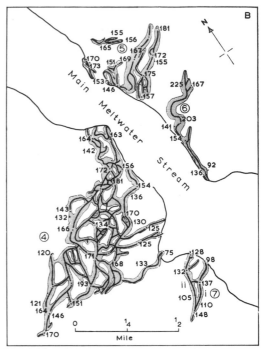

Fig. 12 b. Esker systems 4, 5, 6 and 7 – 1963
Both maps were plotted on a Wild B8 Aviograph Plotter.

plotting machines. The average difference between the heights is 12.8 feet (3.84 m). The
discrepancy between the heights is probably due
to three factors.

1. The difficulty of locating accurately the plane
   table spot heights in the plotting machine.
2. The instrumental error of the plotting machine
   (±3 feet, 0.9 m).
3. The different datums used for the plane table
   and photogrammetric maps.

However, it is significant that the distribution
of the discrepancies between the plane table
heights and the photogrammetric heights does not
indicate a systematic increase in any direction.
This suggests that there was no tilt present on the
1963 photogrammetric models, and since the 1948
photogrammetric map was based on the same
control system it was not produced from tilted
models either.

It is concluded therefore, that it is possible to
use the 1948 and 1963 photogrammetric maps for
the measurement of changes in altitude of landforms so long as it is remembered that there is a
possible error of ±3 feet (±0.9 m) in all altitudes
with reference to a common datum.

Morphological and altitude changes between 1948 and 1963

There were no detectable changes in the position,
altitude and extent of the eskers in systems 1, 2, 3
(Fig. 3). It must be admitted that between 1948
and 1963 there was considerable growth of alders
over each of these eskers and any attempt to
detect height changes in these eskers over this
period would have severe limitations.

Esker system 4 is clearly visible on the 1948
photography (Fig. 9) and when this system was
plotted in the photogrammetric machine it was
seen to have a very different character compared
with its 1963 condition. In 1948, (Figs. 8, 12a,
and 12b) the margin of the Casement Glacier was
in direct contact with this esker system and several
of the eskers actually continued up on to the ice
surface for a distance of one half mile (0.8 km).
The south-western part of the esker system was
covered by a lake presumably dammed by the
Casement Glacier. The altitude of the surface of
this lake was at 173 feet (52 m) and all the eskers
below this altitude were beneath the lake.

An attempt has been made to plot profiles
along the crest and base of segments of the esker

120

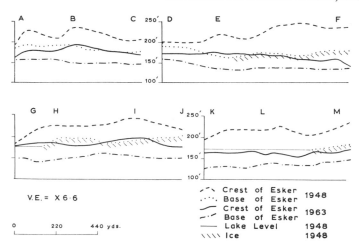

Fig. 13. Profiles of segments of esker system 4 constructed from spot heights obtained photogrammetrically. A—M are locations on Figure 6.

system which appear both on the 1948 and 1963 photogrammetric maps (Fig. 13). It can be seen that each of these esker segments was lowered between 1948 and 1963. The extent of this lowering increased towards the north-eastern end of the system. In general the crests of the eskers had been reduced in altitude by between 20 and 80 feet (6 and 24 m) whereas the bases of the same eskers had been reduced by between 20 and 60 feet (6 and 18 m). These altitude reductions are consistent throughout the system and are far in excess of the probable heighting error in the photogrammetric maps (±3 feet, ±0.9 m). It is suggested that this change in altitude is due to the wastage of buried ice from beneath the eskers during the 15 year period 1948—1963. During this period the margin of the Casement Glacier had retreated over one mile to the north-east of the esker system,

the lake at the south-western end of the system had been drained and a major southward flowing meltwater stream had started to destroy the north-eastern end of the system.

Esker system 6 is one of the most noticeable features of the 1948 photography (Figs. 8 and 12). It is just over one mile (1.6 km) in length and is on the glacier surface throughout its length. At its south-western end the base of the esker is at 162 feet (90 m) and it rises steadily to an altitude of 299 feet (90 m) at its north-eastern end. The ridges forming the system are generally 50 to 80 feet (15 to 24 m) high, have a sinuous pattern and an irregular crest line. Two thirds of this esker system was destroyed between 1948 and 1963 by the southward flowing meltwater stream but the remaining third has survived a considerable lowering in altitude as a result of ice wastage

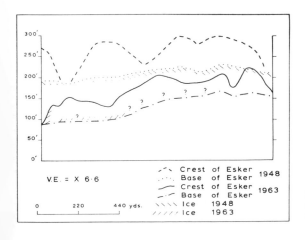

Fig. 14. Profiles of esker 6 based on spot heights obtained from the 1948 and 1963 photography. The extent of the ice under the esker in 1963 is unknown.

from beneath the esker (Fig. 14). The crest line has been lowered by between 40 and 100 feet (12 and 30 m) while the base of the esker has been lowered by between 40 and 110 feet (12 and 33 m).

In 1948 only about one third of esker system 7 existed (Fig. 12a). The north-eastern third of ridge i and ridge ii do not occur on the 1948 photograph. Comparisons between spot heights on that part of the esker system which can be seen on both the 1948 and 1963 photography indicates that there has been a decrease in altitude during this period of between 20 and 40 feet (6 and 12 m). In esker system 8 there has also been a general decrease in height of between 20 and 30 feet (6 and 8 m), since 1948. Only minor changes have taken place in the morphology of the esker system.

It can be concluded from the above evidence that there has been considerable retreat of the 'active' ice margin during the period 1948—1963. Many of the eskers which actually occurred on the glacier surface, in 1948, have subsequently been let down on to the sub-glacial surface, preserving their distinctive forms and in some instances still contain an ice core.

It is interesting to note that measurements of the amount of ice wastage on the snout of the Casement Glacier (Petrie and Price 1966) indicate that the ice on either side of the eskers which extended up on to the glacier surface in 1948 wasted away at an average rate of about 30 feet (9 m) per year between 1948 and 1963. In contrast the ice beneath the eskers melted away much more slowly as some eskers still contain ice cores while the ice on either side has disappeared. It is not possible to state an accurate rate of wastage for the ice beneath the eskers, as it is not known exactly when any given esker became ice free. However, it can be stated that esker system 6 has been lowered at the approximate rate of 5 feet (1.5 m) per year but it is not known how much ice still remains in the core of the esker. In the same way, the general lowering of the base of the eskers in system 4 has been of the order of 20 to 60 feet (6 to 18 m) and although the actual date at which these eskers became ice free is not known, the rate of wastage of the ice beneath the eskers was of the order of 1 to 4 feet (0.3 to 1.2 m) per year.

The origins of the eskers

The deposition of material on the beds of subglacial streams has for a long time been the main explanation for the development of eskers. The fact that meltwater can flow up-hill under hydrostatic pressure in these subglacial tunnels has fitted the evidence that some eskers ascend slopes which are opposite to the general direction of meltwater flow. The englacial and supraglacial hypotheses have been much criticised by many authors who have pointed out that narrow strips of fluvioglacial deposits laid down on the floors of englacial streams would not produce a ridge 10 to 50 feet (3 to 15 m) high after the ice around and beneath it had melted away.

The eskers in front of the Casement Glacier exhibit a wide variety of features which have to be accounted for in an explanation of their formation. The single eskers have uneven crest lines and one (Fig. 4) has an up-hill section in terms of the general movement of meltwater. The up-hill section could have been produced either by deposition under hydrostatic pressure, or by the letting down of the deposit from a tunnel within the ice or from a channel in the ice surface. There is no way of deciding which of these processes actually occurred.

The eskers leading to the fan (Fig. 3) were formed subglacially. The location of the ridges on the floor of a meltwater channel and the ice contact form of the proximal side of the fan indicates a subglacial origin for these particular ridges as it is unlikely that an englacial or supraglacial deposit would have such a well-adjusted relationship with the fan.

It is the complex esker systems (Fig. 2: 4, 5, 6, 7, 8) that provide the most interesting evidence relating to the origins of eskers. The irregular crest lines, the common occurrence of higher altitudes at the distal end of the esker systems, and the presence of ice-cored eskers all present problems in interpretation. The photographic evidence indicates that in 1948 parts of esker system 4 and all of esker system 6 occurred on the glacier's surface. At the same time, the remainder of system 4 and all of systems 7 and 8 were underlain by buried ice while system 5 was englacial.

The greater part of the major esker system (4) has been laid down on a surface which slopes to the north-east. Both ice movement and meltwater movement was towards the south-west. Since the eskers in this system extended up on to the ice surface in 1948 it is unlikely that any part of the system was deposited in a lake. The development of the lake clearly seen on the 1948 photographs took place after the eskers were formed. It is suggested that this complex esker system developed either in

122

englacial or supraglacial channels and was subsequently let down on to the north-easterly sloping surface. This process would account for the higher altitudes at the distal end of the system, and the irregular crest lines of the eskers and the presence of buried ice in the cores, of some of the eskers. It appears that a similar process resulted in the development of systems 5, 6 and 8.

Only a part of esker system 7 was probably formed in the same way as systems 4, 5, 6 and 8. Because only a small part of system 7 is visible on the 1948 photography (Fig. 12) it is rather difficult to envisage how the rest of the system was formed during the period 1948 to 1963. This system is located on the floor of a meltwater channel and it is possible that the present gravel ridges represent the remnants of a surface of gravel which has been eroded by meltwater. This possibility requires some very selective erosion by the meltwater streams. Alternatively this system could have been produced englacially or supraglacially and was then let down on the subglacial floor. In 1948 the buried ice on either side of the esker system had not melted out and therefore most of the esker system was not upstanding. However, by 1963 the buried ice on either side of the system had melted and the system stood above the level of its surroundings. This hypothesis is supported by the fact that kettle topography occurs between the ridges in the system and covers an extensive area on the northern side of the system.

It has been agreed by numerous writers that gravel ridges established on or within an ice mass will be destroyed when the ice melts out. In his paper on the supraglacial debris of the Wolf Creek Glacier, R.P. Sharp (1949) indicates that the differential melting of the ice beneath the debris produces a constant redistribution of the material with the consequent destruction of specific forms. However, in a later paper on eskers in Minnesota, Sharp, (1953) points out that the wastage of ice beneath the spruce trees that have grown along the stagnant margin of the Malaspina Glacier must have been sufficiently uniform so that no marked tilting of the trees or serious damage to their roots has resulted. He further states (p. 879) that: "Preservation of supraglacial eskers underlain by no more than 10 to 20 feet (3 to 6 m) of ice appears to be within the realm of possiblity under favourable conditions."

A recent paper by Jewtuchowicz (1965) includes a description of an esker in Bungebreen, Vest-spitsbergen, which is gradually emerging from the glacier surface as the ice on either side melts. However, the author infers that the ice is not thick and that this particular esker probably extends down to the subglacial ffloor. In an adjacent ice free area the author describes eskers which in cross-section show inward dipping strata. Jewtuchowicz then states (p. 720): "The inward dip of the layers may have resulted from the melting of ice beneath the esker and subsequent collapse of the material, a fact which may indicate the englacial origin of the esker. Observations on the ablation of glaciers show that the sediments resting on the ice surface, even though they may be more than 4 m thick, do not remain stationary during melting but begin to flow as the water supply increases. It is therefore doubtful whether an englacial formation resting on an unstable base of melting ice can possibly preserve its original form and structure after total melting of the ice".

The photogrammetric evidence of the change in altitude of some of the eskers near the Casement Glacier clearly indicates that eskers can be let down from the surface of a glacier on to the subglacial surface without being destroyed. Only small amounts of ice remained in 1962 in the extreme north-eastern end of esker system 4 (Fig. 2) and it can be clearly demonstrated that parts of this esker system were underlain by ice which ranged in thickness between 20 and 60 feet (6 and 18 m) (Fig. 13).

Several examples of ice-cored gravel ridges have been cited in this paper. If the deposits of englacial and supraglacial streams survive the early stages of ice wastage they will prevent the ice beneath them from melting at the same rate as the clean ice on either side (Fig. 15). This will result in an ice-cored gravel ridge. Such a feature will have little of its original stratigraphic character, as slumping will take place down the sides of the ice core. If this process continues long enough either a series of very small ridges or mounds would be the end product, but if the deposits arrive at the subglacial surface at an earlier stage distinct ridge forms will be preserved. Unless the stream deposits are very wide it is also unlikely that flat-topped features will be preserved. The dominance of sharp-crested ridges in the esker systems near the Casement Glacier support this hypothesis. The irregular crest lines of the eskers in these systems could also be the result of the varying thickness of the ice forming the ice cores.

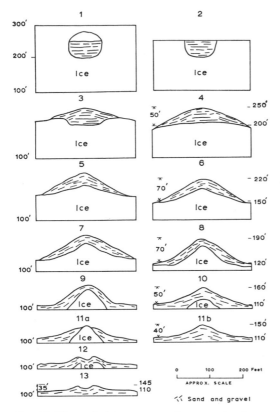

Fig. 15. The probable sequence of events whereby an englacial (1) or supraglacial (2) stream deposit, 50 feet (15 m) thick, can be lowered on to the subglacial floor. The fluvioglacial deposits protect the ice beneath them from wasting as rapidly as the clean ice on either side (3—7). If the ice core remains long enough for sufficient slumping to take place down each side of it (8, 9, 11a) two esker ridges will be produced (12, 13). Alternatively, if the ice core is wasting away rapidly only one ridge about 40 feet (12 m) high will be produced (10, 11b).

In all of the complex esker systems discussed in this paper there are many examples of tributary, distributary and meander patterns. It appears that the meltwater stream system which was responsible for these deposits must have developed on stagnant ice and must have had access to a considerable amount of morainic debris either on or in the ice. It is significant that all the complex esker systems near the Casement Glacier are on the lee side of a series of solid rock ridges which are between 200 and 300 feet (60 and 90 m) high. (Fig. 2). These rock ridges were just beneath the ice surface in 1948 but were completely ice free in 1963. The ice on the lee side of these ridges must have been stagnant for a considerable period of time and permitted the development of a complex drainage system which was not subsequently destroyed by ice movement. The meltwater streams in the system must have been of considerable size as they permitted the accumulation of up to 50 feet (15 m) of sediment on their channel floors. The source of this material is easily found as the surface of the Casement Glacier for some distance up glacier from the esker system is completely covered with ablation moraine.

There are many references in the literature to large meltwater streams occurring on or beneath the surface of temperate glaciers. The author has observed such streams on Breidamerkurjokull, Iceland, and again in a zone of debris covered ice, the streams are seen to wash this material into the englacial environment. On this glacier eskers are also seen to stand up above the ice surface and other eskers along the margin of this glacier are known to have ice cores. Further work is being carried out on those eskers.[1]

On the basis of morphology, sedimentary character and location, it is not possible to state whether an esker originated as a subglacial englacial or supraglacial meltwater stream deposit. Certain of the eskers near the Casement Glacier, however, were deposited by englacial or supraglacial meltwater streams and were subsequently let down on to the subglacial surface.

Acknowledgements

The fieldwork on which this paper is based was carried out as part of a study of the proglacial deposits of the Casement Glacier which was financed by the National Science Foundation, Grant No. G-24149. This grant was administered by the Ohio State University Research Foundation as project No. 1496.

This study would not have been possible without the initial planning and strong encouragement of Dr. R. P. Goldthwait. During the three months of fieldwork Mr. R. Welch and Mr. S. Calmer gave the author much support as field assistants and the staff of the Glacier Bay National Monument provided transportation, radio communication and much encouragement.

[1] Englacially formed eskers have been described at several glaciers in Iceland, particularly by E. Todtman (1960 and other papers).

124

The author is pleased to acknowledge the co-operation of his colleagues Mr. G. Petrie and Mr. D. Nicol, of the University of Glasgow, who carried out the photogrammetric plotting of the esker systems.

A grant was made by the Carnegie Trust for the Universities of Scotland to cover part of the costs of the illustrations.

Bibliography

Flint, R. F., 1957: *Glacial and Pleistocene Geology*. John Wiley and Sons.

Goldthwait, R. P., 1963: Dating the Little Ice Age in Glacier Bay, Alaska. *Int. Geol. Congr.* 21*st Session, Norden* 1960, Part 27 (1963).

Jewtuchowicz, S., 1965: Description of Eskers and Kames in Gåshamnöyra and on Bungebreen, south of Horn-sund, Vestspitsbergen. *J. Glaciol*, Vol. 5, No. 41, pp. 719–725.

Petrie, G. and Price, R. J., 1966: Photogrammetric Measurements of the Ice Wastage and Morphological Changes near the Casement Glacier, Alaska. *Canadian Journal of Earth Sciences* (in press).

Price, R. J., 1964: Land Forms Produced by the Wastage of the Casement Glacier, Southeast Alaska. *Institute of Polar Stud.* (Ohio State Univ.) Rep. No. 9.

Price, R. J., 1965: The Changing Proglacial Environment of the Casement Glacier, Glacier Bay, Alaska. *Trans. Inst. Brit. Geogr.* No. 36, pp. 107–116.

Sharp, R. P., 1949: Studies of the Supraglacial Debris on Valley Glaciers. *Am. Journ. Sci.* Vol. 247, pp. 289–315.

Sharp, R. P., 1953: Glacial Features of Cook County, Minnesota. *Am. Journ. Sci.* Vol. 251, pp. 855–883.

Todtmann, Emmy., 1960: Gletscherforschungen auf Island. *Universität Hamburg, Abhandlungen aus dem Geb. der Auslandskunde, 65*, C, 19.

17

Reprinted from *Fennia*, **52**, 3–7, 13–17, 20–25, 30–41 (1931)

On the Transportation of Materials in the Esker of Laitila

A. HELLAAKOSKI

* * * * * * *

I. Area investigated.

THE BED ROCK. — The investigation outlined on the following pages sets out from the well-known fact that the r a p a k i v i rocks in Southwestern Finland are with respect to their petrography sharply distinguished from the rock formations of the surrounding areas. The rapakivi massif, represented on fig. 1 by two separate spots, was styled in earlier literature the rapakivi area of Uusikaupunki. Recently, in accordance with the example set by ESKOLA (4 & 5), the smaller of these spots has been called the rapakivi area of Vehmaa, and the larger one that of Laitila. The Vehmaa area has already been treated in a monograph (12). But the Laitila area with its more multiform petrographic variations has heretofore been less known in its details, although sufficiently for the purposes of the present study, the more so as P. ESKOLA, who suggested this subject to the writer, proffered his help both in setting up the working plan and in checking the results.

This study is confined to the rapakivi area of Laitila, especially to its southwestern part, which is traversed by an ose chain, the Laitila esker.

The rapakivi area of Laitila is bounded by A r c h a e a n r o c k s, represented by granites, gneisses, amphibolites etc. Only the northeastern and northern borders, where a stratum of J o t n i a n s a n d s t o n e sends out a long tongue indicated on the map between the rapakivi and the Archaean area (see also 7 & 13), form an exception. Within the rapakivi area, as well as outside it, occurs the latest intrusion of this region, the o l i v i n e d i a- b a s e, which in the form of narrow dikes and sills pierces the Laitila rapakivi, the Jotnian sandstone and the Archaean rocks nearest to them in the vicinity of the town of Rauma (7, 13).

LANDSCAPE. — The most salient characteristic feature of the rapakivi landscape are the disintegrated boulders, which also occur frequently at a certain distance southeast of the rapakivi region. In its scenery, however, the Laitila area does not differ very sharply from the hilly landscape of other parts of South Finland. To be accurate, the relief may be said to be more unquiet

Editor's Note: A row of asterisks indicates material deleted owing to lack of space. Tables 2, 3, 4, and 7 have also been omitted for reasons of brevity. All numbered excavations are shown in Plate 1, page 310.

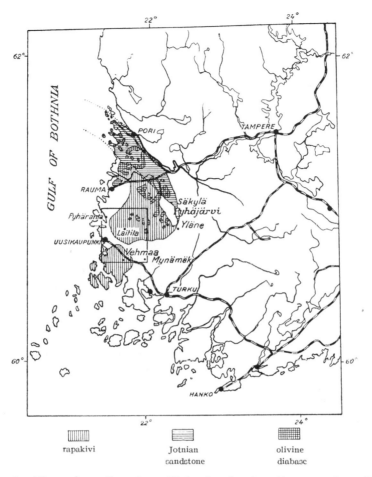

Fig. 1. Map of southwestern Finland, showing the area investigated. After *Atlas of Finland 1925*.

than usually is the case. This is due partly to the scantiness of the morainic mantle, partly to the vertical joint-walls characteristic of rapakivi. The variations in height, expressed in metres, are insignificant. Even the highest »mountains» do not rise as much as 100 m, and in the vicinity of the esker chain the highest summits reach the height of 50 or 60 m above sea-level. The tops of the rocky elevations are very often rather flat, because rapakivi possesses, in addition to the vertical jointings, a nearly horizontal jointing parallel to the surface; in this direction also rock horizons weathering more easily alternate with such weathering less easily (18).

After the disappearance of the glacial inland ice the highest points of the area investigated were covered by water, varying between 80 and 100 m in depth, inasmuch as the 140 to 150m isobases of the highest shore — the Yoldia stage — in all probability cross this area (23). As is generally the case with subaquatic landscapes, the summits are clean-washed. In valley depressions the soil mantle (morainic drift) is so thin that the bed rock may peep from under an arable field. The bed rock is very often exposed in the close vicinity of the esker chain. A peculiar feature of the area is the absence of clay, while postglacial mud and peat beds cover wide areas.

Fig. 2. View along the row of esker hills in the lowland of Laitila. Photo P. Eskola.

THE ESKER OF LAITILA. — In this landscape, with its thin soil mantle, the Laitila esker (ose, harju) is a prominent accumulation, although it slopes gently and although its height, measured from base to top, very seldom exceeds 20 m. In its direction the esker oscillates as a rule, between N30°W and N40°W. In the southeast the former direction dominates, in the northwest the latter. In the vicinity of the Gulf of Bothnia the esker bends towards N45°W. According to the measurements of the author, the figures expressing the direction of the glacial striae range between N27°W and N36°W. The mean directions of the striae in the immediate environment of the rapakivi area vary, according to earlier determinations, between N25°W and N45°W (7). The esker thus runs throughout its course indisputably in the direction of the striae and at right angles to the glacial ice front (20), and consequently is a radial ose. Plateaus are absent, as is generally the case with subaquatic oses (22).

In the parish of Pyhäranta, the ose chain rises from under the water-level of the Gulf of Bothnia, taking the form of small gravel accumulations with plane tops, but towards the southeast it assumes the form of a ridge as early as on its arrival in the district shown on the map plate 1, *i.e.* before it enters the rapakivi area. In the parish of Laitila it continues its southeasterly direction, being very distinct, although fragmentary here and there, then further on it crosses the parish of Mynämäki. In its course the esker climbs the highest elevations of the region and descends to the deepest depressions, as if it took no notice of the variations of the relief. The shape of the esker, however, seems to be influenced by the variations in the altitude of the ground which it traverses.

Among the forms presented by the esker of Laitila two main types may be distinguished: the ridge, and the row of hills. The ridge form predominates in the highest regions, *e.g.* on both borders of the rapakivi area, in which there is the rocky landscape of Neitsenkallio, 49 m high, on the NW side, and on the SE side the 56 m high summit of the elevation of Matoluolanmäki. (At the tops of both summits the esker, if it ever existed here, has been washed away by the sea, to such an extent that only scattered boulder heaps have been left.) The form of a row of hills predominates between these two elevated areas, where the surface of the rapakivi massif slopes gently down to a wide lowland. The church village of Laitila being situated at the centre of this lowland, the latter will be called in the sequel the lowland of Laitila. In its hilly southeastern half the esker is disconnected and capricious in its course, being here and there somewhat displaced from its general trend. Across the more level northwestern half of the lowland the esker runs for some 7 km as an almost straight row of hills between Lake Valkojärvi (11 m above sea-level) and the large peat bog of Laitila. At the bottom of the peat bog and the lake the depression probably descends below the present sea-level. Consequently, at the time when this area was freed from its ice cover, the waters of a sea about 150 m deep rolled above the row of hills.

Southeast of the rapakivi area the esker also tends to break up into hills on the lowlands (*e.g.* Nästi, Peulunsuo, Juva, the church village of Mynämäki), while on the higher lands between the localities mentioned the continuous ridge predominates, although hills are frequent.

The deposit of which the esker consists will be called in the seque lose material, or ose gravel.

AIM OF THE INVESTIGATION. — As the Laitila esker for some 27 km runs
through a continuous rapakivi area, it is particularly well suited to yield
an answer to the following questions:

1) To what extent does the petrographical composition of the ose material
change quantitatively in the distal direction away from both contacts?

2) What is the relation of the ose material, with respect to its petrographical composition, to the morainic drift of the same tract?

With a view to finding a reply to these questions the author in the summer
of 1929 carried out stone countings both at the Laitila esker and in the
morainic drift in its vicinity. This work was financially assisted by The Seth
Sohlberg Fund.

*   *   *   *   *   *   *

COLLECTION OF STONES — From the sides of an excavation the required quantity of stones larger than 6 cm have been collected into a box and set aside for assorting and counting, the boulders affected by disintegration being noted and recorded at the same time. The size class from 6 to 2 cm has been taken with the soil itself and placed in boxes with appropriate precautions. From this sample the stones have been gathered one by one, measured and put aside. The grain classes smaller than 2 cm have not been removed from their sample until after their arrival in the laboratory.

THE TOTAL NUMBER OF THE STONES COUNTED. — The relative proportions of the different rocks of each size class are presented in percentage by the tables 2 and 4. Similarly, the total numbers to which the percentages refer are shown in the tables.

The rule observed in counting stones larger than 6 cm and from 6 to 2 cm in size has been to collect at least 200 stones of either class. For size 6 to 2 cm it has always been possible to collect this minimum number, although sometimes several boxes of gravel have been required both in the esker and the morainic drift, whereas the stones larger than 6 cm, which sometimes are entirely absent in the esker, may be so sparse that in some cases it has been necessary to base the percentages on a less number than 200, never, however, on less than 100 stones.

In all the cases in which stones smaller than 2 cm have been determined, the minimum number has been 100.

POSSIBLE ERROR. — The possibility of error is introduced into the computations by the following three circumstances:

1) T h e   v a r i a t i o n s   i n   t h e   p e t r o g r a p h i c a l   c o m p o-
s i t i o n. — The percentages reached in each place of excavation by the method described above, are entitled to represent the petrographical composition
of the deposits with accuracy only on condition that the different rocks are
fairly evenly mixed in the deposits. This may be the case with respect to the
esker, as will be seen from the percentages derived from the large gravel pit
at Laitila (excavation 9), in which stones from four different beds have been
counted. The author has sometimes counted the stones in two divisions (100
and 100, for instance), and has thereby found out that the esker yields such
regular percentages for the rocks that 100 stones give much the same result
as do 200. But in dealing with morainic drift one soon discovers that its
petrographical composition may indisputably undergo a change in the space
of a couple of metres. The most striking instance of this fact is excavation
II, one end of which yields a percentage of rapakivi nearly 20 % higher than
the other. The range of the possibility of error implied in cases like this is difficult to estimate without special investigations, but an attempt has been made
to reduce it by collecting the stones from excavated walls at a length of 2 or
3 metres, or, in the case of morainic drift, from two pits at a distance of 2 or
4 metres from each another. As to the excavations in the esker, this interval,
as a rule, has been more than the minimum mentioned, and has at the same
time been extended to several cross-beds.

2) P o s s i b l e   e r r o r   d u e   t o   d i s i n t e g r a t i,o n. — In collecting
stones from deposits containing an abundance of loose rapakivi the deposits
may have lost a number of the rapakivi cobbles larger than 6 cm, while the
smaller grain sizes may have been enriched from the disintegrated materials.
The excavation methods described above have, however, reduced the possible error due to this source to such narrow limits that there is no need to
take it into account. But a possibility of error ranging from 2 to 3 % may be
implied in countings involving a greater amount of rapakivi, especially with
respect to morainic drift, in which crumbling is difficult to recognize owing
to the angular shape of many of the boulders. As for countings embracing those
pebbles smaller than 2 cm, which previous to counting have to be washed, the
range of error for pebbles from 2 to 0,6 cm in diameter may not be higher
than that mentioned above, because the numbers of the pebbles have been
noted previous to washing. On the other hand, for pebbles 0,6 to 0,2 cm in
size, which have been sifted from the samples only after having been washed,
an error amounting to about 5 % in favour of the rapakivi is quite possible.

3) E r r o r   i n   d e t e r m i n a t i o n   may not affect the percentage for cobbles
larger than 6 cm, either, although it is probable that some varieties of the rapa-

kivi difficult to recognize megascopically, in the first place the rapakivi aplites, have been classed among the Archaean rocks. But in the case of the stones from 6 to 2 cm in size, this certainly gives rise to a source of error, the range of which is widest, in this case also, where rapakivi is most abundant and where rocks difficult to recognize are found in greatest profusion. By counting the most uncertain boulders apart — those with the question mark in the tables — the author, however, believes himself to have avoided this source of error to such an extent that at most 2 to 3 % of the rapakivi stones may possibly have been referred to the Archaean rocks. Since in such case it is rapakivi which loses through this source of error, the errors due to crumbling and those in determination should neutralize one another. Accordingly, in examining the percentage tables it must be kept in mind, with a view to the possibility of error, that the cobbles measuring from 6 to 2 cm in size entered in the column with the question mark, in all probability for the most part belong to rapakivi, since their number is greatest where the percentage for rapakivi is highest. In fact, this column is of little importance, its maximum being 2,5 and 2,6 %, the esker excavation 19 excepted, where the largest stones measured 2 to 4 cm, and whence in consequence the column under consideration received the exceptional value 3,5 %.

This applies also to stones measuring 2 to 0,6 cm in size, in the column with the question mark 3 % at most, with the reservation that 100 stones never yield so valid a percentage as do 200 stones of the greater size classes. — The only grain class in which the percentages admit of interpretation is that of 0,6 to 0,2 cm. In examining the table dealing with it, perhaps the best means of neutralizing the possibility of error due to disintegration is to transfer the percentages 6 to 9 of the uncertain stones (those with the felspar dimmed, for instance) to the column embracing the Archaean rocks.

III. Countings in Morainic Drift.

EXCAVATIONS. — The scantiness of morainic drift in Laitila is shown by the general thinness of the morainic mantle as well as by the sparseness of local morainic accumulations. This as well as the absence of ready-made excavations is responsible for the fact that the number of excavations and counting places in morainic drift is not more than 11 (map pl. 1 and Table 2, p. 18 & 19).

The perpendicular distance of the excavations from the esker ranges from 1,5 to 3 km, except in the case of one excavation (excavation X), which lies no more than 200 m from the margin of the esker. Three of the excavations

lie on the western, the rest on the eastern side of the esker. The distance between those widest apart from one another is 38 km. The distances between
the excavations vary so greatly that while the minimum distance is not more
than 0,5 km the maximum one is 9 km. The distances are smallest in the
neighbourhood of the northwest boundary of the rapakivi area (= »proximal
contact»), where the computations have to answer the question: — Which
rules are observed by the rapakivi in making its appearance in the morainic
drift? Eight of the excavations remain within the limits of the rapakivi area,
one lies on its southeast boundary (= »distal contact»), and two of them
project towards the southeast, outside the rapakivi area.

Some 4,700 stones have been counted from the excavations in morainic
drift in the immediate vicinity of the esker.

TYPES OF MORAINIC DRIFT. — As a preliminary to the examination of the
tables recording the countings in morainic drift it may be remarked that an
attempt has been made to include in the countings as many different types of
morainic drift as possible, cement-like morainic drift (excav. I, IV, IX, XI) as
well as more gravelly, stratified varieties (excav. VI, VII). Both types are
exhibited in the following mechanical analyses, which do not embrace stones
larger than 2 cm.

Table 1. Mechanical analyses of morainic drift.

N:o	< 0,002 mm	0,002—0,02	0,02—0,2	0,2—0,6	0,6—2,0	2,0—6,0	6,0—20,0 mm	Sum
IV	2,2	13,3	42,0	15,8	13,2	7,0	6,5	100
VII	0,4	1,1	11,9	13,6	26,8	15,8	30,3	99,9
IX	1,3	5,1	26,6	19,4	22,6	9,1	15,7	99,8

Among these the soil of excavations IV and IX has to be described, according to the dominating size of grains, as morainic sand, and, consequently,
represents the commonest type in the Northern countries (1, 3). As to its origin, undoubtedly it has to be characterized as unwashed and unassorted ground
moraine drift. The more assorted deposit of excavation VII (Fig. 3), too, is
shown to be unequivocal morainic material even by its appearance; but as to
its source, this washed subaquatic morainic gravel may have been derived
alike from the bottom and the interior of the land-ice. The common feature in
all the three samples analyzed is the predominance of the grain classes smaller
than 2 mm. Consequently, even though the counting of stones be extended

Fig. 3.   Excavation in morainic drift VII. The largest boulders (at the surface) consist of rapakivi

from stones larger than 6 cm down to those of 2 mm, a very considerable part of the morainic material falls outside the countings, and even, in the case of morainic fine sands, an overwhelming majority.

The soil in excavations II, V and VIII approaches in appearance morainic fine sand, but is poorer in coarser materials than the samples analyzed. The farthest departure from the common types of morainic drift is found in the deposits of excavations III and X. By their distinct stratification and by the lack of the finest grain classes they resemble esker material in a belwidering manner, whereas the angularity of the stones still more closely recalls morainic drift. In the opinion of the author, the soil of excavation III consists of morainic drift washed suddenly and violently; while the original purpose of excavation X, lying at a distance of only 200 metres from the margin of the esker, was to serve as a test to ascertain whether such an uncertain case as this could be solved by means of stone countings. The comparison of the countings later on showed in fact that it was a deposit in its petrographic composition entirely different from the material of the adjacent esker and in this respect was well fitted to be called morainic drift.

*   *   *   *   *   *   *

COMPARISON OF THE COUNTINGS. — The first impression left by a comparison of the countings with one another is that the occurrence of the different rocks is highly irregular, an irregularity which, however, does not seem to depend on the different types of morainic drift, nor on the situation of the excavations with respect to the esker. Nor does the distance from the contacts clearly control the percentages for *e. g.* rapakivi.

This apparent disorder, however, conceals some measure of regularity.

Even a glance at the table reveals to what a limited extent the bed rock is able to influence the petrography of morainic drift. It should be noted in the first place that the boulders from distant sources, *i. e.* Jotnian sandstone, sierakivi and porphyry, occur at every spot excavated, except the last one, in which neither sierakivi nor porphyry were found. Jotnian sandstone is present, and in every grain size and in fairly high percentages at that. The rare granophyre, too, has been encountered in most excavations. Olivine diabase is not absent even in the excavation farthest to the southeast, where it must be referred to boulders of distant derivation.

An examination of the column for rapakivi, on the other hand, shows that the underlying bed rock may sometimes mark the overlying morainic drift very strikingly. Even at a distance of 1,5 km from the proximal contact it is represented in the percentage-figures, and its percentage increases rapidly, as if scattered by exploded bombs, being soon some 50 %. But further away from the contact, excavations IV and V exhibit rapakivi in the morainic drift in a proportion of only 20 to 25 % (smaller sizes, it is true, some 40 %). Should the eleven excavations in morainic drift under consideration have happened to disclose in the freaks of morainic drift something particularly local and accidental in character, these 40 to 50 % for rapakivi at a distance of only 2 to 2,8 km from the contact would be of that kind. This occurrence can hardly be interpreted as fortuitous in the sense that the rapakivi bombardment might have been brought about by a horizon in the bed rock being liable to disintegration in an exceptional degree. This occurrence, corroborated by two excavations (II & III), must possess also the kind of evidence referred to above: a fairly large part of the morainic material may sometimes be derived from so close a vicinity as to have been transported no more than some 2 or 3 km from its parent rock.

The composition of the morainic drift in the Laitila district clearly embraces materials from distant as well as near sources, which, intimately blended together, are rivals for predominance. At a distance of 14 to 15 km from the proximal contact (excavations VI and VII lying at different sides of the esker, at a distance of 4,5 km from each other), rapakivi seems finally to gain the upper hand. But as the distal contact approaches, its percentage distinctly declines (excavations VIII & IX), at the same rate as an increase is shown in the number of boulders of Archaean rocks and of those with distant derivation in the strict sense of the word.

Besides, this case is not unique. According to private information ESKOLA has carried out stone countings about 20 km further east, in the rapakivi area, at Maurunoja in Lellainen, and has arrived at the following results (depth 1 m, 219 stones, size 6 to 2 cm):

Archaean rocks .... 46,6 %
Rapakivi .... 23,8 »
Olivine diabase .... 2,3 »
Diabase porphyry .... 0,9 »
Jotnian sandstone .... 19,6 »

Sierakivi .... 3,2 %
Porphyry .... 0,9 »
Granophyre .... 2,7 »

The rapakivi area is widest in the east, and, therefore, the counting place was further away from the proximal contact than any of the author's counting places in the rapakivi area. The morainic mantle in the counting place under consideration is exceedingly thick, an adjacent well, sunk entirely in morainic drift, being more than 12 m deep.

Even where the rapakivi area is widest, it has evidently not been wide enough to stamp the glacial morainic material with its own mark in any decisive degree. Consequently, it is nothing exceptional, if in Laitila, in spite of local rapakivi bombardments, even at the brink of the distal contact, a material of distant derivation, transported across the whole of the rapakivi area, should gain the upper hand among the stones with 20 to 2 cm size in the morainic drift.

Therefore, as soon as new local materials derived from the Archaean bed rock associate outside the distal contact with this morainic drift of chiefly distant derivation, the percentage of rapakivi declines (excavation X) so far that among the excavations discussed heretofore only excavation I contained less rapakivi. Although, on the basis of only two excavations, it is not permissible to pronounce anything definite about the petrography of the morainic drift outside the distal contact, the increase in the number of the boulders derived from the Archaean rock seems to observe nearly the same rates at which rapakivi increased from the proximal contact onwards. This appears from the following comparison: — If the exceptional percentages for excavations II and III be omitted, the proportions of rapakivi, at a distance of 4,9 km from the proximal contact, in excavation IV, are 18,5 % and 28,3 %. On the other hand, the distance from the distal contact (excavation IX) to excavation X is about 3,5 km, and the increase in the percentage of the Archaean rocks may have to be estimated as $82,5 - 57 = 25,5$, and $75 - 53,5 = 21,5$. This estimate is supported by the decrease in the number of boulders of distant derivation between excavations IX and X, a fact which must be due to the supervention in the morainic drift of new material, derived from the Archaean rocks southeast of the rapakivi. Farther to the southeast rapakivi does not, however, wholly disappear from the morainic drift, but persists in small proportions: excavation XI 5,5 % and 4,5 %.

As the rapakivi and the Archaean rock landscape do not differ appreciably in altitude, the rapakivi bed rock has been almost as resistant against the glacial erosion as the Archaean rocks. The stone countings show that no distinct difference is discernible in the degree of wearing out of the boulders of the rapakivi and of the Archaean rocks. But Jotnian sandstone has evidently been the most resistant rock, particularly when compared with the rapakivi. In excavation XI the rates of both are almost equal, and farther to the southeast the proportion of rapakivi would obviously fall below that of the Jotnian sandstone.

In the above comparisons no attention has been paid to the different size classes of the morainic boulders. There seems to be no rule prevailing in the percentages for stones larger than 6 cm and those measuring 6 to 2 cm throughout the whole series of the countings. In most cases, however, the percentage for cobbles of 6 to 2 cm in diameter is somewhat higher, for which reason rapakivi tends to occur in morainic drift preferably in small sizes. This rule, which finds its explanation in the liability of rapakivi to disintegration, is corroborated still more clearly by stones smaller than 2 cm: in excavations IV and IX the percentage for stones measuring 2 to 0,6 cm and 0,6 to 0,2 cm is manifestly higher than for stones 6 to 2 cm in diameter, and twice as high as for stones larger than 6 cm. It is, however, to be noted that this rule does not hold good for excavation VII as indisputably as in the cases discussed above.

IV.   Countings in Esker Gravel.

EXCAVATIONS — Thanks to the numerous gravel pits in the esker it has been possible to carry out stone countings in the esker at shorter and more regular intervals than in the morainic drift. Within a stretch of 39,5 km excavations together with countings were carried out at 23 spots, of which 15 lie within the rapakivi area and the rest outside the distal contact. The widest intervals (between excavations 5 and 6, and 15 and 16) are both less than 4 km, all the others are 3 km at most. The succession of excavations is closest in the row of hills on the plain of the lowland of Laitila, where countings have been performed in the separate hills, although the distances between them fell short of 0,5 km. However, one of the two wide intervals mentioned above (between 5 and 6) is situated on this same hill row, because here is the widest gap in the portion of the esker falling within the area investigated.

The total number of stones counted in esker material is about 10,300.

TABLES 4a and 4b record the stone countings in the esker in the same manner as the countings in morainic drift, running from the proximal contact to-

Fig. 4. Large gravel pit in the church village of Laitila, excavation 9.
Gradient of beds 40° SSE.

wards the southeast. The distances from both contacts indicated in the tables
are measured in the direction of N35°W. For the reasons specified in connection with Table 2, it has been possible here, too, to record the distances
from the distal contact only within 0,5 km.

ESKER MATERIAL. — First of all, as mentioned above, all the countings have
been carried out in the cross-bedded strata typical of the esker, 11a excepted,
which has deliberately been chosen to represent the stones measuring 6 to 2 cm
in diameter in the surface mantle of the esker. The strata often overlap one
another regularly, and in all probability mostly slope towards the southeast,
in the sides of the esker also outwards. In the large gravel pit in Laitila (Fig.
4), in which the author exposed the surface of a stratum, this had a steep,
40°, slope towards SSE.

The degree of coarseness remains nearly the same in the same stratum,
but varies irregularly from one stratum to another, as is stated in the literature
on eskers to be the case in many other tracts of the country. The largest stones
very seldom exceed the size of 10 or 12 cm in diameter. These stones larger
than 6 cm occur — when at all — at most places of excavation only in a few
strata, in which there is always an admixture of finer grain sizes. The author
has encountered only two exceptions to these rules: in the vicinity of excavation 16 an esker hill consists at least in its surface exclusively of boulders 6

to 40 cm in size; in a gravel pit lying about 1 km south of excavation 23, the finer gravel is interspersed with a considerable number of stones 10 to 50 cm in diameter.

Whether stones larger than 6 cm are present or not, stones 6 to 2 cm in size are common everywhere. Even in excavations where sand predominates they are met with, either scattered here and there (excavations 2 and 9a), or as coarser intercalated layers (excavations 19, 22a), sometimes a thick stratum consists almost entirely of stones measuring 6 to 2 cm (excavation 9c). As to their form, the stones, both large and small, are of the roundish type familiar from the literature on eskers.

Since the finest-grained esker material falls outside the countings, some sand-dominated pits (4, 12, 19, 22a) are represented in the counting by one or two stony layers. On the other hand, excavation 15, consisting almost exclusively of gravel from 6 to 0,2 cm in size, in which counting has been extended down to stones of 0,2 cm in diameter, is represented in the counting pretty nearly by the whole of its materials. In a single place, *viz.* in the large gravel pit of Laitila (excavation 9) the countings have been made from several layers separately. The countings in excavation 22 have tentatively been made in two pits lying close to each other: in one of them stones larger than 6 cm were counted, in the other those 6 to 2 cm in size.

* * * * * * *

COMPARISON OF THE COUNTINGS. — The figures of the tables 4a and 4b show that the material of distant derivation is represented by a minor but constant percentage in all the excavations. The Archaean rocks occur in all places of excavation, rapakivi everywhere else but in two excavations next to the proximal contact. The percentages shown in each column for the different strata of the same pit (excavation 9) and of the neighbouring pits are on the whole very much the same. The petrographical character of esker material does not, accordingly, seem to depend on the degree of washing nor on the position of a stratum in eskers. The general impression made by the table is, that in addition to its assortedness esker gravel appears in its petrography to be a soil very evenly mixed.

R a p a k i v i makes its appearance in the esker[1]) some 5 to 8 kilometres from the proximal contact (excavations 3—5) with a steadily rising percentage. As at the same distance from the distal contact rapakivi and the boulders of distant derivation begin decisively to fall in number (excavations 18—20), this must be due to the supervention of Archaean rock material derived from a near by source. The boulders of the esker have thus been transported at least 5 or 8 km from their source in the bed-rock.

The percentage of rapakivi is at its highest, *i.e.* 80 to 90 %, in the southeastern part of the rapakivi area (excavations 13—15). The reason why this maximum does not continue beyond the distal contact is revealed by the increase of the material of distant derivation in excavations 16 and 17. With a proportion higher than 50 % rapakivi dominates without interruption for some 20 km from the church village of Laitila in the southeastern direction (excavations 11—19). The only indisputable deviations from the smooth run of the numbers of the column for rapakivi take place along the row of hills crossing the lowland of Laitila, where the number of rapakivi boulders in

---

[1]) Among the surface boulders rapakivi occurs on the esker at a distance less than 1 km from the proximal contact.

excavations 7 and 8 exhibits a sudden fall and so abrupt a rise in the next excavation 9 that the percentage here is somewhat higher than in excavation 10. As leaps and bounds like these have not been met with elsewhere, in spite of numerous excavations, it will suffice to state in this connection that we are dealing with an exceptional case, the explanation of which must be postponed. However, it may be pointed out that in these leaps and bounds of the percentage only the stones larger than 6 cm are involved.

During the rise of the percentage for rapakivi towards its maximum the proportions in which the different grain classes stand to one another differ from those prevailing during the anticlimax. Rapakivi occurs for the first time as boulders larger than 6 cm (excavation 3), which in excavations 4—6 continue to be ahead of the smaller sizes in percentage. The smaller sizes in their turn seem to be on a par with one another in their percentages (excavation 5 and further 9d). Meanwhile the sizes larger than 6 cm have sustained the local defeat mentioned above, but in excavations 9—14 they are again increasing in number *pari passu* with the smaller sizes until the rapakivi maximum is reached. During the course of the maximum (excavations 15—16) the sizes larger than 6 cm are definitely distanced by the sizes smaller than 6 cm. But the mutual relations of the sizes smaller than 6 cm are undergoing changes: the stones 6 to 2 cm in size are a little distanced by the sizes 2 to 0,6 cm and 0,6 to 0,2 cm (excavation 15), of which at least the former size ousts the Archaean rocks almost entirely from the countings. As the percentages of rapakivi soon after this, beyond the distal contact, begin to decline sharply in number (excavations 18—21), the order mentioned above is preserved with the exception that the disappearance of stones smaller than 2 cm takes place far more slowly than that of the sizes larger than 6 cm and the size 6 to 2 cm. The first size to disappear from the esker is that larger than 6 cm (excavation 23).

The boulders of distant derivation being so sparse, the percentages of the A r c h a e a n r o c k s are pretty nearly the complementary values of the percentages of rapakivi and do not require any special account.

As to the b o u l d e r s o f d i s t a n t d e r i v a t i o n, it is their uniform sparseness which deserves to be specially noted. Granophyre, porphyry and sierakivi enter into all the countings, the last-mentioned being the most frequent, but sometimes they may be entirely absent. Only Jotnian sandstone occurs in every place excavated, its maximal rate being 7,5 %, but the most common rate is only 2 to 3 %, in the southeast still less.

V. Comparison between Morainic Drift and Esker Gravel.

DIAGRAM. — The comparison of the stone countings carried out in morainic drift and esker gravel is illustrated by the diagram (pl. 2). The chief results of the countings have been projected in the diagram in the following manner: The distances of the excavations from the contacts in the direction of the movement of the inland ice, and in the direction of the esker (N35°W), have been plotted on the abscissa, and the percentages of the different rocks on the ordinate. The direction of the abscissa is marked on map pl. 1.

In order to give a clearer general synopsis only two grain classes have been entered in the diagram, *viz.* stones larger than 6 cm and those from 6 to 2 cm in size, and only two rocks, *viz.* rapakivi and sandstone. To what extent the curves for the Archaean rocks would deviate, if actually drawn, from the reflected image of the curves for the rapakivi, is shown by the curves for sandstone, which represent the sums of the percentages of the Jotnian sandstone and the sierakivi. Other rocks of distant derivation (porphyry and granophyre) occur in percentages so small, that they cannot be presented in the same system of co-ordinates as those mentioned above. The same is the case with olivine diabase. In examining the diagram it is advisable to bear in mind that the stones of the size class from 6 to 2 cm in diameter, shown in the countings with the question mark, probably ought to be referred to rapakivi (cf. p. 15), for which reason the 6—2 cm curve for rapakivi in excavations 12, 15, 17, and 19 in the esker as well as the corresponding curve for morainic drift in excavations III & VII should perhaps represent proportions 2 to 3 % higher. In other countings there occur in the question mark column of the tables only low percentages (<2 %) which in no way affect the course of the curves.

THE DIFFERENCES IN THE CURVES FOR RAPAKIVI. — In the comparison of the curves for rapakivi in esker gravel on the one hand, and in morainic drift on the other, attention is attracted to four striking disagreements:

1) Rapakivi makes its appearance in morainic drift at a distance of 1,5 km, and in the esker at a distance of 5,3 km, from the proximal contact. Similarly the new Archaean rock material makes its appearance in morainic drift beyond the distal contact at a distance of 1 to 3,5 km from the latter (cf. p. 22), in the esker at a distance of 5 to 8 km from the same (cf. p. 30). In so far as samples of the bed rock in the morainic drift present themselves 4 km nearer to their source than in the esker material, the former may be said to be derived from nearer sources than the latter.

2) The high pillar in the curve representing the percentage of the rapakivi in the morainic drift at a distance of only 1 km from the spot where rapakivi first appears in morainic drift, is a striking illustration (cf. p. 21) of the petrographical heterogeneity of the morainic drift, especially when compared with the esker material, in which rapakivi appears very evenly, increasing percentage after percentage. This difference — in morainic drift the percentage leaps to and fro, in the esker it is more regular — is certainly one of the chief dissimilarities between these two deposits. The morainic drift can be characterized as a deposit of poorly mixed petrographical composition, while the esker material is well mixed. The esker material could be called exceptionally poorly mixed in the esker hills between excavations 7 to 9, where within a distance of 1,3 km the curve for esker rapakivi stones larger than 6 cm traces the local minimum pointed out above, and immediately thereafter a minor maximum pillar. These oscillations will be neutralized, if the mean percentage is deduced for the three excavations.

3) Especially striking is the high and even arch of both the curves for the rapakivi of the esker within the interval between about 15 and 35 km from the proximal contact (excavations 11 to 19). Within this space these curves keep all the while above 50 %, while the curves for morainic drift remain for the most part below that percentage; in the esker the percentage of rapakivi is twice as high as in the morainic drift. In five successive excavations (12 to 16) the highest maximum of the esker surpasses the highest maximum of rapakivi encountered in morainic drift (excavation VII). It is possible that more numerous excavations in the morainic drift of this tract, or the taking into account of the volume and weight of the stones might have transferred the climax of rapakivi to the side of the morainic drift; in that case the course of the curve for morainic drift would again exhibit steep pillars, which would still more clearly evidence the petrographic variability of the morainic drift. Whether this be the case or not, the indisputable fact remains that neither of the curves for rapakivi in the esker (= neither grain size) in the course of their common maximum (about 70 to 85 %) descends for a moment to such percentages as have been encountered in the morainic drift of the same tract (30 to 35 % in excavations VIII and IX). The overwhelmingly greater part of the esker material is derived from near by sources. This is true even in the event that in the adjacent morainic drift rocks derived from distant sources should happen to dominate.

4) The relations of both curves for rapakivi in morainic drift seem to vary irregularly. In connection with the comparison of all the four size classes of the table for morainic drift (cf. Chapter III), it was found, however, that the

highest percentages are yielded by the smallest sizes, *i. e.* the countings in morainic drift prove that the rapakivi is an easily disintegrated rock; whereas the curves for rapakivi in the esker repeat the rule deduced from the tables for the esker (cf. p. 31): *viz.* rapakivi makes its first appearance in the esker as stones larger than 6 cm, the percentage of which remains for a while higher than for other sizes; in the course of the maximum the smallest sizes definitely gain the lead; in the final catastrophe the largest size disappears first, the smaller ones less rapidly. Consequently, the morainic drift reveals by its rock material that during the glacial transportation wholesale crushing and crumbling were the dominant processes, while the esker material discloses that the varying size of its stones is an expression of the varying degrees of grinding and rubbing, or the wear and tear, of the boulders, the relations of which to one another depend in each place of excavation on its situation in the ose chain.

AGREEMENT BETWEEN THE CURVES. — Besides these strikingly different features, the curves for rapakivi in morainic drift on the one hand and in the esker on the other exhibit also some measure of agreement.

The curves for rapakivi in both deposits begin to decline from their maximum before the distal contact (morainic drift), or from the distal contact onwards (the esker), *i. e.* in both deposits at points where new Archaean rock material from the distal side of the rapakivi area has not yet joined company. Both features (cf. p. 22 & 30) must be due to Archaean rock material of distant derivation which, for some as yet unknown reason, has been enriched in the moraines of the southeastern part of the rapakivi area of Laitila.

It would seem a legitimate conclusion to say that the same cause produces the same effect with respect to both deposits, and to add that in the morainic drift the rapakivi seems to disappear more readily than in the esker, in which it more stubbornly resists the material of distant derivation. It would be equally legitimate to conclude that the enrichment of materials of distant derivation in the moraines is a factor the influence of which is reflected rather sensitively in the curves for the esker, but it must be added that the stone asset furnished by the morainic drift is, however, only one of the factors which have determined the petrographical composition of the esker material.

Which of these two interpretations is the more correct? The curves have to answer this question, and they do so decidedly in favour of the latter one. Let us examine more closely at what distance from one another rising and falling occur in the curves for rapakivi in the morainic drift and in the esker. For facilitating the comparison let us imagine two curves for morainic drift (taken together) on the one hand and two curves for esker material on the other (taken together) as separate graphs, which may be moved in the direction of the abscissa over each other.

The average interval between the enrichment in materials of distant derivation in the morainic drift and in the esker (see Diagram) appears to be about 3 to 5 km. As stated above, the rapakivi makes its appearance in the morainic drift at a point some 3,8 km (size larger than 6 cm) to 4,4 km (size from 6 to 2 cm) nearer to the proximal contact than in the esker, while the new Archaean rock material appears beyond the distal contact in morainic drift at a distance of 1 to 3,5 km, and in the esker at that of 5 to 8 km from the contact; accordingly the average difference in the last-mentioned cases is 3 to 5 km.

Supposing that the curves for the esker are moved along the abscissa axis 4 km nearer to the proximal contact, the curves for the esker and those for morainic drift exhibit a considerable measure of agreement at the same distances from both contacts. The starting point of the curve in the northwest, the enrichment in material of distant derivation and the final loss of the rapakivi in the southeast must, of course, coincide, since they have determined the replacement of the curves. Moreover, the curves for morainic drift and the esker on the lowland of Laitila almost coincide, between points IV and V and 5 and 6, for instance, similarly the rapakivi minimums V and 7 fall almost one upon the other. Hence onwards towards the southeast the curves for the esker rise high above those for morainic drift, while e. g. the maximums remain at the same place, and the final stages of the curves agree to a great extent.

It may be pointed out already in this connection that if the curves for sandstone are moved some 4 km towards one another, their ascents and descents become fairly well coincident. The sandstone maximum in the esker (excavations 8 & 9), for instance, falls close to the maximum of morainic drift.

THE CAUSES OF DISAGREEMENT. — The position of the curves with respect to one another supports the interpretation that the morainic drift is the source of the esker material: the morainic drift is the reservoir that furnished the rock material which, after having, so to say, rushed a few kilometres — 4 km on an average, the small pebbles even more — in the direction of the motion of the inland ice, stopped as a new deposit with a very homogeneous petrography, viz. the esker gravel. This rush forwards and intermixing was accompanied by a powerful friction destructive to the stones, as is shown by the different percentages for the different stages of wear and tear of the rapakivi in the different tracts, pointed out in paragraph 4 (p. 33) above. The same acceleration of friction during the genesis of the esker further appears from the arch of the curve for rapakivi in the esker (see paragraph 3, p. 33), rising high above that for morainic drift: as the morainic drift changes into esker gravel, the boulders of the Archaean rocks derived from distant sources are continually worn and torn, and therefore diminish rapidly, whereas the rapakivi boulders seem to withstand the wear thanks to the continual arrival of reinforcements. The contrary case, *i. e.* the cessation of the rise of the percentage of rapakivi between excavations 5 to 8 and even the decline in the size larger than 6 cm in excavations 7 & 8 may be due in a decisive degree to the exhaustion of reinforcement in the deepest depression of the lowland of Laitila, although other factors in this case were the poor intermixing of the esker material between excavations 7 to 9, as assumed above, and of the local morainic material of distant derivation recorded in excavation V. Beyond the distal contact, where rapakivi certainly looses its reinforcement entirely, while the percentage for the Archaean rocks is constantly augmented by the bed rock of the ground, the disappearance of rapakivi from the esker takes place catastrophically: within an interval of 6 kilometres (excavations 18 to 21) the rate of stones larger than 6 cm falls from 57,5 % to 3 %, and that of the size 6 to 2 cm from 68 % to 5,5 %.

Figuratively and relatively it could be said that the esker is the destroyer of stones, the morainic drift their preserver.

MATERIALS DERIVED FROM DISTANT SOURCES. — The same record of the conditions which prevailed when an esker arose is yielded by materials indisputably derived from distant sources in the strict sense of the word, *i. e.* the Jotnian sandstone, sierakivi, porphyry and granophyre.

The curves of the diagram for the s a n d s t o n e, representing the sums of its two varieties, represent so considerable a percentage of the material derived from distant sources in both deposits that it is worth while to compare the curves. It has been pointed out above in several connections how sensitively the curve for sandstone reflects the oscillations in the materials of distant derivation, both in the morainic drift and in the esker. The curves further show that the percentage for sandstone in morainic drift is everywhere (with the exception of excavation X) strikingly higher than in the esker. In morainic drift the minimum is 3,5 % and the maximum 17,5 %, the average 8 or 10 %. In the esker the minimum is 0,5 %, the maximum 8 %, in most cases below 5 %. This must have been due to violent wear and tear during the genesis of the esker. The representation of the different grain sizes in the esker is marvellously equal: in the diagram only the curve for the size larger than 6 cm has been drawn, but the same curve is almost capable of combining also the small black circles representing the size 6 to 2 cm; whereas with respect to the morainic drift, the small rings representing the size 6 to 2 cm may diverge far from the curve for the size larger than 6 cm, mostly above it. If further the combined percentages for the sandstones smaller than 2 cm be taken from the counting tables and compared with the stones larger than 2 cm, the same proportion will be found: the proportion of the size 2 to 0,6 cm is in the morainic drift 3 to 8 %, in the esker 1 to 4 %, that of the size 0,6 to 0,2 cm 3 to 5 % in morainic drift, and 0 to 2 % in the esker.

The distinctive feature of the curves for sandstone in the diagram is in the first place the J o t n i a n   s a n d s t o n e, which in percentage is everywhere far ahead of the sierakivi. In the examination of Jotnian sandstone in the tables attention is arrested by the exceptionally high capacity of resistance to wear and tear. The countings in morainic drift embrace this rock at a rate of 2 to 3,5 % even far to the southeast, where rapakivi begins to disappear. This durability of the Jotnian sandstone is borne out still more unequivocally by the countings in esker gravel, in which this distant wanderer occurs in quite small percentages (only twice more than 5 %), but still is found so persistently that it is never absent in the percentages showing stones larger than 2 cm, not even within the maximum of the rapakivi, where the stones of the Archaean rocks have dwindled to the rate of about 10 %. The most striking instance is excavation 15, in which the proportion of the Archaean rocks among the stones from 2 to 0,6 cm in diameter amounts to 1 % (to which possibly some of 3 uncertain stones should be added), while that of sandstone similarly amounts to 1 %. Among the stones from 6 to 2 cm in size, taken from the same excavation, the proportion of the Archaean rocks is 4 times that of the Jotnian sandstone; in both of the sizes counted in excavation 16 the proportion of the Archean rocks is 4,3 to 5,2 times that of the Jotnian sandstone. In like manner this wanderer from afar persists, while the rapakivi in its turn is on the wane: in excavations 22 and 23 the percentage both for rapakivi and for Jotnian sandstone is at last nearly the same, and sandstone persists even among the largest stones of excavation 23, while rapakivi has disappeared from the percentages.

In view of the capacity of the Jotnian sandstone to withstand the heavy wear and tear of an esker to such an exceptional degree, it is no wonder that it occurs among the boulders of morainic landscapes everywhere in Southwestern Finland (9, 13).

The less common s i e r a k i v i is seen to occur in every excavation in morainic drift, (XI alone excepted) although it does not occur in every size. The three maximum percentages are 5,0, 3,9 and 3,5 %; while in the esker its maximums are 2,5, 1,5 and 1,5 %. It further appears from the number of stones counted that the percentage for sierakivi in the esker is in most cases based on a single specimen in each counting. Sierakivi is entirely absent in excavations 2, 4, 13, 23. Sierakivi with globular grains of quartz, in which calcite has cemented the round quartz globules whose diameters vary between 1 and 3 millimetres, has been encountered only in morainic drift.

The p o r p h y r y seems to observe the same rule as the sandstone. In the excavations in morainic drift porphyry has occurred among the stones counted everywhere save in excavation XI. The maximums were 3,0, 2,6 and 2,5 %. In the esker the maximums were 3,0, 1,5 and 1,0 %, and nine excavations yielded no percentage whatever for porphyry.

And finally as to the g r a n o p h y r e: while it entered into the countings in morainic drift as often as 19 times, it appears only 8 times in the twice as numerous countings in the esker.

Morainic drift is the preserver also of materials from distant sources, the esker is their destroyer.

THE SHAPE AND SIZE OF STONES. — The stone countings have revealed so profound a petrographic discrepancy in character between morainic drift and esker material that it must perforce presuppose entirely different conditions, nay, entirely different mechanics in action at the time when these two soils were deposited from the land-ice. Additional light is thrown upon these conditions by the shape and size of the stones.

Among the common formulations of the differences between morainic drift and esker gravel is the following: the bulk of the stones of morainic drift are angular, those of an esker rounded. This rule holds good of the majority of the stones embraced by all the stone countings; excavation 21, in which most of the stones were angular as in morainic drift, being an exception. Thus, rapakivi stones larger than 6 cm occur in morainic drift in most cases in so angular a form that it has often been difficult to make out whether the rapakivi stones in morainic drift had perhaps not been broken up during excavation. Unless the form of such an angular rapakivi boulder is not wholly abnormal, it resembles in its outline a brick or a thick book. In the esker, on the other hand, the rapakivi boulders larger than 6 cm are in their dominant type rounded from their first percentages onwards, *e. g.* like balls (excavations 4, 7, 18, 22), or like discs (excavation 9).

Among the smaller rapakivi stones of special interest are the o r t h o-c l a s e g r a i n s, which are common in the sizes 6 to 2 cm and 2 to 0,6 cm.

In morainic drift they have generally preserved their smooth crystalline faces or cleavage planes and their sharp edges, in fact, sometimes they are quite idiomorphic, while in the esker they are generally rounded at the edges and dimpled on the surfaces. Similarly the q u a r t z g r a i n s, which occur especially among the stones from 0,6 to 0,2 cm in size, are in morainic drift almost always recognizable as rapakivi quartz owing to their crystal faces. Sometimes unmistakable double pyramids are discernible. In the esker these grains have always lost their crystalline faces, so much so that it has been necessary to enter the quartz grains in the column with the question mark, unless it has been possible to recognize them as rapakivi quartz by some orthoclase inclusion determined by the microscope. Even the b i o t i t e of the rapakivi invariably occurs in different forms in the morainic drift and in the esker. Among the morainic stones from 0,6 to 0,2 cm in size loose biotite is always present in the form of knobs (often accompanied by a trifling quantity of quartz and orthoclase from which it may be identified as rapakivi biotite). In the same size class of esker gravel these knobs do not occur at all. Evidently the loose biotite in the esker has been worn and torn into scales, which abound in the size class smaller than 0,2 cm to such an extent that they constitute the most distinctive characteristics of the grain size 0,2 to 0,002 cm (coarse and fine sand) in all the samples of esker gravel analyzed.

Although these facts clearly reflect the effect of wear and tear upon the stones of the esker gravel, wear and tear alone does not suffice to explain the general small size of esker stones when compared with the stones 0,2 to 1 m in size present in fairly high proportion in the morainic drift. This disagreement is of course one of the features of the different assortedness of these two deposits: esker gravel, which, besides being worn and torn, is characterized by assortedness, displays the last mentioned character in the scarcity or complete absence of both the smallest and the largest grain sizes. This being the case, it is, however, legitimate to conclude that the agency which drove the esker material to its actual site was not capable of dragging the large boulders, lying, perhaps in plenty, at a short distance, to the line of the ose chain.

VI. Formation of the Esker of Laitila.

The comparison of morainic drift and esker gravel brings out that the esker material possesses several highly characteristic features, which distinguish it from the different types of morainic drift even in the immediate vicinity of the esker. The stone countings especially constantly force one to acknowledge that, in comparison with morainic drift, the rock material of the esker is heavily worn and torn, mixed and washed to a considerable degree, and has been transported about four kilometres on an average in the distal direction from its parent drift.

The gravel of which the ose chain consists may originally have been set in motion and rushed forwards in ice laden with drift along the line of the ose chain as well as at the side of the same, but it has certainly been transported for a considerable distance in a lineal direction and thus become mixed before its final accumulation. The direction of the esker must reflect the direction of the agency which wore, washed and accumulated the esker material.

Consequently, with respect to the esker of Laitila, the older or so-called glacial or moraine theories may safely be eliminated from the number of the theories concerning the formation of eskers. Outside the range of possibilities in this case falls among them also the explanation advanced by LEIVISKÄ (14), according to which the radial oses are in most cases englacial drift loads concentrated along certain stripes parallel to the ice movement, uncovered by ablation and washed by melt-waters. In their stead, with respect to the esker of Laitila, recourse must be had to some of the more recent explanations, the so-called fluvial theories (11, 25, 2), which conceive the esker material to be accumulated and the ose forms shaped by glacial drainage waters, running in channels cut in the ice- sheet.

The stone countings do not warrant any definite choice among the different fluvial theories. If one takes into account the fact that at the time of the formation of the esker in front of the retreating debouchure of the glacial river there existed a sea nearly 100 metres deep even above the elevations of the bed rock, it becomes, undoubtedly, difficult to understand that an open s u r f a c e   r i v e r on the land-ice should have been capable of gathering in its channel such large quantities of stones derived from near by sources in the underlying bed rock, as shown by the stone countings. The testimony of the stone countings seems to be more in favour of b o t t o m   r i v e r s, as presupposed by the theories of STRANDMARK and DE GEER, than of surface rivers. Such bottom rivers, with tunnel passages in the ice, are supposed to have been able to penetrate so far below the general base level, that it is assumed that even in the subaquatic regions the rivers which accumulated eskers ran at the bottom of the ice-sheet along the surface of the bed rock.

Although we dispense in this connection with a more precise definition of our standpoint with respect to the different fluvial theories, it is in any case evident that the genetic term f l u v i o-g l a c i a l   m a t e r i a l employed in connection with the fluvial theories applies well to the material of the Laitila ose chain. To the definition of fluvio- glacial material, moreover, has been added a description of the petrographical composition more accurate than heretofore. Future investigation may show, whether the rules arrived at for the area now investigated are valid also elsewhere.

Literature.

Bull. = Bulletin de la Commission Géologique de Finlande.
G. F. F. = Geologiska föreningens i Stockholm förhandlingar.

1. B. AARNIO: Etelä-Pohjanmaa. Agrogeologisia karttoja N:o 5 (Valtion maa-tutkimuslaitos, 1927).
2. GERARD DE GEER: Om rullstensåsarnas bildningssätt (G. F. F. 19, 1897).
3. GUNNAR EKSTRÖM: Klassifikation av svenska åkerjordar (Sveriges Geol. Unders. 20, 1927).
4. P. ESKOLA: Petrographische Charakteristik der kristallinen Gesteine von Finnland (Fortschr. der Mineralogie, Kristallogr. und Petrogr. 1927).
5. —»— On rapakivi rocks from the bottom of the Gulf of Bothnia (Fennia 50: 27, 1928).
6. —»— On the occurrence of orthoclase and microcline in the Finnish granites and pegmatites (Bull. 85, 1929).
7. HJ. GYLLING: Beskrifning till kartbladet N:o 12 (Finl. geol. unders., 1888).
8. H. HAUSEN: Stenräkningar på Åland (G. F. F. 33, 1911).
9. —»— Data beträffande frekvensen af jotniska sandstensblock i de mellan-baltiska trakternas istidsaflagringar (G. F. F. 34, 1912).
10. —»— Undersökning af porfyrblock från sydvästra Finlands glaciala af-lagringar (Fennia 32: 2, 1912 & Bull. 31, 1912).
11. N. O. HOLST: Om de glaciala rullstensåsarne (G. F. F. 3, 1876).
12. ILMARI KANERVA: Das Rapakivigebiet von Vehmaa (Fennia 50: 40, 1928).
13. AARNE LAITAKARI: Über das jotnische Gebiet von Satakunta (Fennia 45: 8, 1925 & Bull. 73, 1925).
14. I. LEIVISKÄ: Über die Ose Mittelfinnlands (Fennia 51: 4, 1928).
15. FRITZ MENDE: Typengesteine kristalliner Diluvialgeschiebe aus Südfinn-land und Åland (Zeitschr. für Geschiebeforschung I: 3, 1925).
16. V. MILTHERS: Scandinavian indicator-Boulders (Danmarks. geol. Unders. II: 23, 1909).
17. —»— Ledeblokke i de skandinaviske Nedisningers sydvestlige Grænseegne (Meddelelser fra dansk geol. Forening 4, 1913).
18. AD. MOBERG: Beskrifning till kartbladet N:o 10 (Finl. geol. unders. 1887).
19. E. MÄKINEN: Über die Alkalifeldspäte (G. F. F. 39, 1917).
20. MATTI SAURAMO: Studies on the Quaternary varve sediments in southern Finland (Fennia 44: 1, 1923 & Bull. 60, 1923).
21. —»— Tampere, Maalajikartan selitys (Suomen geol. yleiskartta B 2, 1924).
22. —»— Über das Verhältnis der Ose zum höchsten Strande (Fennia 51: 6, 1928 & Bull. 84, 1928).
23. —»— The Quaternary geology of Finland (Bull. 86, 1929).
24. J. J. SEDERHOLM: Beskrifning till bergartskartan B 2 (Geol. öfversiktskarta öfver Finland, 1911).
25. P. V. STRANDMARK: Om jökelelfvar och rullstensåsar (G. F. F. 11, 1889).

MAP OF THE AREA INVESTIGATED.

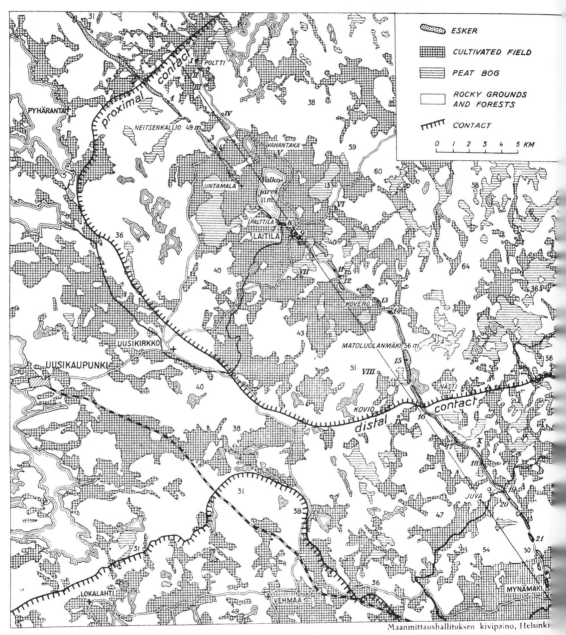

I—XI excavations in morainic drift.   *1 — 23* excavations in the esker.

The map gives an idea of the relative altitudes of the area: cultivated fields and peat bogs lie lower, rocky grounds and forests higher. The altitudes of some of the highest points in metres are designated.

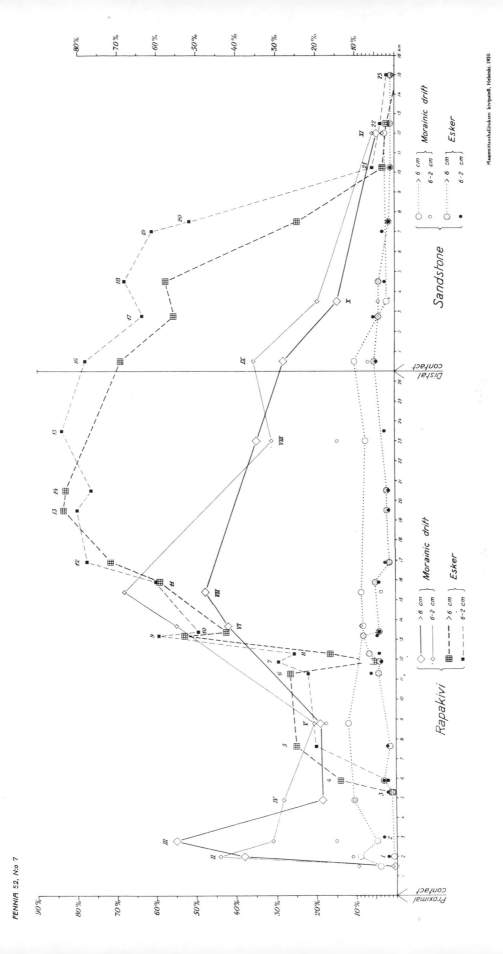

Maanmittaushallituksen kivipaino, Helsinki 1930.

18

Reprinted from *Jour. Glaciol.*, **8**(54), 413–425 (1969)

OBSERVATIONS ON A COLLAPSING KAME TERRACE IN GLACIER BAY NATIONAL MONUMENT, SOUTH-EASTERN ALASKA*

By GARRY D. McKENZIE

(Institute of Polar Studies, Ohio State University, Columbus, Ohio 43210, U.S.A.)

ABSTRACT. Detailed observations on a collapsing kame terrace indicate that the terrace is being reshaped by: slumping and sliding of debris into depressions, melt-water erosion on the side of the terrace, debris flows in the gullies, and stagnant-ice bursts, a phenomenon analogous to a glacier burst except in the mode of formation of the water. Temperatures in the gravel over the ice, where the gravel is about 4 m thick, indicate that the rate of melting of the upper surface of the ice due to conduction may be as high as 24 cm year^{-1}. Highest temperatures in the gravel were recorded during periods of heavy rainfall.

RÉSUMÉ. *Observations sur une terrasse affaissée dans le Glacier Bay National Monument, sud-est Alaska.* Des observations détaillées sur une terrasse affaissée d'un "Kame" indiquent que la terrasse est reformée par: effondrement et glissement de débris dans des dépressions, érosion par l'eau de fonte du bord de la terrasse, débris coulant dans les ravins et résurgence d'eau de fonte de la glace stagnante, phénomène analogue à celle des glaciers sauf dans le mode de formation de l'eau. Les températures dans les débris au-dessus de la glace où ils atteignent 4 m d'épaisseur, indiquent que la vitesse de fonte de la surface supérieure de la glace due à la conduction thermique, peut atteindre jusqu'à 24 cm par an. Les plus hautes températures dans les débris furent enregistrées durant les périodes de fortes pluies.

ZUSAMMENFASSUNG. *Beobachtungen an einer zerfallenden Kames-Terrasse im Glacier Bay National Monument, Südost-Alaska.* Eingehende Beobachtungen an einer zerfallenden Kames-Terrasse zeigen, dass die Terrasse umgeformt wird durch: Stürzen und Gleiten von Schutt in Vertiefungen, Schmelzwassererosion an der Terrassenflanke, Schuttfluss in die Gräben und Ausbrüche stagnierten Eises, eine Erscheinung analog einem Gletscherausbruch bis auf die Art der Bildung von Wasser. Temperaturen im Schutt über dem Eis zeigen bei einer Schuttmächtigkeit von etwa 4 m, dass die Abschmelzung durch Wärmeleitung an der Eisoberfläche bis zu 24 cm pro Jahr betragen kann. Die höchsten Temperaturen im Schutt wurden in Perioden heftigen Regenfalls beobachtet.

INTRODUCTION

During the 1966 and 1967 summer field seasons, the writer made observations on a collapsing kame terrace in Adams Inlet, Glacier Bay National Monument, Alaska (Fig. 1). From these observations the mechanisms and rate of collapse of the ice-cored kame terrace were determined.

The mechanisms of formation of kames, eskers and other ice-contact deposits have been summarized by Charlesworth (1957), Flint ([1957]) and Embleton and King (1968). Field observations (Sharp, 1947, 1949; Clayton, 1964) on the arrangement of supraglacial deposits have helped to clarify problems associated with ice-contact deposits existing in areas long since deglaciated. Generally, it is concluded that an irregularly distributed surface cover promotes differential ablation, forming a hummocky surface. Movement of debris from mounds or kames into depressions, followed by a change in the ablation rate on the mounds and in the depressions, results in an inversion of the topography.

Observations have been made on the rate of collapse of ice-cored deposits and the various factors affecting this rate. Sharp (1949, p. 298) noted that Philipp had observed as early as 1912 that a thin mantle of debris promoted ablation. Lister (1953) made a quantitative study of the rate of melting of debris-covered ice and found that in the Icelandic climate a debris cover of 4–7 mm was required to provide protection to the ice. Thinner debris covers increased the ablation rate. The grain-size of the debris was observed by Sharp (1949) to be an important factor in the ablation rate with a fine grain-size being a more effective insulator because fine-grained debris prevented the circulation of the air to a greater degree than coarse-grained material.

---

* Contribution No. 149, Institute of Polar Studies, Ohio State University, Columbus, Ohio, U.S.A.

Swithinbank (1950), in a study of debris cones, concluded that the amount of heat transmitted to buried snow or ice depended on the thermal properties (conductivity and radiation absorption coefficient), and the mechanical and structural properties (thickness and grain-size) of the debris cover.

Location and setting

Adams Inlet (lat. 58° 52' N., long. 136° 50' W.) is 108 km north-west of Juneau, Alaska, and it occupies an area of 27 km² (Fig. 1). It is bordered by mountains ranging in elevation from 1 200 to 2 200 m on the east and south, and on the north and west by low hills and outwash plains. In 1890–92, when the area was visited by Reid (1896), ice was at an elevation of 380 m in Adams Inlet, and between 1890 and 1940 the surface of Adams Inlet Glacier was

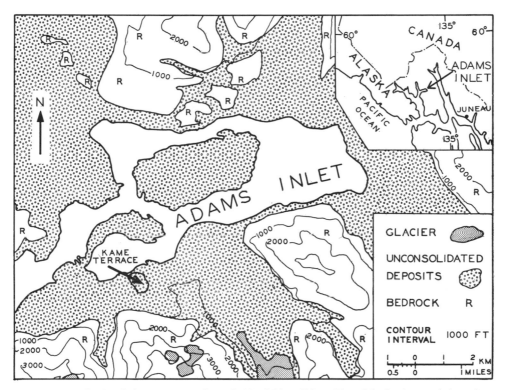

Fig. 1. Location map showing the kame terrace on the south side of Adams Inlet. Adams Inlet (on inset map) is on the eastern side of Glacier Bay.

lowered by ablation at the rate of 7.9 m year⁻¹ (Field, 1947). During this period melt water from glaciers in tributary valleys and cirques deposited outwash against the sides of and over Adams Inlet Glacier. A remnant of this glacier, covered by debris deposited about 1930, was investigated in this study.

The deposit of gravel overlying stagnant ice occurs on the south side of Adams Inlet and is classified as a kame terrace (Fig. 1). According to Flint ([1957], p. 149), a kame terrace is ". . . an accumulation of stratified drift laid down chiefly by streams between a glacier and an adjacent valley wall and left as a constructional terrace after disappearance of the glacier. . . ." The deposit in question exhibits the terrace form only in those parts nearest the mountainside —the other parts, distal from the mountain, were originally deposited on ice and are collapsing because of the melting of this ice. The deposit in Adams Inlet is bounded on the north by alluvial fans; on the south, the undisturbed parts of the terrace abut other unconsolidated deposits. Gullies are being formed in the collapsing east and west slopes of the terrace. The elevation near the north edge of the terrace is about 60 m. Only the outer collapsing part of the terrace was studied and mapped (Fig. 2).

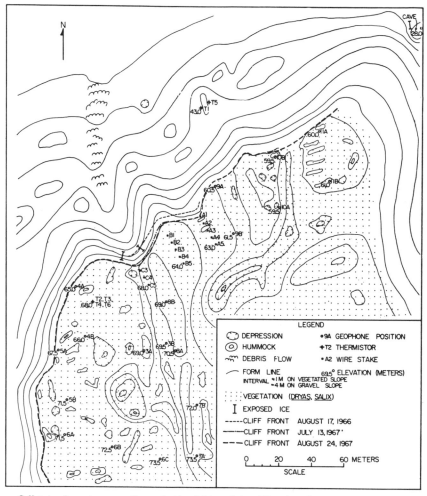

Fig. 2. Collapsing kame terrace on the south side of Adams Inlet. Elevation of vegetated surface is about 65 m.

Climate

Most of Glacier Bay National Monument lies within Köppen's Dfc climatic types (cool snow-forest climate, moist in all seasons, with cool summers). January, the coolest month, probably has a mean temperature between −3°C and −5°C in the vicinity of Adams Inlet. The July average temperature for the two field seasons in Adams Inlet was 12.2°C. The average precipitation for the same period was 0.24 cm d⁻¹; the average for the latter part of June was lower and for August it was higher. In August 1966 there were 25 days with more than 0.025 cm of precipitation; the maximum precipitation in a 24 h period during that month was 5.49 cm.

Mean relative humidities were generally high. In July 1966 and 1967 they were 87 and 89%, respectively. Winds were weak at observation times (07.00 and 21.00 h) with speeds generally less than 0.8 m s⁻¹.

Average cloudiness in tenths for the two summers, determined from the two observation times, was 8.7 for July and 8.5 for August.

No data are available on the date of the disappearance of the last snow from the ground or on the commencement of continuous winter snow cover for this area.

NATURE OF THE DEBRIS COVER

The ice beneath the collapsing deposit is covered with a sandy, coarse cobbly gravel with little remnant stratification. The gravel cover was first determined by seismic survey and

Fig. 3. North side of the kame terrace showing exposed stagnant ice. The gravel overlying the ice at the top of the terrace is about 3.5 m thick. Debris flow in gully is 0.6 m thick at the terminus.

ranges from 3 to 6 m in thickness with an average of about 4.2 m. A 60 m long exposure of ice (Fig. 3) showed that the gravel cover near the edge of the terrace averaged 3.5 m in thickness. The surface of the kame terrace has a kame-and-kettle topography with depressions as deep as 5 m. Most of the hummocks are less than 1 m high and circular in outline with a basal diameter of approximately 2 m, although some hummocks near the collapsing edge of the terrace are elongated (12 by 2 m) parallel to the wasting slope (Fig. 2).

The surface vegetation is mainly *Dryas* with some alder and poplar (Fig. 4). This type of cover, a *Dryas* mat with scattered shrubs, is stage III of the plant succession established in Glacier Bay by Decker (Goldthwait and others, 1966) and it indicates a time span of 20–25 years since the area became available for vegetation. This cover of vegetation and gravel has slowed the ablation of the underlying ice compared with the exposed glacier that occupied

Adams Inlet until approximately 1950. Tilted shrubs and trees, and solifluction lobes, covered with *Dryas* indicate that the surface form of the terrace is constantly changing. At several spots on the kame terrace, hummocks show open patches of gravel where a recent change in the shape of the kame has resulted in an opening in the vegetation cover.

Fig. 4. Hummocky vegetated surface of kame terrace. Hammer seismograph is in operation to determine depth to stagnant ice. View is north-east.

NATURE OF THE STAGNANT ICE

The stagnant core was viewed in two places. One was a gully on the north side of the terrace where the gravel cover had slumped and slid from the face of the ice and flowed out as a debris flow (Figs. 2 and 3). The exposed thickness of ice here was about 5 m, and the slope of this face was 39–40°. The other was the north-east corner of the terrace (cave in Figure 2), where the source of a spring was dug out to expose about 1 m of the basal ice. At this locality, the only place where the base was seen, the ice rested on sand and silt with pebbles.

A conservative estimate of the thickness of ice remaining in the terrace in 1967 is 35 m. The estimated width of the ice core is about 380 m and the length is about 400 m. The surface of the ice, as seen in Figure 3, is slightly undulatory, at least near the outer parts of the terrace. Water was observed flowing from the ice–gravel interface, from fractures within the ice and from springs at the base of the ice. At the north-east corner of the terrace at the base of the ice, several 2 to 4 cm thick layers of silt with some sand and pebbles were found separated by 30–60 cm of ice. These layers may have been incorporated debris in or near the base of the former glacier.

Several ice crystals as long as 22 cm were found, indicating a relatively long period of stagnation.

Cave in stagnant ice

At 09.55 h on 13 August 1967 a large volume of water burst from beneath the north-east side of the kame terrace where a small spring had been found a month earlier. The outrush of water was audible from a distance of at least 1.5 km. Two blocks of ice that apparently had been part of the stagnant ice mass were found on the alluvial fan adjacent to the kame terrace (Fig. 5). One of the blocks had dimensions of 2.2 by 1.4 by 0.8 m and was discovered 96 m from the cave; the other block, with dimensions 0.9 by 0.6 by 0.4 m, was discovered about 100 m from the cave.

Fig. 5. North-east corner of the kame terrace showing the entrance to the stagnant-ice cave (arrow). A man is standing beside a large ice block expelled from the entrance to the cave.

The cave was visited less than 10 h after it had been opened and at that time it had an entrance 6 m wide and 2 m high. This width continued for several meters in from the entrance. At a distance of 26 m from the entrance the cave had narrowed to 1 m across, and it extended for another 10 m at this diameter. The top of the entrance to the cave was rounded, and sand and water were flowing over the 35° ice slope above the entrance. A stream occupied a 2 m wide channel in the unstable silt and clay floor. The roof of the cave consisted of debris-free ice with a scalloped surface. It is evident from its size that the cave had been formed before the release of water. Frozen silt at the side of the entrance to the cave, below the ice, and the slight flow of water that came from the spring at this location before the opening of the cave suggest that this part of the terrace at least was saturated with water. The cave apparently had been melted in the ice by the flow of rain and melt water through and beneath the ice. The impermeable silt and clay beneath the terrace also helped to hold the water in channels and fractures within the ice. Much of the terrace may have been saturated with water, because water was observed to flow over the surface of the large exposure of ice in the gully

317

(Fig. 3). The pressure and quantity of water at the time of the outflow must have been considerable in order to move the ice blocks more than 90 m from the terrace. No water-level lines were seen on the walls of the cave and it was probably filled with water prior to the cave opening.

The sudden release of water with corresponding erosion and transport of debris and blocks of ice from buried stagnant ice is here termed a "stagnant-ice burst". This release of large volumes of water and accompanying erosion is, on a small scale, analogous to the glacier bursts or jökullhlaups that occur during a subglacial eruption of a volcano. In the case of the stagnant-ice burst, melt water and rain water are dammed by stagnant ice and an impermeable bed; the water is released when the pressure becomes too great or the ice dam is weakened by ablation. The stagnant-ice burst opens large areas in the ice to rapid ablation by warm air and thus speeds collapse of the ice-contact deposit.

Mechanisms of Collapse

As the ice melts beneath the gravel of the kame terrace, the gravel cover shifts and takes on a different topographic form. The fastest melting takes place where the ice is closest to the surface. This results in a mound or kame being formed where there is a thick cover of gravel, as the more rapidly melting ice around it subsides. The cover of gravel over the mound then slides, or moves as a lobe into the depression surrounding the kame. In this manner the terrace changes its shape and mixing of the previously stratified gravel occurs. In places where the terrace is completely underlain by ice all traces of the terrace form may be lost as the feature changes to an accumulation of kames. At this point the feature may more properly be termed a collapsed terrace or collapsed outwash.

In addition to the slumping and sliding of material that occurs on the vegetated part of the kame, there is re-arrangement of material on the unvegetated slopes by the action of running water. Springs emanating from within and beneath the ice carry debris that forms alluvial fans on the slopes of the collapsed part of the terrace. These streams also tend to undermine the loose gravel, causing further collapse and mixing of the deposit.

Debris flows are another mechanism in the formation of kames. The unvegetated slopes where the ice is closest to the surface may give way and slide into a gully. Then or several days later when sufficient moisture has accumulated in the gravel at the top of the gully, the mass of gravel becomes quick and flows down the gully (Fig. 3). Continued melting of the buried ice in this part of the terrace forms kettle holes in the debris flows and in the alluvial fans.

Still another factor in the movement of material in the collapsing kame terrace is the so-called stagnant-ice burst described previously. In Adams Inlet the stagnant-ice burst is apparently responsible for forming or enlarging gullies in the side of the kame and transport of debris away from the kame terrace. How often this process occurs is not known. The winter freeze may be enough to seal the outer parts of the kame that previously were open channels. In the spring, melt water and run-off from the mountain may be trapped in the stagnant ice and the whole process may recur. It may also be a unique occurrence that will not be repeated at the kames in Adams Inlet. The importance of stagnant-ice bursts as factors in reshaping ice-contact deposits is not known, for none are reported in the literature.

Soil Temperature Variations in the Kame Terrace

Soil temperatures were measured during the second summer using a YSI Tele-thermometer model 42SC. This model is readable to ±0.2°C in the range of operation and is accurate to ±0.5°C. Comparison with a standardized thermometer in an ice–water bath gave readings ranging from 0.1 to 0.2°C for the six thermistors.

Four thermistors were set near the top of the kame terrace at depths of 0.5, 0.9, 1.1 and 1.3 m as shown in Figure 6, and two thermistors were set at depths of 0.8 and 0.55 m (Fig. 6), on the west and east flanks respectively of an unvegetated kame near the bottom of the collapsing north slope. Thermistor readings were taken once daily between 19.00 and 21.30 h. These results, including the cloudiness and precipitation determinations, are given in Figure 7.

The results of the temperatures for the thermistors on top of the terrace indicate that the most significant temperature rise occurs during heavy rain. This is best illustrated by the readings for 8 August (Fig. 7) for thermistors on top of the terrace. Between 07.00 and 21.00 h on 8 August, 2.13 cm of rain were recorded. A rise in temperature of 2°C at 0.5 m in the kame between the evening readings of 7 and 8 August was noted. Even at 1.3 m depth a slight increase in the temperature was recorded. Although the maximum increase in temperature was recorded at the highest thermistor on the evening of 8 August, the maximum was not noted at the lower thermistors until the evening of 9 August. By 13 August most of the

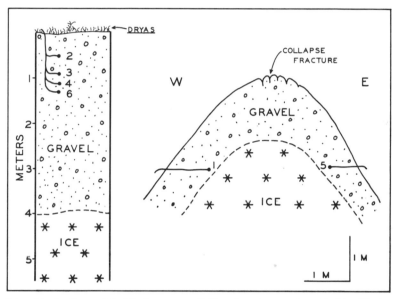

Fig. 6. Sketch of thermistor locations on top of the kame terrace (left) and on a kame on the north slope of the kame terrace (right)

thermistors indicated temperatures about the same as before the heavy rain. The temperature in the upper part of the kame continued to decrease and fell lower than it had been before the rain until it reached a minimum on 14 August. By then the gravel at the surface, without continued high input of solar energy on the overcast and foggy days that prevailed, could no longer act as an effective heat source for the percolating water. Thus the temperature of the upper layers continued to fall. The total amount of precipitation for the period from 21.00 h on 7 August to 07.00 h on 11 August was 8.25 cm. The greatest input of heat into the kame occurred on 8 August when the rain was the heaviest and the upper layers of the gravel warmest. The temperature impulse caused by the heavy rains attenuated as it traveled into the kame (Fig. 7). Temperature data following other heavy rains are not available.

The effectiveness of rain as an agent of ablation on bare glacier ice is not completely known (Marcus, 1964). Part of the increase in ablation of a glacier during rain is due to the kinetic energy produced by the rain drops; most of the energy comes from the rain that may be warmer than the surrounding air at the surface of the ice. Although many studies suggest that the effect may be negligible, Marangunic (unpublished) noted a two- to three-fold increase in ablation on the debris-free glacier surface during periods of heavy rain.

In the climate of Adams Inlet, solar radiation is ineffective alone in raising the temperature at depth in the soil. It is through a combination of a sunny period, in which the surface temperatures are raised, followed by heavy rains that transfer surface heat to the lower part of the kame, that the temperature at depth is raised. Near the surface, daily fluctuations in temperature reflect the changes in solar radiation. This is particularly evident in the case of the thermistors buried at 10 and 12 cm in the unvegetated part of the kame (Fig. 7).

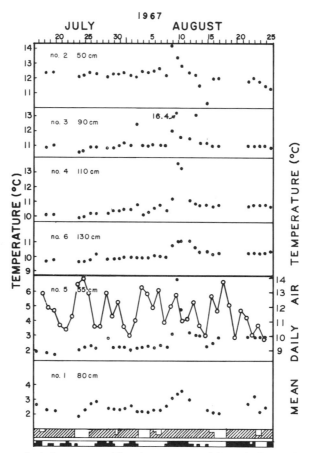

Fig. 7. *Daily soil temperatures in the kame terrace. Thermistors 2, 3, 4 and 6 were in the top of the terrace (Fig. 6) with No. 6 thermistor 2.7 m above the ice. Thermistors 1 and 5 were on the west and east sides, respectively, of a kame on the unvegetated slope. Cloudiness given in the upper set of shaded blocks: solid is overcast, 2/3 shaded is mostly cloudy, 1/3 shaded is partly cloudy and unshaded is clear. Precipitation code: total shaded is > 0.2 cm, 1/2 shaded is < 0.2 cm.*

Hourly readings were taken for 29 h on all thermistors on 30 and 31 July (Fig. 8). Both of these days were overcast with only 1 h of sun on 30 July. 31 July was foggy with drizzle and a total of 0.03 cm of precipitation was recorded. Over this period most of the thermistors, including those at shallow depth, showed only 0.2–0.7 deg change. This is to be expected because of the minor change in solar radiation and little precipitation during this period. Two thermistors showed increased temperatures over a 5 h period between 07.00 and 12.00 h. At 90 cm a rise of 0.7 deg was recorded and at 110 cm a rise of 1.4 deg was noted. These changes do not appear to be related to fluctuations in radiation or precipitation and were not recorded in the levels below or above. These temperature changes might be due to phase changes. Water vapor might be condensing in this layer and thus supplying heat to this part of the kame. Horizontal mass transfer in this layer could be the source of the moisture.

Over the 39 d observation period in 1967 the temperature at 1.3 m in the kame terrace rose 1.4 deg. This increase in temperature is part of an annual sinusoidal heat wave that penetrates into the kame. It reaches a maximum in late summer and then decreases over the winter; the depth of frost in this area is unknown. Other minor fluctuations in the 39 d record may also be due to percolating precipitation; however, the effectiveness of slight rainfalls seems to be minor.

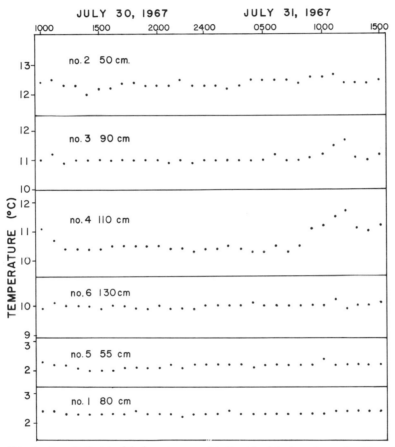

Fig. 8. Hourly soil temperatures in the top of the kame terrace and in the side of the kame terrace (see Figure 6 for location of thermistors).

Rate of collapse

The rate of melting of the stagnant ice determines the rate of collapse of the kame terrace. An estimate of the rate of transfer of heat to the ice by conduction can be obtained by the temperature-gradient method (Staley and Gerhardt, 1958). Heat may also be transferred by percolation of rain water, change of state, radiation and convection within the gravel. No estimate of the amount of heat transferred to the ice by these methods is included here. As seen by the increase in temperature during rainy periods, the percolation factor may be very important in this climate. Basal melting by terrestrial heat flow also reduces the volume of the buried ice. For this area the supply of heat is about 6.49×10^{-6} J cm^{-2} s^{-1} (Lee and MacDonald, 1963). This would be sufficient to melt 0.61 cm year^{-1}.

The heat flow by conduction to the ice can be determined by the formula

$$S = \lambda \frac{\mathrm{d}T}{\mathrm{d}\mathcal{Z}}$$

where S is the soil heat-flux, λ is the thermal conductivity, and $\mathrm{d}T/\mathrm{d}\mathcal{Z}$ is the temperature gradient with depth. For the vegetated area on top of the kame terrace (Figs. 2 and 6), the lowest thermistor (No. 6) is 2.7 m above the ice. During the period of observation the temperature increased from 9.7°C to 10.4°C not including the pulses caused by percolation (Fig. 7). Taking 10.0°C as an average during this interval, the writer determined the temperature gradient, from a depth of 1.3 m to the surface of the ice, to be 0.037 deg cm^{-1}.

An estimate of the thermal conductivity of the gravel has been obtained by comparison of moisture contents and grain-size distributions for two samples with values for similar material given in the literature. The grain-size distribution and conductivities for several unconsolidated materials are given in Table I along with the characteristics of the two samples (67–23–1, 67–23–2) from the unvegetated part of the kame terrace.

TABLE I. GRAIN-SIZE ANALYSES AND SOME THERMAL CONDUCTIVITIES FOR SELECTED UNCONSOLIDATED MATERIALS

Sample	Grain-size analysis weight %			Moisture weight %	Conductivity J cm^{-1} s^{-1} deg^{-1}
	>2.0 mm	2.0 to 0.05 mm	<0.05 mm		
Gravel*	60.0	32.5	7.5	5.0	1.25×10^{-2}
Chena gravel†	80.0	19.4	0.6	3.0	1.57×10^{-2}‡
Fairbanks sand†	27.5	70.0	2.5	5.0	1.48×10^{-2}‡
67–23–1	34.0	62.0	4.0	5.4	—
67–23–2	34.8	60.0	5.2	5.2	—

* Wechsler (1966).
† Kersten (1966, unpublished).
‡ Average for several densities.

According to Kersten (1966), the conductivity of a soil varies with the grain-size, moisture content, density, composition, temperature and the state of the moisture (solid or liquid). Texture and moisture content are probably the most important, and from a comparison of the values in Table I an estimated conductivity of 1.5×10^{-2} J cm^{-1} s^{-1} deg^{-1} for the gravel in the kame terrace is reasonable.

From the gradient and assumed conductivity value for this gravel the supply of heat to the surface of the stagnant ice block during July and August averages about 5.5×10^{-4} J cm^{-2} s^{-1}. Assuming that 0.90 g of ice occupy 1 cm^3 and that the heat of fusion of water is 335 J g^{-1}, then during these 2 months approximately 9.6 cm of ice would melt from the upper surface of the stagnant ice block. This rate of melting probably decreases during the spring and fall, and melting ceases during the winter. If we assume that this rate lasts for 5 months of the year where the debris cover is about 4 m thick, the rate of surface lowering of the ice due to ablation by conduction would be about 24 cm year^{-1}.

The rate of melting of ice beneath the unvegetated part of the kame, where the gravel cover was only 0.5–0.8 m thick, averages about 0.8 cm d^{-1} during July and August, and occasionally it may be as high as 1.0 cm d^{-1} during rainy periods. Thus the back-wasting of the thinly covered (at times uncovered, as in Figure 3) part of the kame terrace is faster than the down-wasting of the upper vegetated and more thickly covered area.

If it is assumed that the average annual ablation of the ice is about 24 cm, and future changes in the vegetated surface that might affect the ablation rate are neglected, then the

35 m thick ice block would melt in 146 years. The actual length of time that the ice remains will in fact be much less because of factors not included in the estimation. Back-wasting of the kame terrace sides, where the ablation rate is much higher, will also contribute to the much earlier removal of the ice. This ice will melt more quickly than those buried ice blocks that are completely surrounded by unconsolidated material.

SUMMARY

A collapsing kame terrace, consisting of 4 m of sandy, coarse cobbly gravel with a vegetated surface of *Dryas* and some alder and poplar, was studied to determine the mode and rate of collapse. The hummocky terrace with depressions as deep as 5 m is underlain by ice estimated to be 35 m thick.

Re-arrangement of the debris cover on the ice is seen to take place by: slumping and sliding of debris to low areas, melt-water erosion with deposition in alluvial fans on the sides of the terrace, debris flows in gullies, and the stagnant-ice burst phenomenon in which water under pressure breaks through an ice dam causing much erosion of the surrounding gravel.

Temperatures (at several depths) in the gravel on top of the terrace indicate that the flow of heat by conduction to the surface of the buried ice may be as much as 5.5×10^{-4} J cm^{-2} s^{-1} during the months of July and August. This rate of heat flow is sufficient to melt 9.6 cm of ice from the buried ice block during this period. Assuming this rate for 5 months of the year (although the heat flow is lower at the beginning and end of the summer, the ablation season may be longer than 5 months at lower rates), the lowering of the ice surface may be as much as 24 cm year^{-1}. As expected, this is much lower than the ablation rate for the debris-free stagnant ice of Adams Inlet Glacier that ablated at the average rate of 7.9 m year^{-1} between 1890 and 1941. The estimated length of time for survival of ice in this debris-covered kame terrace, that is thinly debris-covered and at times uncovered, is less than other ice blocks that are completely surrounded by unconsolidated material.

ACKNOWLEDGEMENTS

Support for this research was supplied by a National Science Foundation grant (GA-403) awarded to The Ohio State University Research Foundation and administered by the Institute of Polar Studies. Logistic support was provided by the National Park Service. W. C. Lefler, T. R. Merrell III and S. Goldthwait assisted in the field investigations. G. Dewart, M. Hoyer, J. Lindsay, J. Mercer and J. Splettstoesser read the manuscript and made valuable suggestions on its improvement.

MS. received 12 February 1969

REFERENCES

Charlesworth, J. K. 1957. *The Quaternary era, with special reference to its glaciation.* London, Edward Arnold Ltd. 2 vols.
Clayton, L. 1964. Karst topography on stagnant glaciers. *Journal of Glaciology,* Vol. 5, No. 37, p. 107–12.
Embleton, C., *and* King, C. A. M. 1968. *Glacial and periglacial geomorphology.* London, Edward Arnold Ltd.
Field, W. O., *jr.* 1947. Glacier recession in Muir Inlet, Glacier Bay, Alaska. *Geographical Review,* Vol. 37, No. 3, p. 369–99.
Flint, R. F. [1957.] *Glacial and Pleistocene geology.* New York, John Wiley and Sons, Inc.
Goldthwait, R. P., *and others.* 1966. Soil development and ecological succession in a deglaciated area of Muir Inlet, southeast Alaska, by R. P. Goldthwait, F. Loewe, F. C. Ugolini, H. F. Decker, D. M. DeLong, M. B. Trautman, E. E. Good, T. R. Merrell III, and E. D. Rudolph. *Ohio State University. Institute of Polar Studies. Report* No. 20.
Kersten, M. S. 1966. Thermal properties of frozen ground. (*In* U.S. National Research Council. Building Research Advisory Board. *Proceedings of an international conference on permafrost.* Washington, D.C., p. 301–05. ([U.S.] National Academy of Sciences—National Research Council Publication 1287.))

Kersten, M. S. Unpublished. Final report, laboratory research for the determination of the thermal properties of soils. [Written for U.S. Army Corps of Engineers, 1949.]

Lee, W. H. K., *and* MacDonald, G. J. F. 1963. The global variation of terrestrial heat flow. *Journal of Geophysical Research*, Vol. 68, No. 24, p. 6481–92.

Lister, H. 1953. Report on glaciology at Breiðamerkurjökull 1951. *Jökull*, [Vol. 1], Ár 3, p. 23–31.

Marangunic, C. Unpublished. Effects of a landslide on Sherman Glacier, Alaska. [Ph.D. thesis, Ohio State University, 1968.]

Marcus, M. G. 1964. Climate–glacier studies in the Juneau Ice Field region, Alaska. *University of Chicago. Dept. of Geography. Research Paper* No. 88.

Reid, H. F. 1896. Glacier Bay and its glaciers. *U.S. Geological Survey.* 16th *Annual Report*, [for the year] 1894–95, Pt. 1, p. 415–61.

Sharp, R. P. 1947. The Wolf Creek glaciers, St. Elias Range, Yukon Territory. *Geographical Review*, Vol. 37, No. 1, p. 26–52.

Sharp, R. P. 1949. Studies of superglacial debris on valley glaciers. *American Journal of Science*, Vol. 247, No. 5 p. 289–315.

Staley, R. C., *and* Gerhardt, J. R. 1958. Soil heat flux measurements. (*In* Lettau, H. H., *and* Davidson, B., *ed. Exploring the atmosphere's first mile*. New York, Pergamon Press, p. 58–63.)

Swithinbank, C. W. M. 1950. The origin of dirt cones on glaciers. *Journal of Glaciology*, Vol. 1, No. 8, p. 461–65.

Wechsler, A. E. 1966. Development of thermal conductivity probes for soils and insulations. *U.S. Cold Regions Research and Engineering Laboratory. Technical Report* 182.

19

Kame Structure West of Zieleniew

STEFAN JEWTUCHOWICZ

*This article was translated expressly for
this Benchmark volume by Sandi and
Paul Mayewski, from "Struktura Kemu
W Zieleniewie," in* Folia Quaternaira,
30, pp. 59–68 (1969)

Within the ground moraine west of Kutno, in the village of Zieleniew, there is a small hill. It measures 600 m in length, 500 m in width, is approximately 10 m in relative height, and has a long axis which stretches NW–SE. The hill is composed of gravel, sand, silt, and clay. Measurements have shown that the dip directions of these layered sediments are widely dispersed and vary from several to as much as 90°. Another structural aspect of the described hill includes frequent vertical displacements. These displacements are composed of clumps of fluvioglacial debris and rock debris.

The fact that this hill is a kame is substantiated by its location on a ground moraine and by the well-sorted fluvioglacial material of which it is composed. The analysis of its internal structure testifies to the fact that during deglaciation numerous structural disturbances occurred here. Kame sediments with a disturbed structure are divided by sediments whose structure has been undisturbed. For this reason bed deformations cannot be correlated with the pressure of an active continental glacier. No traces of renewed activity of the receding continental glacier whose sediments cover the described surface are evident in the terrain adjacent to the kame. Thus, the structural deformations of the kame must be connected with a process occurring during dead ice conditions.

Keller's (1954) hypothesis referring to cores squeezed up from the glacier foundation is well known. The mechanism of this occurrence according to Keller is as follows. In an open crevasse of dead ice in which a kame is being formed, there exist variations in basal pressure. Great pressure exists under the ice and it cannot be equilibrated by the pressure of fluvioglacial deposits at the bottom of the crevasse. The greatest difference in pressure exists at the edge of the crevasse. Consequently, this is the place of up-squeezing of basal deposits, most often clay, which forms a nucleus for the feature.

A similar, although somewhat modified concept of the genesis of an up-squeezed core in eskers is accepted by H. Bramer (1961). He asserts that the eskers which have an up-squeezed clay core originated between two ice masses. The weight of the ice causes a squeezing out at the foundation. Marginal forms with an up-squeezed core were noted by W. Niewiarowski (1964), who observed that in the central part of some hill, beds are up-squeezed in the shape of a cupola. These up-squeezings occurred during a period of glacier stagnation.

In view of the above-mentioned hypotheses on the origin of fluvioglacial accumulations with a squeezed-up core, bed deformations in the Zieleniew kame call for a discussion concerning their classification into deposits caused by stagnating ice, up-squeezing, or as structures originating on the surface.

A bed of clay, sections of which are tilted almost perpendicularly, can be observed in the Zieleniew kame (Fig. 1). Measurements have shown that their orientation is

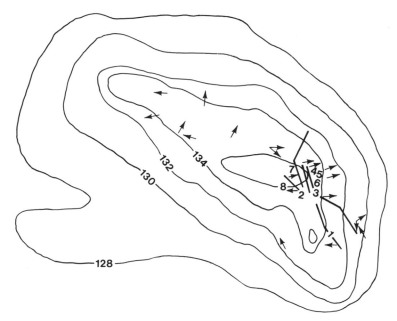

Figure 1. Morphology of the kame. Arrows show the dip directions of the kame deposits; thick lines, vertically situated fragments of boulder clay beds in the kame; numbers, the places where the photographs discussed in the text were taken.

always concurrent with the longer axis of the hill. Distances between the perpendicular sections of clay range from 5 to 25 m, and the space between them is filled with sand and gravel. These deformations do not occur everywhere and are not symmetrically located within the kame. A substantial curving of the beds can be observed commonly in the SE section of the feature (Fig. 1).

One of these structural disturbances is shown in Fig. 2. Here a layer of clay is curved up to a 60° dip. This clay is covered by segmented stratified sands. In the eastern section of this exposure, masses labeled b, c, and d are displaced in relation to each other and to their laminae and are offset by faults in which the movement was downward. Thus, individual masses of sand were settling. As a result of the settling of the eastern section of the structure, fracturing of the sand series occurred in the central part and a fissure appeared along which intensive vertical displacement of material took place. The western part of the deposit (Fig. 2) has a structure similar to that of the eastern. The offset of the beds as well as the traces of the faults testify that sediment displacement also took place in this section and caused fracturing and settling of the sand series. The following relationships support the fact the the displacement was downward: (1) the shape of the terminal portions of the sand layers in contact with the clay, (2) the shape of the crevasse walls, and (3) faults noted by the nonsynchroneity of thinly deposited layers. On the other hand, there is no evidence that points to the movement being upward.

If the clay squeezed up from the foundation into the fluvioglacial deposits, diaprism-type deformations should be noticeable. The structure of the kame (Fig. 2) shows no curvature of the layers at the clay and sand contact in response to upward pressure. Thus, the hypothesis that the occurrence of clay deposits amidst fluvioglacial deposits is due to clay squeezing up from the base appears not to hold true in this case. In view of these facts, one must conclude that the clay layers were deposited concurrently with other materials forming the kame. The position of the clay between the sands testifies to the changeable depositional environment during ice ablation. The clay layer represents a phase of low meltwater activity. Sand fracturing into slices, settling of these masses, and the great distortion of clay layers are all the result of a settling process probably caused by drying and shrinking of the deposits. An important role in this process was played by the melting of the stagnant ice on which the kame was built.

In many deformations one can observe a complete lack of evidence of an upward movement of clay which should have been marked at the contact of the clay and the fluvioglacial material. In the exposure presented in Fig. 3, layers of sand under the clay exhibit traces of water erosion and the sand above has eroded the clay surface. These deposits have also been subject to settling. Owing to the great cohesiveness of the clay, this settling process caused relatively few changes in the clay structure; fluvioglacial features under uneven settling are most often subject to numerous crackings and slumpings.

In the southeast section of the kame, gravel and sand have been exploited, but the clay remains intact and appears as if sculpted from the other deposits. The bottom of the gravel pit is also composed of clay whose surface is slightly inclined and was settled in various directions. Sections of clay that are curved upward at a steep angle crop out above the bottom of the gravel pit to a height of 4 m. Figure 4 presents a cross section of one of these steep features.

Most often, these sections are composed of not only clay, but also a curved series of various other deposits. For example, Fig. 4 shows a 1.5-m-thick layer of clay overlying a 0.5-m layer of silt which overlies sand masses. These lumps have various orientations, with some being stratified in the same direction as the clay and silt surface and others being stratified vertically. The clay layer is also stratified and consists of thin silt layers which, owing to drying processes, lessen the clay's cohesiveness. Consequently, the clay splits into plates. In places, instead of silt layers, one can often find thin layers of gravel in the clay (Fig. 5).

According to G. Keller (1954) and H. Bramer (1961), the up-squeezing process is the cause of clay splitting in the kame and esker cores. During the up-squeezing movement clay lumps are believed to form within the adjacent fluvioglacial material.

Figure 2. Zieleniew. Structure of the kame. a, boulder clay; b, c, d, e, f, g, lumps of sand.

Figure 3. Zieleniew. The erosive character of the contact phane between the boulder clay and the sand series in the kame. a, boulder clay; b, sand.

Figure 4. Zieleniew. Transversal section through a vertical layer of kame deposits. a, boulder clay; b, silt; c, d, lumps of sand.

Figure 5. Zieleniew. Bedding of the clay in the kame. a, beds of clay; b, beds of gravel.

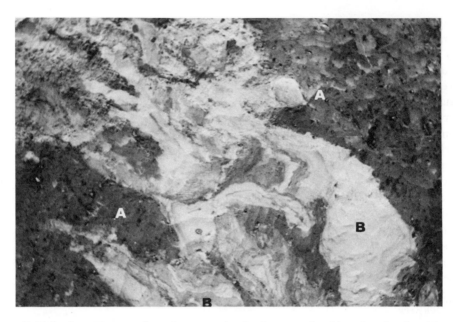

Figure 7. Zieleniew. Traces of mud streams in the kame. a, boulder clay; b, silt with sand.

These lumps are then subject to slumping and overthrusting. In light of the hypotheses of these authors, the question of clay flaking in the Zieleniew kame calls for a discussion. Stiff clay would not be subject to up-squeezing, therefore, one may conclude that the clay was in a plastic state. Had the fluvioglacial material also been plastic, structures other than blocks should have formed. Therefore, it has to be accepted that the fluvioglacial material was stiff and cracked under pressure. Consequently, a question arises as to whether a 0.3- to 1.5-m-thick layer of clay could have pushed its way all the way to the top, through fluvioglacial material several meters thick, causing it to split into masses. Also, is it possible for plastic clay squeezing up through a resistant fluvioglacial material to preserve its stratified structure as seen in Fig. 5? The above-mentioned facts do not present a satisfactory answer to the questions raised and force one to seek an interpretation other than up-squeezing through stagnant ice to interpret the cause of structural deformation in the Zieleniew kame (Fig. 4).

The position and structure of this kame allows for the hypothesis that it was formed in an open crevasse on the ice surface. As a result of ablation, the upper part of the crevasse was enlarged. This, in turn, lessened the steepness of its walls. Debris newly melted out was deposited not only on the bottom of the crevasse but also on its sloping walls. As a result, part of the kame was built on ice. Melting of the ice caused morphological changes on the surface of the forming feature. The uneven melting rate of the stagnant ice was dependent on the thickness of the deposits covering it and on the meltwater activity. This uneven melting influenced surface settling, which in turn caused slumping downward and sideward of previously deposited material. Places of settling were also locations for small debris deposits during periods of slow ablation and low-energy meltwater activity.

The ice melting determined the superposition of various geological events. Here one can single out the material-slump phase marked by sand blocks which were later covered by clay. Short periods of increased water activity occurred during clay accumulation. Within these periods coarser material was deposited, causing stratification. Water increased the plasticity of deposits. The pressure of large sand and gravel accumulations over a small water-saturated surface caused the deposit to curve without disturbing its stratification. The hypothesis that the clay beds were so steeply curved by pressure is based on the fact that the lower portion, at the bottom of the gravel pit, spreads almost horizontally over a large surface, and only certain sections are folded. The fact that the vertical position of clay beds is a result of settling and pressure by deposits during accumulation is further substantiated by the feature's stratification presented in Fig. 1. G. Keller's (1954) hypothesis of parallel dragging of the cores of kames forming behind a retreating edge of an ice crevasse cannot be applied to the Zieleniew kame. As previously noted, the crevasse edges in which the described kame formed did not retreat in parallel fashion but as a result of ice melting where lowered from the surface. Thus, conditions for uniform crevasse retreat and for up-squeezing processes were not present in Zieleniew.

In the southeast section of the hill described, one can observe a collection of sand lumps presently near the surface (Fig. 6).* The stratification of the lumps points to the fact that they have various orientations, some being vertical and others slanted. The pressure of newly slumping sections and loose sand and gravel caused folding of beds and faults and movement of already slumping sections. In another part of the kame one can note traces of debris which had flowed into the depressions of a former surface (Fig. 7). Melting of stagnant ice caused vertical slumping (Fig. 8). As a result of settling, large sections of the hill were also subject to slumping (Fig. 9).

The structural examples noted in Zieleniew testify to the great influence of glacial ablation on the internal structure of the kame. Frequent changes in the intensity of melting and the resulting changing rate of meltwater activity caused differential movement and consequent fracturing within these deposits. One can find stagnation deposits, mud flows, and slump deposits adjacent to well-sorted materials. Melting of ice, which was covered by rock debris, caused several structural deformations among which are numerous vertical displacements, large-scale tilting, stretching, cracking of beds, and pressure deformations. As a result of these disturbances, the morphology of the kame in certain sections is not in accord with its structure.

It is also worth noting that the structural deformations in the Zieleniew kame were caused by surface processes having nothing in common with squeezing up at the base of a stagnant glacier. Therefore, this kame cannot be classified with accumulation features derived by the up-squeezed core hypothesis.

*The original print for Fig. 6 is unobtainable; the text accurately describes this section of the terrain, however.

Figure 8. Zieleniew. Traces of settling caused by melting of stagnant ice in the kame.

Figure 9. Slumping of a kame fragment caused by melting away of stagnant ice.

Summary

The kame at Zieleniew is composed of sand, gravel, and silt deposits. Boulder clay is less abundant. The dip directions of layers are widely dispersed (Fig. 1), the dip values varying from several degrees to as much as 90°. In the kame two to four nearly vertical boulder clay beds can be seen (Fig. 1). The strikes of these beds do not always coincide with the morpholoical long axis of the hill. The distance between the boulder clay beds varies from 5 to 25 m, these beds being separated by sands and gravels. Other features of the structure of this kame are: numerous vertical displacements of beds, aggregations of lumps of the fluvioglacial material, and traces of downflow in the material.

The kame deposits which reveal a disturbed structure are separated by unperturbed ones. That is why the disturbance cannot be connected with pressure exerted by an active glacier. Keller's hypothesis referring the up-squeezing of the kame cores to the dead ice pressure does not explain the complicated structure of the form discussed. The structural evidence presented in Figs. 1–9 shows that the material was moving down at the time of the displacement of the fluvioglacial lumps. There is no trace which might indicate an upward movement of the boulder clay layers and of the sand and gravel lumps. The presented features of the Zieleniew kame show that the cause of disturbance was the melting away of dead ice covered by kame deposits. As a result of melting, the settling of the material took place. An irregular accumulation of great masses of gravel and sand over a small surface, hence unequally distributed pressures, made the water-saturated layer curve over.

References

Bramer, H. 1961. Bemerkungen zum Problem der Aufpressungs-Oser. Geologie, Berlin, 4/5.
Keller, G. 1954. Drucktexturen in Eiszeitlichen Sedimenten. Eiszeitalter u. Gegenwart, Öhringen, 4/5.
Niewiarowski, W. 1964. Formy marginalne z wycisnietym jadrem z okolicy Jaworza kolo, Wabrzezna. Zeszyty Nauk. UMK w Toruniu, Nauki mat.-pryr., Torun, 10.

V
Disintegration Deposits and Outwash

Editor's Comments on Papers 20 Through 22

The use of the term "disintegration deposits" is a recent addition to the terminology of glacial geology. In many senses they have always been recognized in the form of kame fields and pitted or collapsed outwashes (Thwaites, 1926), but those who published most (Clayton, 1967; Parizek, 1969) have shown that forms, origins, and complications such as doughnut shapes and high lacustrine platforms involve far more than simple kame deposition. Furthermore, many of these features are composed of ablation till. All around the glaciated world, examples of ice stagnation and disintegration fell into place in the 1957–1967 decade (Hoppe, 1957; Holmssen, 1963; Niewiarowski, 1963; Stalker, 1960; Winters, 1961). Because they were first fully recognized in print by Gravenor and Kupsch in 1959, their work is reproduced (Paper 20). The impetus came from deposits in the Missouri Coteau region of North Dakota as well as Alberta. Living glacier models were sought in Alaska by Clayton and others.

The background for development of "disintegration" ideas probably came from the 1929 to 1944 arguments about the nature of ice-sheet retreat. "Normal retreat," as it was called, and believed by most workers until 1930, meant active, moving ice right to the glacier's edge, emplacing neat recessional moraines whenever net ablation equaled annual motion, and generating kames only in a narrow marginal belt over hilly topography. Flint (1929) drove home the "stagnation" concept, and others soon agreed. Hills and barrier mountains shut off ice flow early in the thinning of an ice sheet, creating a long period of downwastage. This turning point in interpretation is not included here. As always, a few (Lougee, 1940) held out for the old "normal" concept as the only explanation. In the postwar era, with recognition of glacial surges even in continental glaciers, the ice sheet moved out over previously dead, wasting, and dirt-covered ice. This easily explains collapsed outwashes and high lake-bed plains in areas of disintegration like Missouri Coteau.

As with eskers, outwash plains or sandur (Scandinavian) have been described by dozens of authors over more than a century (Jewtuchowicz, 1953). Other than for the basic deduction that a large glacial stream came directly off former ice onto the apex of each outwash fan and spread in shifting unrestricted channels, little attention was paid to origin. For example, the outwash aprons forming southern Cape Cod were well displayed in Woodworth and Wigglesworth's extensive volume and maps (1934) but not until 1939–1940 did a U.S. Geological Survey party under Mather deduce some details of their origin (Paper 21). Chronological terrace studies were done all over the world (e.g., New Zealand by Gage, 1958, and Suggate, 1963).

Until 1950 diagnostic studies of the slow-developing outwashes in front of actual glaciers were neglected. Early investigation began in Iceland (Hjulström, 1954;

Krigstrom, 1962). My studies in the Glacier Bay, Alaska, region (Goldthwait, 1963) produced the first log-dated record of the rate of outwash accumulation. Price's (1971) studies south of Vatnajökull and other Iceland glaciers provided the first measured sequence of changing channels and buried ice masses. Hjulström's hydrologic deductions (1954) was followed by Fahnestock's detailed hydrologic study of a mountain glacier valley train on Mt. Rainier (Paper 22). As with other long professional papers, the following omissions were made here with the author's knowledge, to get it into this volume: sections and illustrations on antidunes, tables supporting some graphs, and duplicating photographic examples.

References Cited

Clayton, L. (1967) Stagnant ice features of the Missouri Coteau in North Dakota: North Dakota Geol. Survey Misc. Ser., v.30, p. 25–46.

Flint, R. F. (1929) The stagnation and dissipation of the last ice sheet: Geograph. Rev., v. 19, p. 256–289.

Gage, M. (1958) Late Pleistocene glaciation of the Waimakariri Valley, Canterbury, New Zealand: N.Z. Jour. Geol. Geophys., v. 1, p. 123–155.

Goldthwait, R. P. (1963) Dating the Little Ice Age in Glacier Bay, Alaska: 21st Internat. Geol. Congr., Norden, Report, pt. 27, p. 37–46.

Hjulström, F. (1954) The Hoffellsandur: Geografiska Annaler, v. 36, p. 135–145.

Holmssen, G. (1963) Glacial deposits of southeastern Norway: Amer. Jour. Sci., v. 261, p. 880–889.

Hoppe, G. (1957) Problems of glacial morphology and the Ice Age: Geografiska Annaler, v. 39, p. 1–18.

Jewtuchowicz, S. (1953) La structure de sandre: Lødz Soc. des Sciences et des Lettres, classe 3, v. 4, no. 4, 23 p.

Krigstrom, A. (1962) Geomorphological studies of sandar plains and their braided rivers in Iceland: Geografiska Annaler, v. 44, p. 328–346.

Lougee, R. J. (1940) Deglaciation of New England: Jour. Geomorphol., v. 3, p. 189–217.

Nicwiarowski, W. (1963) Some problems concerning deglaciation by stagnation and wastage of large portions of the ice-sheet within the area of the last glaciation in Poland: 6th Conf. Internat. Assoc. Quaternary Res., Warsaw, 1961, Rept. 3, p. 245–256.

Parizek, R. R. (1969) Glacial ice-contact rings and ridges: Geol. Soc. America, Spec. Paper 123, p. 49–102.

Price, R. J. (1971) The development and destruction of a sandur, Breidamerkurjökull, Iceland: Arctic Alpine Jour., v. 3, no. 3, p. 225–237.

Stalker, A. M. (1960) Ice-pressed drift forms and associated deposits in Alberta: Canada Geol. Survey, v. 57, 38 p.

Suggate, R. P. (1963) The fan surfaces of the central Canterbury Plains: N. Z. Jour. Geol. Geophys., v. 6, p. 281–287.

Thwaites, F. T. (1926) The origin and significance of pitted outwash: Jour. Geol., v. 34, p. 308–319.

Winters, H. A. (1961) Landforms associated with stagnant ice: Professional Geographer, v. 13, p. 19–23.

Woodworth, J. B., and E. Wigglesworth (1934) Geography and geology of the region including Cape Cod, the Elizabeth Islands, Nantucket, Martha's Vineyard, No Man's Land, and Block Island: Harvard Coll. Mus. Comp. Zool. Mem., v. 52, 338 p.

20

Reprinted from *Jour. Geol.*, **67**, 48–64 (1959)

ICE-DISINTEGRATION FEATURES IN WESTERN CANADA[1]

C. P. GRAVENOR[2] AND W. O. KUPSCH[3]

Research Council of Alberta and University of Saskatchewan

ABSTRACT

The wasting Wisconsin glacier left predominantly till deposits in western Canada and only subordinate amounts of stratified drift. In the final phases of wasting, the ice separated in places into a large number of small, dead, ice blocks: it disintegrated. This disintegration caused the preservation of many different land forms, some of which were initiated during the time of ice flow, others originated after flow ceased. Those features that show the influence of the previous live ice are said to be "controlled." Such control may be exerted by crevasses and thrust planes which are the response to stresses operative in a living glacier. Uncontrolled deposits do not reveal the influence of former flow. All gradations between controlled and uncontrolled disintegration can be observed.

The depositional disintegration features include hummocks, moraine plateaus, round and irregularly shaped closed ridges, linear ridges, and washboard moraines. Ice-walled channels are an erosional form. The hummocky terrain and the closed ridges are regarded as the dominant product of uncontrolled disintegration. The linear and washboard ridges developed along inherited lines of weakness in the disintegrating ice and are regarded as controlled disintegration features.

Both uncontrolled and controlled deposits resulted from the sloughing of ablation material into cracks and cavities in the ice and from the squeezing of till upward into openings at the base of the ice.

INTRODUCTION

Although glacial features associated with stagnant ice of late Wisconsin age have been described from other regions, the writers feel justified in calling attention to those of western Canada for the following reasons:

1. Gradations have been observed between several glacial features generally believed to be of different origin but which are all preserved because of final stagnation, ice wastage, and consequent disintegration.

2. Unlike eastern North America, where stagnant ice features are largely glaciofluvial in origin, those in western Canada are composed mainly of till.

3. They are well preserved as a result of the dry climate and consequent scant postglacial erosion. They are also well displayed both on the ground and on air photos because of the lack of trees and, furthermore, have been little altered by human activities.

In view of the fact that the terminology of till features associated with stagnant ice is at present not clearly defined or universally accepted, the writers take this opportunity to propose a new descriptive classification of these features, supported by a dis-

cussion of their characteristics and possible origin.

DEFINITION OF ICE DISINTEGRATION

The term "wastage" applied to a glacier is usually considered as including melting, wind erosion, evaporation, and calving (Howell, 1957, p. 318). A wasting glacier may maintain a distinct receding terminus while flow continues actively, or it may, because of widespread thinning, stagnate and separate (Flint, 1957, p. 33). Along some parts of an ice margin, stagnation extended for many miles back from the front, whereas elsewhere along the same margin the ice was actively flowing and deposited or pushed up end moraines. In still other places the ice was receding at a high rate, thus leaving no separate blocks of ice. As a result, only a sheet of ground moraine without any patterned relief remained.

Although the writers use the term "separation," they prefer *ice disintegration* to describe the process of breaking up into numerous small blocks, which finally comes about in a stagnant and thus wasting glacier. It best describes the falling-to-pieces of the ice as a result of physical and chemical action and thus conforms to the definition of the geological meaning of "disintegration" as given in Webster's *Dictionary* (Neilson, 1956): "the falling to pieces of

[1] Manuscript received February 21, 1958.

[2] Geology Division, Research Council of Alberta, Edmonton, Alberta.

[3] Department of Geology, University of Saskatchewan, Saskatoon, Saskatchewan.

rocks as a result of chemical or physical action, weathering, frost, etc."

Disintegration may give rise to a variety of land forms, depending upon a great number of factors, such as the amount of debris carried by the ice, the position of the debris on, in, or under the ice, the amount of meltwater, and the resultant erosion and deposition. They all have in common, however, that they are preserved because they represent the last phase of glacial deposition and were not destroyed by later glacier advances.

When the forces that operate to break up an ice sheet are equal in all directions, the disintegration may be said to be *uncontrolled*, and the result is a field of round, oval, rudely hexagonal or polygonal features, and a general lack of dominant linear elements. Where the ice separated along fractures or other lines of weakness, the disintegration may be said to be *controlled*, and the result is a field of linear or lobate land forms. In places the ice broke along open crevasses or along thrust planes, both of which formed when the ice was still flowing, and the disintegration thus shows *inherited flow control* (pl. 3). The linear elements then bear a direct and understandable relationship to the preceding flow directions and are usually parallel, perpendicular, or at 45° to the direction of flow. In some areas, however, the fractures may have resulted from causes other than earlier ice flow, such as the settling of ice into bedrock joints. Gradations between controlled and uncontrolled disintegration features are to be expected and can be found in many places.

In places disintegration features are *superimposed* on previously formed live-ice features, such as drumlins and flutings parallel to the direction of ice movement or end moraines perpendicular to it.

Relief of ice-disintegration deposits may vary from pronounced to hardly noticeable in the field. In the latter case it may be evident only from the study of air photographs. *High-relief deposits* have local relief greater than 25 feet; *intermediate-relief deposits* from 10 to 25 feet; *low-relief deposits* less than 10 feet.

Ice-contact stratified drift deposits are generally believed to be associated with the thinning of an ice sheet. Deposits of this type in western Canada are subordinate to till features that resulted from stagnation. Because they have been described as stagnation features from elsewhere and because they can be independent of disintegration, they will not be considered here.

RECOGNITION OF ICE-DISINTEGRATION FEATURES

The distinction between end moraines and ground moraines on some of the older glacial maps of western Canada (Johnston *et al.*, 1948) is based largely on relief. Areas of drift having a local relief of more than 15 feet and consisting predominantly of till were regarded as end moraines. Areas of low relief were mapped as ground moraine. The following characteristics have now been found:

1. Some uplands mapped as end moraines have only a very thin till cover, and both their relief and their trend are due primarily to topographically high underlying bedrock.

2. Till ridges representing true end moraines cut across the trend of the high-relief uplands formerly mapped as end moraines with a different trend.

3. End moraines of low relief occur in areas formerly mapped as ground moraine.

4. Some areas mapped as end moraine appear as "blobs" on maps and show no trend, either as a whole or in detail.

These findings necessitate criteria for the recognition of areas of predominantly stagnant-ice deposition. The distinction between end moraine and ground moraine should be based on observable *trends*, in addition to *relief*. The writers regard a true end moraine as a ridge or narrow elongated area consisting of glacial material produced at the front of actively flowing ice or as a ridge of pre-existing material structurally disturbed by ice-push, with or without a covering of glacial material. This definition attempts to embody the concepts generally accepted by European geologists (Woldstedt, 1954, p. 101–110).

Some of the high-relief areas formerly regarded as end moraines and believed to have been deposited by live ice are interpreted by the writers as the result of ice disintegration; others appear to be complexes of end-moraine ridges. In some places low-relief ground moraine may show the effects of wasting of unoriented dead-ice blocks. In others the effects of the living ice are still visible, and the position of the former ice front can be reconstructed from the controlled disintegration pattern.

UNCONTROLLED DISINTEGRATION FEATURES

HUMMOCKY DISINTEGRATION MORAINE

Description.—Broad tracts of rough morainal topography are common features over much of the western plains. Many of these areas are irregular in outline and show no pronounced elongation. They consist of a nondescript jumble of knolls and mounds of glacial debris separated by irregular depressions. The knolls do not align into ridges, and no dominant trends are discernible. These areas have the characteristic "knob-and-kettle" topography. They are tracts of high local relief and were mapped as end moraines by early workers, even though clearly defined ridges may be absent and the area itself lacks any definite elongation. Thus the Moose Mountain upland in southeastern Saskatchewan was regarded as a northwest-trending end moraine (Johnston and Wickenden, 1931, p. 40). Recently, however, Christiansen (1956, p. 31) found that this area is devoid of any trends and shows no visible traces of live-ice deposition. It is therefore considered to be deposited by stagnant ice. Washboard moraines, indicating the former position of the ice front, are found encircling the Moose Mountain upland but not on it.

The high-relief disintegration moraine may cover a preglacial upland, a condition which appears to hold true for the Max Moraine of the Missouri Coteau (Townsend and Jenke, 1951, p. 842), for the Turtle Mountain area in Manitoba, and for the Moose Mountain area in Saskatchewan. Dead-ice moraine of pronounced relief may also occur in areas that were topographically low in preglacial times. Limited drilling carried out on the Viking moraine (which the writers believe is a disintegration moraine and not a true end moraine) of east-central Alberta suggests that the depth to bedrock is greater there than in the surrounding ground moraine.

A careful analysis of a hummocky moraine generally reveals the presence of several distinct elements besides the dominant knobs and kettles. Among these are (1) moraine plateaus, (2) closed and linear ridges, (3) ice-walled channels (pls. 1 and 6). The last two features also appear outside the high-relief areas and will therefore be treated separately.

Moraine plateaus are relatively flat areas in the hummocky moraine (pl. 1, *A*). Their level surface is generally at the same elevation or slightly higher than the summits of the surrounding knobs. In east-central Al-

---

PLATE 1

High-relief disintegration moraine. (Location of pl. 1 shown on fig. 4.) Location: T. 46, R. 12, W. 4th mer., Alberta. (Air photograph reproduced by permission of government of Alberta, Department of Lands and Forests.)

A, Moraine plateau.

B, Closed till disintegration ridge on moraine plateau (rim ridge).

C, Linear till disintegration ridge, sinuous.

D, Linear ridge, beaded.

E, Moraine plateau which forms the headwater region of an ice-walled channel leading off to the south.

F, Closed till disintegration ridges, surrounding depressions (rimmed kettles).

PLATE 2

Closed disintegration ridges in low-relief till area. (Location of pl. 2 shown on fig. 4.) Location: Sec. 33, T. 3, R. 3, W. 2d mer., Saskatchewan. (Air photo reproduced by permission of Royal Canadian Air Force.)

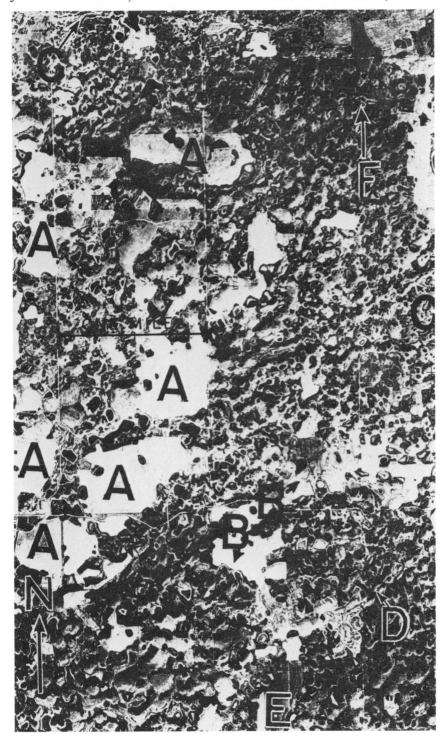

High relief disintegration moraine.

Closed disintegration ridges in low relief till area.

Linear disintegration ridges.

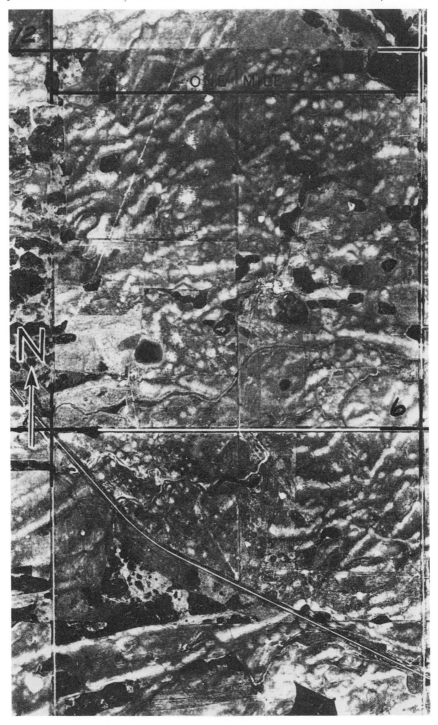

Linear disintegration ridges.

berta drilling operations show that the moraine plateaus are composed of clayey till. In some places there is a thin cover—2–10 feet thick—of lacustrine silts and clays at the surface overlying the till. The moraine plateau are roughly circular, elongate, or irregular in plan. They may be pitted with kettles, which in rare instances have poorly developed rims. The edges of the moraine plateaus may be defined by minor till ridges, referred to as "rim ridges," but in the plateaus of east-central Alberta such rim ridges are the exception rather than the rule. In some parts of the Missouri Coteau in Saskatchewan, on the other hand, where the plateaus are distinctly "pock marked" with keetles, rim ridges are common. In places meltwater channels may lead away from the plateaus (pl. 1, E). Similar situations are also recorded from till moraine plateaus in Sweden (Hoppe, 1952, p. 9).

Moraine plateaus composed chiefly of stratified silt and clay, of probable lacustrine origin, have been noted on Moose Mountain, Saskatchewan (Christiansen, 1956, p. 11). They appear to be similar to the plateau hills described from Denmark by Schou (1949, p. 10).

The knobs of hummocky dead-ice moraine in western Canada consist mainly of till, which in places is quite compact. Sand and gravel knobs, however, are common and generally show the characteristic "collapse" structures of ice-contact deposits. Some excavations revealed knobs composed of crudely stratified drift overlain by a layer of loose "washed" drift. Some of the knobs formerly believed to be composed of till may

therefore have sand and gravel in their cores. In others, layers of sorted sand and gravel cover till cores. The knobs may occur as isolated mounds, in clusters, or connected by ridges.

Circular undrained depressions are present at the tops of many mounds. They also occur between the knobs, but irregular hollows are more common in that position.

Distinctive features.—It is likely that the hummocky, knob-and-kettle topography has various origins (Woldstedt, 1954, p. 96). An end-moraine complex of hummocky topography deposited by the live-ice front may appear very similar to a hummocky moraine deposited by dead ice. Inasmuch as blocks of ice may be left in the end moraine on retreat, some dead-ice forms can occur in the end moraine. Therefore, it becomes necessary to distinguish between a hummocky moraine which still shows the former position of the ice front from a hummocky moraine which does not. Detailed field work supported by study of air photographs and large-scale topographic maps generally provides the answer. If no trends, such as aligned knobs or hollows, are apparent from the air photographs and if the moraine itself is not elongated and narrow but round and broad, the moraine is regarded as deposited by stagnant ice. The hummocky disintegration moraine generally has a rounded outline, its borders are indistinct, and they may grade almost imperceptibly into the surrounding low-relief ground moraine (pl. 5).

In hummocky moraines deposited as true end moraines, the knobs are aligned as more or less clearly defined ridges, which are emphasized by intervening aligned depressions.

PLATE 3

Linear disintegration ridges. (Location of pl. 3 shown on fig. 4.) Location: T. 48, R. 1, W. 4th mer., Alberta. (Air photograph reproduced by permission of government of Alberta, Department of Lands and Forests.) The ice movement in this area was slightly east of south, which is at right angles to the curvilinear or lobate ridges. On the eastern side of the photograph the intersecting ridges form a box pattern.

PLATE 4

Linear disintegration ridges. (Location of pl. 4 shown on fig. 4.) Location: T. 50, R. 27, W. 3d mer., Saskatchewan. (Air photograph reproduced by permission of Royal Canadian Air Force.)

These trends define successive positions of the ice front. Generally they are gently curved in outline and are regularly spaced.

Some hummocky moraines show alignment of knobs or ridges in two directions, which is indicative of ice-fracture control. These fractures were developed at the time when the ice was still active, and their trends do not necessarily bear a simple relationship to the trend of the ice front.

Terminology.—The purely descriptive term *hummocky moraine* is recommended to designate areas of knob-and-kettle topography. Where an origin along a live-ice front is indicated, the term *hummocky end moraine* is recommended. Where stagnant ice was responsible for deposition, the deposit may be referred to as *hummocky disintegration moraine* or *hummocky dead-ice moraine*. German authors often use *kuppige Grundmoräne* ("hummocky ground moraine") to designate such areas (Woldstedt, 1954, p. 95–96). Hoppe (1952) extended the term *hummocky moraine* to low-relief features, including some long ridges that resulted from dead-ice deposition. Such an extension does not appear desirable because the term *hummock* to most geologists means a hillock of more or less equidimensional shape and not a ridge.

CLOSED DISINTEGRATION RIDGES

Description.—At many localities on the Western Plains, circular, oval, or irregular closed ridges of glacial material are present. They are most noticeable on air photographs, but some larger, more regularly shaped ones are also obvious in the field. The closed ridges may be one of the elements of a hummocky moraine (pl. 1, *F*), or they may occur in low-relief areas where they occur in large numbers alone or together with linear ridges. Gradations between closed and linear ridges are not uncommon.

The most symmetrically shaped type of closed ridge resembles a giant doughnut from the air. Perfectly circular ridges range in height from a few to 20 feet and in diameter from 20 to 1,000 feet. The ridge, which surrounds a central depression, is commonly referred to as a "rim." It is generally unbroken and of the same height around the depression. Concentric ridges have been noted, and in places a low circular knoll takes the place of the more common central depression. The circular ridges are similar to one type of "prairie mounds" described by Gravenor (1955), but in "prairie mounds" the base of the central depression lies well above the general ground level, whereas the base of the central depression in the circular ridges lies at or below that level. Intermediate forms, however, exist. In general, the low circular ridges surround a depression that is shallow, but the floor is somewhat below the general level of the till plain (pl. 2).

Locally the circular ridges occur with ridges that are oval or irregular in plan. Such forms are also described by Deane (1950, p. 14). They surround depressions which may vary in shape from circular to irregular. All such ridges are referred to as *closed ridges* by the authors, even though they may show some minor breaks resulting from irregular deposition or later erosion.

Some closed ridges do not surround depressions but mounds of glacial material or moraine plateaus in hummocky moraine (pl. 1, *B*). They occur on the outside edge or within the confines of the plateau. Hoppe (1952, p. 5) has described similar ridges from moraine plateaus in Sweden. In western Canada they are not common, and many moraine plateaus are devoid of ridges of any type.

Although most closed ridges in the areas studied are composed of till, some are of stratified materials and are usually associated with eskers, kames, and crevasse fillings. The sand and gravel ridges vary in diameter from 50 to 800 feet and from 5 to 30 feet in height. In general, these ridges are higher and have steeper sides than the till ridges, and the base of the central depression lies at or above the level of the surrounding stratified drift.

Terminology.—Well-developed circular ridges of glacial material are colloquially referred to as "doughnuts"; mounds with round depressions are often referred to as "humpies." The depressions and the surrounding rim have been more formally designated as "rimmed kettles" (Christiansen, 1956, p. 11), the ridges alone as "ice-block ridges" (Deane, 1950, p. 14). More irregular ridges may produce a "brainlike" or "vermicelli" pattern on air photographs. Ridges on the edges of moraine plateaus are called "rim ridges" by Hoppe (1952, p. 5).

Closed ridge is proposed as a purely descriptive term to designate both the regular and the irregular forms. The adjective *circular, elliptical,* or *oval* can be applied to the more regular forms of closed ridges. Most authors seem to agree that the closed ridges resulted from stagnant ice that separated into individual dead-ice blocks, even though they do not agree as to the mechanics of deposition. If this origin is accepted, the term *closed disintegration ridge* may be used. *Ice-block ridges* and *rimmed kettles* are alternatives, provided that they do not imply construction of the ridge from material on the ice alone but leave room for the concept that till was squeezed upward from beneath the block. Generally, fields of closed ridges show no evidence of depositional control, even on a regional scale. In places, however, distinct end moraines are composed of hummocky moraine in which closed ridges are noticeable. The trends indicate a depositional control oriented transverse to the direction of ice flow. In other places the closed disintegration ridges are aligned because they are superimposed on previously formed trends (pl. 6). This is the case in some parts of the Missouri Coteau in Alberta and Saskatchewan, where structural end moraines of ice-shoved bedrock form ridges which caused the alignment of rimmed kettles in the overlying moraine.

CONTROLLED DISINTEGRATION FEATURES

LINEAR DISINTEGRATION RIDGES

Description.—Linear ridges resulting from ice disintegration are common features over much of the Plains region of western Canada and the United States. They have been described by many workers (e.g., Colton, 1955; Gravenor, 1956) and have been observed over large areas as undescribed.

Linear disintegration ridges are composed chiefly of till and may or may not have included pockets of stratified materials. A thin layer of gravel commonly lies on top and on the flanks of the ridges. In other places the ridges may consist predominantly of stratified material. They vary in height from 3 to 35 feet, in width from 25 to 300 feet, and in length from a few yards to 8 or more miles. They are straight or slightly arcuate. An important characteristic is that, in general, two sets intersect at acute or right angles. Such intersection of ridges forms a "waffle," "diamond," or "box pattern." These patterns indicate controlled deposition; orientation with regard to the direction of ice movement, however, may vary in different localities. In the Wolf Point area of Montana (Colton, personal communication) the majority of the longer and better-developed ridges lie at 45° to the direction of ice movement. In the Dollard district of Saskatchewan, on the other hand, the more distinctive ridges lie parallel to the ice flow (Kupsch, 1955, p. 329). In the Vermilion area of Alberta, the ridges show distinct parallelism to the lobate ice front in the center of the lobe, but with a more confused box pattern along its margin (pl. 3). Therefore, although the ridges are not of themselves reliable indicators of ice-movement direction at all localities, it is nevertheless significant that the most prominent ridges commonly lie normal, parallel, or at 45° to the direction of flow and hence show flow control.

In some areas three or more sets of linear disintegration ridges are present (pl. 4), and a complex pattern of linear elements results. Where two or more ridges meet, they produce forms resembling hairpins, wishbones, crosses, and many other angular shapes. Especially interesting are those which represent gradations between closed and linear disintegration ridges, with the resultant shape of a shepherd's crook.

The junction of two ridges may be at the same level, or one ridge may be superimposed on the other (fig. 1). Superimposition is not common but has been observed (Kupsch, 1956, fig. 1), and no distortion of the lower ridge is apparent. The material of the upper ridge was apparently let down without disturbing the lower ridge.

The most striking, easily recognizable, linear disintegration ridges occur in till areas of low to intermediate relief. They are distinct because of their straightness and field pattern. In high-relief moraine they may be straight but are more commonly irregular or slightly sinuous (pl. 1, *C*). In some places they connect a series of closed till ridges, thus forming a feature like a string of beads (pl. 1, *D*). Linear ridges in high-relief areas

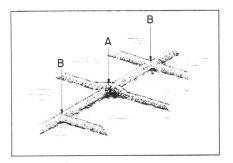

Fig. 1.—Diagram to illustrate different ridge junctions. *A*, superposed junction; *B*, confluent junction.

may have parallel trends, as in the Viking disintegration moraine of east-central Alberta, where they are oriented northeast at right angles to the direction of ice movement.

Although linear ridges composed of till predominate in the areas studied, some consist of sand and gravel and thus suggest ice-contact deposition; the proportion of till and stratified materials varies between wide limits. Gravenor (1956, p. 14) describes such a ridge of stratified drift from east-central Alberta, where it terminates in a broad kame.

Linear disintegration ridges of "sandy moraine" are described from the Cree Lake area, Saskatchewan, by Sproule (1939, p. 104) as "ice-crack moraine." The material composing these ridges is a very sandy

till and not a stratified drift. The area lies south and west of a large exposure of the pre-Cambrian Athabasca sandstone that lay in the path of glacier advance. It is interesting to note that some of the linear disintegration ridges of the Cree Lake area occur on bare bedrock.

Terminology.—Deane (1950, p. 14) used "ice-block ridge" for both closed and linear till ridges. The most widely used term for the ridges is perhaps "crevasse filling." It was introduced by Flint (1928, p. 415) and is applicable to the till ridges of western Canada, which are thought to have originated in crevasses in the ice. The crevasse fillings described by Flint, however, are composed of stratified material, and, to indicate the different composition of the ridges in the plains region, Gravenor (1956, p. 10) has suggested the term "till crevasse filling." A more general term is *linear disintegration ridge*, which implies less about the origin than any other term. It merely indicates that the ridge originated during stagnation or near-stagnation of the ice. Most geologists apparently agree to this general origin but differ as to the specific mode of formation.

WASHBOARD RIDGES

Gwynne (1942) described minor transverse till ridges from Iowa, where they form what he referred to as a "swell and swale pattern." Such patterns are common in many parts of the northern Great Plains and are now generally referred to as "washboard moraines." Gwynne (1942, p. 206) visualized them as being formed by periodic retreat and readvance of the live glacier front pushing previously deposited ground moraine into ridges. They would thus be minor end moraines or recessional ridges. The washboard moraines of Alberta and Saskatchewan were deposited on land; indications of a subaqueous origin are lacking. They are therefore different from those other minor end moraines, the "winter moraines" described from Sweden which apparently were formed under water (Woldstedt, 1954, p. 149).

Elson (1957, p. 1721) examined washboard moraines in Manitoba, where he found them to be composed of parallel, discontinuous, sandy-silty till ridges as high as 15 feet and as long as a mile, spaced 300–500 feet apart and forming lobate patterns. He concluded that the washboard ridges were deposited subglacially at the base of thrust planes in the ice.

Although it is realized that geologists are not in agreement as to the origin of washboard moraine, the writers tend to favor Elson's hypothesis and regard the preservation of the ridges as the result of ultimate stagnation. The washboard ridges were preserved because, finally, the glacier separated along the thrust planes. They are therefore to be regarded as disintegration features exhibiting inherited flow control. Of all the various disintegration features, they show the most obvious flow control, and they are thus helpful in reconstructing former ice-front positions if it is supposed that the thrust planes are everywhere parallel to the margin.

The linear disintegration ridges are in places different from washboard ridges, in that the linear ridges may be even-crested and have a fairly consistent width (pl. 3), whereas the washboard ridges are almost everywhere discontinuous and of variable height and width, forming a pattern of light "dapples" on air photographs (Gwynne, 1942, p. 202), like some of the disintegration ridges shown on plate 4. This difference may be due to the general open nature of crevasses into which till was squeezed to form continuous linear ridges and the narrow thrust planes along which the discontinuous till masses of the washboard ridges accumulated.

The washboard pattern is characteristically lobate, outlining the position of the thrust planes in the live ice, locally modified by small to large re-entrants formerly occupied by meltwater streams, which may or may not have deposited eskers. Irregular or branching ridges occur in places, but, in general, the washboard "waves" or parallel troughs and ridges are well developed.

Gradations from washboard moraines to patterns indicating linear disintegration along crevasses have been observed. Lawrence and Elson (1953) noted a second direction of trends of acute or right angles to some washboard moraines. The presence of two or three directions is an outstanding characteristic of the field pattern of linear disintegration ridges, indicating a system of fractures composed of several sets. One of these directions may be parallel to the former ice front and show a lobate outline (pl. 3). The ridges are then similar to washboard ridges except that they were probably deposited in continuous open crevasses. Locally the development of a pattern controlled by several sets of fractures (as along the margins of the lobe shown in pl. 3) reveals the linear disintegration origin and distinguishes the fillings of transverse crevasses from washboard ridges caused by squeezing along transverse thrust planes.

EROSIONAL DISINTEGRATION FEATURES

ICE-WALLED CHANNELS

Description.—Meltwater channels in areas of stagnant ice are among the more spectacular glacial features of the Plains region (pl. 5, *A*, *B*, *C*). They are especially well developed in east-central Alberta, where they can be divided into two types: (1) those filled with till, which are recognizable on air photographs by a chain of kettles in the bottom of the valley, and (2) those which are not filled with till but are broad, open troughs with some sand and gravel on their floors. Both types are generally marked by a topographic depression which outlines the channel, but the first type may be completely filled and thus blend in with the surrounding moraine. Gradations between the two extreme types are encountered in many places. The first type is most common in high-relief disintegration moraine, and the bedrock is at a considerable depth below the base of the channel. The second type is most common in areas of thin, low-relief ground moraine, and in many places the bedrock is exposed in the sides of the channel. Some of the channels have been observed to feed into esker systems.

In map view, the channels present a complex pattern composed of parallel and intersecting elements. In east-central Alberta the major channels are roughly parallel to the ice-movement direction, and the intersecting ones are at right angles. That the two directions were not everywhere used at the same time is indicated by the fact that the floor of one channel commonly truncates the deposits of another. The pattern is only roughly controlled, and over small areas control is not very obvious.

Terminology.—Erosional channels occupied by subglacial meltwater streams are referred to as *Rinnentaler* in the German literature. It may be that some of these channels were not eroded by meltwater alone but that they were initiated by glacial erosion and later occupied by meltwater (Woldstedt, 1954, p. 38). The term *Rinnentaler* ("channel valleys") would seem suitable for the ice-walled channels of western Canada, but it is difficult to translate with retention of any descriptive meaning. In Denmark the trenches cut by streams flowing beneath the ice are called *tunneldale* ("tunnel valleys"), and an origin in closed tunnels is implied (Schou, 1949, p. 20). This term is applicable where such an origin is indicated, but the writers prefer *ice-walled channels* to include streams in open trenches as well as in closed tunnels.

ORIGIN OF DISINTEGRATION FEATURES

GENERAL REMARKS

All studies made by the writers on ice-disintegration features suggest that the features were formed late in the existence of a glacier. The features resulted from the letting-down of till due to ablation, from the squeezing-up of till into openings at the base of the ice, or from a combination of both causes.

Vertical air photographs of the Cree Lake area reveal that some of the linear disintegration ridges cross over drumlins without any apparent disturbance of the older deposits. Near North Battleford, Saskatchewan, linear ridges cross such streamlined live-ice features as elongated drumlins and intervening grooves, collectively referred to as "fluting." In both areas the disintegration features are clearly younger than the live-ice features.

Closed ridges that are aligned into end moraines may owe their alignment to *inherited flow control* because the hollows and surrounding rims were formed by blocks of dead ice left in the constructional end moraine after retreat of the ice. The possibility that the development of trends in a hummocky moraine is later than ice disintegration has to be considered if the origin of dead-ice thrust moraines, as described by Seifert (1954, p. 132–133), is accepted. He visualizes the thinning of ice along the margin and the accumulation of morainal material on it as being followed by a thickening of the glacier. The regenerated ice would then override the stagnant ice with its moraine and push it into thrust ridges. If this process caused trends in hummocky disintegration moraine, it could be said to exhibit *regenerated flow control*.

ORIGIN BY ABLATION

In a wasting glacier, debris which was originally in the ice gradually accumulates

---

PLATE 5

Ice-walled channels. (Location of pl. 5 shown on fig. 4.) Location: T. 45, R. 10, W. 4th mer., Alberta. (Air photograph reproduced by permission of government of Alberta, Department of Lands and Forests.)

A, Terrace of the Kinsella ice-walled channel, now covered by dead-ice moraine. To the west this channel is cut into bedrock.

B, Ice-walled channel, almost completely filled with drift.

C, Broad, open type of ice-walled channel, filled with sand, gravel, and some till, showing circular disintegration ridges.

D, Poorly developed flutings (?) in lodgment ground moraine, possibly indicating live-ice deposition, covered by **thin** veneer of dead-ice (ablation) moraine.

F, Dead-ice moraine, intermediate relief.

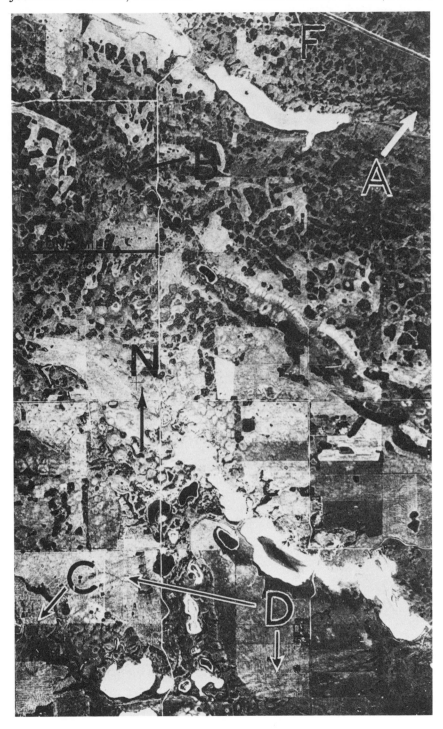

Ice-walled channels.

Closed disintegration ridges and hummocky dead-ice knobs.

on the surface of the ice. As ablation progresses, this surface debris is let down on the basal-ice deposits. While the material is still on the ice, it forms chaotic ridges, knolls, and depressions which reflect mainly relief of the ice surface itself. The drift let down irregularly from the surface of the glacier to the ground gives rise to ablation moraine (Flint, 1957, p. 120). This letting-down process can be used to explain ice-disintegration features (fig. 2). Hummocky moraine may be a reflection of irregular accumulations of debris on the ice (Gravenor, 1955, p. 477). Ablation material is moved by mass wasting into openings in the ice. Although the circular (or polygonal) shape is common for a dead-ice mass and the ablation material thus sloughed off into round closed ridges, in many places the blocks have an angular outline and represent remnants of a glacier that disintegrated along pre-existing fracture systems. When final melting of the ice blocks occurs, round or linear disintegration ridges remain, depending on the shape of the ice block (fig. 2).

It seems obvious from several considerations that many depositional disintegration features owe their origin, at least in part, to material that fell into crevasses, subglacial channels, and irregularly shaped hollows under and in the ice.

Observations on the Kinsella ice-walled channel in Alberta by Gravenor and Bayrock (1956, p. 6–8) show that at one point gravel outwash was deposited which lies *above* the level of the surrounding high-relief disintegration moraine. This outwash is not covered by till and shows no evidence of having been overridden by ice. Downstream from the outwash the channel is cut into bedrock and is floored with till in the form of a high-relief disintegration moraine (this

part of the channel is shown on pl. 5). The fact that the outwash lies above the surrounding high-relief disintegration moraine and has not been overridden suggests that the outwash was deposited between ice walls and by the last ice to cover the area. The downstream portion of the ice-walled channel must have been infilled by debris which slumped in from the ice walls. That the

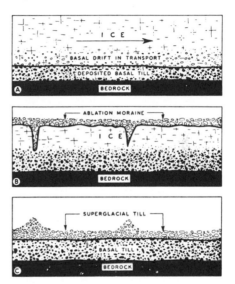

Fig. 2.—Disintegration features formed by ablation (modified after Flint, 1947, p. 112).

channel was not used to any great extent by water after deposition of the ablation moraine is evident from the absence of any notable erosion of the ablation deposits by running water.

Whereas many smaller ice-walled channels may have been true ice tunnels, the width of the larger trenches—over 1 mile—and the fact that they contain little till (which may have slumped in from the ice

PLATE 6

Closed disintegration ridges and hummocky dead-ice knobs superimposed on morainal ridges. (Location of pl. 6 shown on fig. 4.) Location: T. 19, R. 26, W. 4th mer. This photograph shows a series of parallel, curved end-morainal ridges (*A*) (much like "washboard" moraine) on which has been superimposed depositional features of ice disintegration (*B*). Note that the ice-disintegration features show definite alignment which is interpreted as control derived from lines of weakness in the active ice. This photograph suggests that the end-moraine ridges were developed under the ice and at a later date—but along the same lines of weakness—than the dead-ice features were developed.

walls along the sides) would suggest that the larger channels were open to the atmosphere at the time of final ice disintegration.

The superposed junction of linear disintegration ridges (fig. 1) also reveals that some material was let down from the stagnant ice. This type of junction suggests that first one crevasse was open, that material was dumped in it, that another crevasse opened up crossing the first, and that drift was deposited in this second crevasse without disturbing the older deposits.

In hummocky moraine the moraine plateaus of stratified material are generally explained as meltwater or lake sediments in a depression surrounded by ice. On final melting, the fill may then be higher than the surrounding ablation moraine (Schou, 1949, p. 10). If deposition of the water-laid sediments takes place on ice, the melting of this floor will cause "collapse structures" in the sediments. Such "collapse structures" are indeed common in the hummocky dead-ice moraine of western Canada (Christiansen, 1956, p. 12), and Flint (1955, p. 114) noticed them in eastern South Dakota.

Although there are many indications that ablation moraine was being deposited during disintegration of an ice sheet, the hypothesis that all material of a dead-ice origin was let down from above fails to explain certain observed relationships. Ablation material is characteristically loose, non-compact, and non-fissile and contains abundant gravel and larger stones (Flint, 1957, p. 120). Melting is active as ablation progresses, and trickles of water cause crude sorting and a general "washed" appearance. Meltwater deposits, such as kames, eskers, and crevasse fillings of stratified drift, form during the disintegration phase. Such deposits are common in many areas in the eastern part of North America, where they have been interpreted as stagnation features, as, for example, the crevasse fillings described by Flint (1928, p. 411–416). In western Canada, however, many features that apparently resulted from disintegration are composed of till which may be very compact. This material may be the only constituent of the disinte-

gration feature, or it may be overlain by a layer of non-compact till of varying thickness, or the latter deposit may make up the whole land form. The presence of the compact till is difficult to account for if all material is thought to have been deposited by ablation.

Another objection is raised by many workers, who point out that even in Alaska, where thick ablation moraine is common on the lower reaches of living valley glaciers, clear glacier ice is the normal condition and moraine-covered ice the exception. It is held that in ice sheets a complete covering by ablation moraine is impossible (Hoppe, 1952, p. 26). If this is so and if it is believed that dead-ice features result from drift that was let down, it becomes necessary to find a mechanism to get large quantities of drift on top of the ice. Flint (1955, p. 114) therefore assumes a readvance of the glacier which overrode thin residual ice and deposited basal till on top of it. With renewed deglaciation the buried ice masses melted out, and the overlying till was let down. Such conditions are known to have occurred, but it appears that in many cases they are assumed without any corroborative evidence for the assumption of a readvance. Drift can also be brought up from the basal portion of the ice to its surface along thrust planes which curve obliquely upward in the terminal zone of the glacier. The drift melts out onto the ice surface along the thrust line and forms ridges on the ice. This process occurs in live ice, but, when the ice subsequently becomes stagnant, the till ridges on the ice may be let down onto the underlying surface when the supporting ice melts. Investigations of glaciers in Spitzbergen (Hoppe, 1952, p. 28–29; Woldstedt, 1954, p. 54) revealed the presence of ridges parallel to the ice margin and, in addition, others transverse to it. Basal drift is apparently also brought to the surface of the ice along open crevasses oriented transversely to the ice margin probably by squeezing due to ice flow or pressure from the weight of ice blocks. The resulting complex of till ridges (*Lehmmauern*, or "till walls") is strikingly similar to the disintegration pattern of linear ridges encountered in many places in western Canada.

ORIGIN BY SQUEEZING

Although it has been demonstrated that ablation moraine may accumulate on the ice during wasting and that it may become concentrated in ridges and heaps as a result of many different processes, some investigators feel that the quantity of ablation material is not sufficient to account for the disintegration features which can be observed. Hoppe (1952) is a proponent of this school of thought. He believes that the dead-ice features were formed underneath the ice through squeezing of debris into basal cavities such as crevasses and irregular cavities caused by meltwater (fig. 3). The drift could be squeezed up into these cavities because it was soaked with water and therefore in a plastic state. The weight of the ice itself exerted the necessary pressure, but some slight ice-flow movement may have contributed in certain places. Hoppe recognizes that some drift may have been squeezed all the way to the surface of the ice along crevasses (*Lehmmauern*), but he believes that it is quantitatively insufficient to account for all dead-ice moraines features. Hoppe's arguments for assuming that dead-ice features resulted from basal till squeezed into openings, closed to the sky, on the underside of the ice, may be summarized as follows:

1. The till does not show evidence of washing, as would be the case with superglacial material that had fallen from the ice surface or from the side walls into an open crevasse.

2. The till is compact and has all the characteristics of basal till.

3. The till contains pebbles and cobbles that show a distinctive fabric with their long axes oriented at right angles to the long dimension of the ridge. This fabric is regarded as a primary characteristic of the till induced by the lateral pressure of the ice blocks which squeezed the till up into the crevasse.

Although Hoppe stresses the third observation and presents a great number of orientation studies on various types of disintegration ridges in Sweden, it is not clear why a similar orientation could not result from lateral outward flow of the plastic material by soil creep or solifluction when the debris moved slowly down a stagnant ice-block surface. Hoppe's investigations demonstrate, however, the important relationship between minor, or even micro-, relief features and till fabric. Many fabric

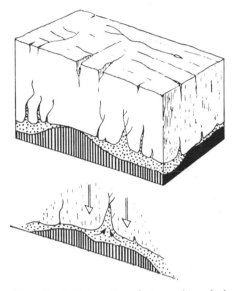

Fig. 3.—Disintegration features formed by squeezing (after Hoppe, 1952, p. 55).

studies in high-relief disintegration moraine previously considered as "end moraine" may have yielded incomprehensible or conflicting results because due account was not taken of this relationship.

The process envisaged by Hoppe hinges essentially on three assumptions: (1) the ice was relatively clear, and most of the debris was underneath the ice; (2) the debris under the ice was not frozen but in a water-saturated state and hence susceptible to plastic deformation; and (3) the fabric of the till in the ridges is the result of squeeze. The first assumption is made by many students of glacial geology. However, in view of the various processes that can bring debris to the top of the ice, it is doubtful that this condition held over the entire continental glacier. It seems likely that locally the margin of the glacier was covered by large quantities of ablation material. The second assumption appears valid under certain conditions, inasmuch as permafrost is generally absent under thick masses of "warm" ice (Woldstedt, 1954, p. 60). In "cold" glaciers, however, it is common for the ground under the ice to be in a permanently frozen state. The third assumption appears to be unproved, as the fabric may be produced during the final melting of the supporting ice, regardless of whether the till ridge was formed by "squeezing-up" or "slumping-in."

Perhaps the main merit of the squeezing theory is that it accounts for the character of the till in the ridges. It should be pointed out, however, that, once the ice had a thin layer of ablation debris on its surface, further down-melting would be extremely slow, and the clayey nature of the till might well be preserved. Recently, Harrison (1957) has argued that the bulk of till in ground moraine originated as englacial material and was deposited by the slow melting-out of debris from the basal zones of the ice. The compact nature or "toughness" of till may, therefore, not be a result of "plastering-on" or squeezing but may simply be due to the till's original texture, structure, and fabric. Hence the character of till in ridges or in hummocky moraine cannot be used as strong evidence in the determination of the origin of the feature.

Arguments against the squeezing theory (such as the superimposed crossing of two ridges, the presence of linear disintegration ridges on bare bedrock, observable "collapse structures" which indicate deposition on ice masses and subsequent lowering, and the characteristics of ice-walled channel fills) have already been presented as favoring the ablation theory.

COMBINED PROCESSES

The hypotheses of ablation and squeezing for the origin of disintegration features are not mutually exclusive, and the writers hold that, in the formation of many of the features observed, both processes may have been operative to varying degrees. In the marginal zone of the glacier, till may be squeezed upward in fractures to the surface of the ice and subsequently let down to the ground by ablation. The ridges thus formed on the ice may lose their identity during ablation and by slump action, eventually becoming disintegration moraine (Woldstedt, 1954, p. 54). Some fields of clearly recognizable linear till ridges may owe their origin to these combined processes, but with the preservation of the till ridges more or less as they existed on the ice.

Material squeezed into subglacial crevasses and other openings may, in addition, receive material from above during ablation. Some disintegration features appear to be composed of cores of clayey till squeezed upward and overlain by a covering of loose material dropped from above. Such cores may have been squeezed upward, the covering dropped from above. The amount of meltwater involved in ablation varies between wide limits, and all gradations between till disintegration features and those composed of stratified drift are to be expected. In places, meltwater erosion, instead of deposition, took place, and subglacial channels were cut in tunnels or in trenches open to the sky. Such channels are therefore the erosional counterparts of eskers.

Because combinations of several depositional processes may be responsible for dead-ice land forms, the writers prefer *disintegration features* to the term "ablation moraine," as the latter implies not only an origin from stagnant ice but also a certain process of deposition.

TERMINOLOGY OF MORAINES

Flint (1955, p. 111–120) reviewed the history of the terms "moraine," "ground moraine," and "end moraine," whose definitions vary widely with different authors. The recognition of ice-disintegration moraines further complicates any classification adopted for mapping. Dead-ice land forms may occur as high-relief features in "moraines" or as low-relief forms in "ground moraine." The term "ground moraine" has always implied deposition of material from the base of the ice, back from its margin (Flint, 1955, p. 111). If, however, the high-relief disintegration features owe their origin to squeezing-up of basal material, as Hoppe (1954, p. 8) and other investigators state, then some areas previously considered "moraine" or even "end moraine" are more properly "high-relief ground moraine." Townsend and Jenke (1951, p. 857) suggested this origin for the Max Moraine in North Dakota when they wrote: "The Max Moraine may be a special type more nearly related in extent and mode of deposition to ground moraine than to end moraine."

The German term *kuppige Grundmoräne* also implies a ground-, not end-, moraine origin for hummocky moraines. Such moraine may be deposited along the ice front at localities where no true end-moraine ridges are formed.

In reconnaissance mapping of glacial deposits in western Canada, where little is known about the direction of ice movement, it is of importance to distinguish between areas of predominantly live-ice deposition and those of stagnation. Features that indicate live-ice deposition are either parallel to the direction of ice movement or transverse to it. Parallel features are streamlined and range from striae to drumlins. Transverse features consist of lobate systems of true end moraines. Dead-ice features may show control inherited from live ice, or they may show no control whatever.

Not only do live-ice features grade into those of stagnant ice, but the superimposition of disintegration features on live-ice features can be noted in many places. The linear disintegration ridges that cross flutings near North Battleford, Saskatchewan,

are but one example. Commonly an almost continuous blanket of disintegration moraine hides any underlying streamline features of the live-ice phase of the glacier. This blanketing layer may show dead-ice features that would be scarcely recognizable, were it not for the gradation into areas with higher relief (pl. 5, gradation from E to F). Areas generally considered ground moraine because of their low relief therefore do not necessarily show basal till at the surface but may include some ablation till if the disintegration originated by the combined processes of squeezing-up and lowering. Locally the disintegration blanket of the stagnant phase may be so thin that it does not effectively hide the streamlined "megafabric" of the underlying basal till deposited by lodgment from live ice (pl. 5, D). In other places the disintegration cover may be removed by erosion after final retreat of the ice, and the "megafabric" of the basal till may be secondarily brought out by that erosion. This may be one reason why some flutings are especially well developed in areas that at one time were occupied by streams (Christiansen, 1956, p. 16).

OTHER SIMILAR FEATURES

GENERAL REMARKS

Some land forms may appear morphologically similar to disintegration features but may have originated in a different way. Hummocky moraine may result from live-ice deposition as well as from dead-ice disintegration and is thus polygenetic. Patterns indicating deposition from live ice, such as alignment of the hummocks into lobate trends, may be hardly discernible, or the disintegration moraine may still vaguely show some inherited flow control. Thus it may be difficult or impossible to assign a definite origin to a particular feature. On the other hand, the end members of the live-ice–dead-ice sequences are recognizable and should be differentiated in mapping. If there is any doubt, the feature should be considered with regard to the following three main characteristics:

1. *Individual characteristics.*—What differences exist between the individual feature and others similar in form but known to be of different origin? These characteristics are best studied on the ground with the aid of large-scale vertical air photographs.

2. *Pattern characteristics.*—What differences exist in the pattern of a group of individual features with respect to the pattern of other groups of features of different origin? The pattern characteristics are most clearly seen on air-photograph mosaics. They are generally more revealing than the individual characteristics.

3. *Positional characteristics.*—Is the postulated origin compatible with other glacial features in the region? In other words, do the supposed live-ice features, disintegration moraines, or frozen-ground features occur where they are understandable with respect to the particular geological setting? These characteristics will generally be clear only after a large area or region has been mapped. Accordingly, subsequent reassignment of some features to a different origin may be necessary as mapping progresses. In general, the positional characteristics provide the strongest evidence for a particular origin.

LIVE-ICE FEATURES

Live-ice features oriented in the direction of ice flow, such as ice-abraded bosses, crag and tail, and elliptical drumlins, are not readily confused with disintegration features. Extremely elongated drumlin ridges may, however, be similar to long crevasse fillings. The long linear ridges near Velva, North Dakota, were first considered to be crevasse fillings but are now reinterpreted as drumlins (Lemke, 1958). The streamlined shape of these elongated drumlins with a steeper stoss side and a long, gently tapering tail, the field pattern of the strikingly straight and parallel ridges with intervening grooves giving the country a "scratched" or "fluted" appearance, and their orientation in the direction of ice movement, as indi-

cated by other evidence, will generally help to identify the ridges as live-ice features.

Large constructional or structural (push) end-moraine ridges with a distinctive lobate pattern are generally clearly evident as live-ice features. Similar, but smaller, transverse ridges formed by live ice may not be so easily distinguishable from some disintegration ridges.

FROZEN-GROUND FEATURES

Land forms created by thawing ground ice may be very similar to those formed by melting and evaporation of dead glacier ice buried by debris. Some features may actually have the same shape and composition, so that only the relationship to other land forms of known origin can provide some hints as to their genesis. The writers prefer the inclusive term *frozen-ground features* (Flint, 1957, p. 195–196) for all features that result from the presence and final thawing of ground ice.

Rimmed depressions in the northern Netherlands previously regarded as kettles have recently been reinterpreted as pingo-remnants (Maarleveld and Van Den Toorn, 1955). Their similarity to true kettles is so great that they are referred to as *pseudo-sölle* ("pseudo-kettles"). They are distinguishable from kettles only because stratigraphic relationships indicate formation during a cold climate long after the final melting of the glacier ice. Such detailed stratigraphic information is not available in western Canada. It appears unlikely, however, that the round disintegration ridges surrounding depressions in low-relief areas (pl. 4) are formed by pingo-collapse. As far as the writers are aware, pingos do not occur in large fields of many square miles. Nor are they known to be so closely spaced that they touch each other and form a continuous pattern of round or modified polygonal features.

Elson (1955, p. 140) regarded features such as those shown in plate 4 as possible thaw depressions which resulted from subsidence following the thawing of perennially frozen ground (Hopkins, 1949). Thaw depressions, like pingos, occur in predominantly silty soils, whereas the disintegration features of western Canada are notably well developed in rather impervious till which has a high clay content.

Any alternative explanation for the origin of features which are here interpreted as ice-disintegration land forms will have to take into account the regional setting in till and their classification can be briefly summarized as follows:

A. *Uncontrolled disintegration*

Generally uncontrolled, but gradations to controlled disintegration exist, which is most commonly *inherited flow control*. In places *regenerated flow control* may have been operative.

Features: *Hummocks* (knob and kettle),

FIG. 4.—Index map showing location of plates

relation to other glacial evidence. Development of frozen-ground features is a function of sediment characteristics, not of glacier motion. The regional setting of disintegration features, on the other hand, shows their dependence on glaciation. Disintegration is a marginal phenomenon of an ice sheet and operates irrespective of the sediment type.

SUMMARY OF CLASSIFICATION

The various depositional ice-disintegration features composed predominately of

moraine plateaus, closed ridges (rimmed kettles; ice-block ridges).

B. *Controlled disintegration*

1. Disintegration along open crevasses

Generally well controlled, but gradations to uncontrolled disintegration can be observed. Control commonly inherited from previous ice flow but could be caused by diastrophic movements, jointing in bedrock, and other causes.

Features: *Linear ridges* (crevasse fillings; ice-crack ridges; ice-block ridges).

2. Disintegration along thrust planes

Distinctive inherited flow control evident from lobate pattern.

Features: *Washboard ridges* (swell and swale topography; minor recessional ridges).

C. *Superimposed disintegration*

1. Superimposition on live-ice features parallel to ice flow, such as drumlins, flutings.

2. Superimposition on live-ice features transverse to ice flow, such as depositional and structural end moraines.

In addition to the above-mentioned till features, land forms composed of stratified drift are part of the disintegration moraine. Of the erosional features, *ice-walled channels* appear to be the most prominent.

ACKNOWLEDGMENTS.—The writers are indebted to V. K. Prest, R. Lemke, and J. A. Elson, who read the manuscript and offered valuable comments and criticisms, many of which have been incorporated in this paper. They would also like to thank the Research Councils of Alberta and Saskatchewan for financial support which made this study possible.

REFERENCES CITED

CHRISTIANSEN, E. A., 1956, Glacial geology of the Moose Mountain area, Saskatchewan: Sask. Dept. Mineral Resources, Rept. 21.

COLTON, R. B., 1955, Geology of the Wolf Point quadrangle, Montana: U.S. Geol. Survey, Geol. Quadrangle Map.

DEANE, R. E., 1950, Pleistocene geology of the Lake Simcoe district, Ontario: Geol. Survey Canada Mem. 256.

ELSON, J. A., 1955, Unpublished Ph.D., thesis: Yale University.

———— 1957, Origin of washboard moraines: Geol. Soc. America Bull., v. 68, p. 1721.

FLINT, R. F., 1928, Eskers and crevasse fillings: Am. Jour. Sci., v. 15, p. 410–416.

———— 1955, Pleistocene geology of eastern South Dakota: U.S. Geol. Survey Prof. Paper 262.

———— 1957, Glacial and Pleistocene geology: New York, John Wiley & Sons.

GRAVENOR, C. P., 1955, The origin and significance of prairie mounds: Am. Jour. Sci., v. 253, p. 475–481.

———— 1956, Air photographs of the plains region of Alberta: Research Council of Alberta, Prelim. Rept. 56–5.

———— and BAYROCK, L. A., 1956, Stream-trench systems in east-central Alberta: Research Council of Alberta, Prelim. Rept. 56–4.

GWYNNE, C. S., 1942, Swell and swale pattern of the Mankato lobe of the Wisconsin drift plain in Iowa: Jour. Geology, v. 50, p. 200–208.

HARRISON, P. W., 1957, A clay-till fabric: its character and origin: Jour. Geology, v. 65, p. 275–308.

HOPKINS, D. M., 1949, Thaw lakes and thaw sinks in the Imbruk Lake area, Seward Penninsula, Alaska: Jour. Geology, v. 57, p. 119–131.

HOPPE, GUNNAR, 1952, Hummocky moraine regions, with special reference to the interior of Norbotton: Geog. Annaler, v. 34, p. 1–71.

HOWELL, J. V. (ed.), 1957, Glossary of geology and related sciences: Washington, D.C., N.A.S.-N.R.C. Pub. 501.

JOHNSTON, W. A., *et al.*, 1948, Surface deposits, southern Saskatchewan: Geol. Survey Canada, Paper 48–18, map.

———— and WICKENDEN, R. T. D., 1931, Moraines and glacial lakes in southern Saskatchewan and southern Alberta, Canada: Royal Soc. Canada Trans., 3d ser., v. 25, sec. 4, p. 29–44.

KUPSCH, W. O., 1955, Drumlins with jointed boulders near Dollard, Saskatchewan: Geol. Soc. America Bull., v. 66, p. 327–338.

———— 1956, Crevasse fillings in southwestern Saskatchewan, Canada: Verh. K. Nederlandsch Geologisch-Mijabouwkundig Genoot., geol. ser., v. 16, p. 236–241.

LAWRENCE, D. B., and ELSON, J. A., 1953, Periodicity of deglaciation in North America. Pt. II. Lake Wisconsin recession: Geog. Annaler, v. 35, p. 96.

LEMKE, R. W., 1958, Narrow linear drumlins near Velva, North Dakota: Am. Jour. Sci., v. 256, p. 270–283.

MAARLEVELD, G. C., and VAN DEN TOORN, J. C., 1955, Pseudo-sölle in Noord-Nederland (Dutch, with English summary): Tijdschr. K. Nederlandsch Aardrijksk. Genoot., v. 72, no. 4, p. 344–360.

NEILSON, W. A. (ed.), 1956, Webster's new international dictionary of the English language: 3d ed., unabridged, Springfield, Mass.

SCHOU, AXEL, 1949, Atlas of Denmark: Copenhagen, Hagerup.

SEIFERT, GERHARD, 1954, Das mikroskopische Korngefüge des Geschiebemergels: Meyniana, v. 2, p. 124–184.

SPROULE, J. C., 1939, The Pleistocene geology of the Cree Lake region, Saskatchewan: Royal Soc. Canada Trans., 3d ser., v. 33, sec. 4, p. 101–109.

TOWNSEND, R. C., and JENKE, A. L., 1951, The problem of the origin of the Max Moraine of North Dakota and Canada: Am. Jour. Sci., v. 249, p. 843–858.

WOLDSTEDT, PAUL, 1954, Das Eiszeitalter: Stuttgart, Enke.

21

Reprinted from *Geol. Soc. America Bull.*, **53**, 1150–1163, 1173–1174 (Aug. 1942)

Pleistocene Geology of Western Cape Cod, Massachusetts

KIRTLEY F. MATHER, RICHARD P. GOLDTHWAIT, AND LINCOLN R. THIESMEYER

* * * * * * *

SUMMARY OF FIRST SECTION

Three distinct types of recessional moraines are present in the general vicinity of Buzzards Bay near the west end of Cape Cod. The Buzzards Bay and Sandwich moraines were constructed along the margins of the Buzzards Bay lobe and the Cape Cod Bay lobe of the Laurentian ice sheet during the Wisconsin glacial stage by ice that readvanced from a position north and northwest of these moraines to which it had earlier retreated. Thus the morainic debris was superimposed upon the margins

Editor's Note: A row of asterisks indicates material deleted owing to lack of space.

of the Mashpee pitted plain and represents in part the result of ice push as well as the deposit made by melting ice.

The Scorton moraine is a discontinuous series of till-veneered hills marking the ragged edge of the Cape Cod Bay lobe at a time of ice wastage that caused the irregular withdrawal of the ice front from the smooth line of the Sandwich moraine. Slight oscillations of the active ice within the Cape Cod Bay lobe resulted in the overriding of the stagnant ice along its frayed margin and left the morainic hills standing in the midst of glacio-fluvial deposits.

The Ellisville moraine is the least complex of recessional deposits. Continuous northward retreat of the margin of the Cape Cod Bay lobe of ice was temporarily halted. The recessional moraine and the bordering apron of glacio-fluvial debris, the Wareham pitted plain, were contemporaneously deposited.

PART II. MASHPEE PITTED PLAIN
SURFACE FEATURES AND COMPOSITION

The Mashpee pitted plain lies within the angle between the Sandwich and Buzzards Bay moraines and occupies about 100 square miles of land surface on Cape Cod (Fig. 3). Several square miles of sea floor beneath Nantucket and Vineyard Sounds represent its submerged southern margin. The plain is in essence a slightly dissected, subaerially constructed, alluvial fan with its apex at an elevation of 220 feet, close to the place of overlap of one moraine upon the other, and its surface sloping gently to the curving shore line, 10 to 14 miles distant toward the south and southeast. At the east, beyond the Pocasset and Falmouth quadrangles, it coalesces with a similar fan which according to Chute (1939, p. 12) may have developed simultaneously.

The surface of the plain is modified by two types of depressions: (1) irregular pits or kettle holes of all sizes, some of which are as much as a mile in diameter and 100 feet in depth, distributed at random, and (2) long furrows increasing in breadth from a few feet to 1000 feet and in depth from a few inches to more than 30 feet as they radiate southward and southeastward across the lower half of the plain. Woodworth (1934, p. 271) called these linear depressions "creases," and others have referred to them as "valleys," "arroyos," or "channels." The designation "furrow" is, however, adopted here because it is a noncommittal term without any implication concerning their mode of origin.

The surface slope of the entire plain, restored by filling in, in imagination, the pits and furrows, is very gently concave (Fig. 4). From the apex south and southeast for 6 miles, the average gradient is 15 to 20 feet per mile. Indeed, close to the apex some nonpitted portions of the plain incline as much as 25 feet in one mile. The outer and lower half of the plain, on the other hand, has a surface slope of only 12 to 15 feet per mile.

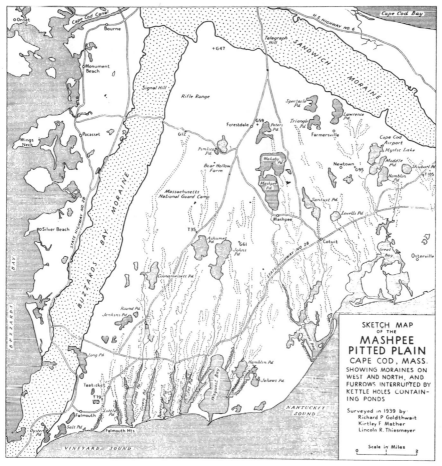

FIGURE 3.—*Sketch map of western Cape Cod*

Showing Mashpee pitted plain, bounded on north and west by Sandwich and Buzzards Bay moraines. Furrows traversing plain are delineated by broken lines. Localities at which stone counts recorded in Table 3 were made are indicated by letter and number, *e. g.*, G59.

The material underlying the plain is bedded sand and gravel, accumulated to a thickness of at least 120 feet, the depth of the deepest pits. The average components vary from well-rounded gravel near the apex to subangular sand near the distal edge. Associated with the gravel in the higher portion of the plain are abundant, small, rounded or subangular boulders, 6 to 24 inches in diameter. One pit along the Pocasset

FIGURE 1. COARSE GRAVEL EXPOSED IN LAWRENCE COMPANY PIT
West of Sols Pond in Falmouth.

FIGURE 2. LARGE CLUMPS OF TILL EMBEDDED IN STRATIFIED SAND AND
GRAVEL
In this exposure at southeast side of gravel pit near Teaticket, both the till
clump in middle of vertical face and the one directly above the man are irregularly stained with iron oxides.

FIGURE 3. HORIZONTALLY BEDDED SAND AND GRAVEL IN ROAD CUT ½ MILE
SOUTHEAST OF JENKINS POND ON ASHAMET-GRASSY POND ROAD
Bedding is truncated by gently sloping wall of a furrow trending nearly at
right angles to road cut.

MATERIALS COMPOSING MASHPEE PITTED PLAIN

364

road, 1000 feet north of a road intersection, the altitude of which is given on the topographic map of the Pocasset quadrangle as 182 feet, disclosed 4 to 6 feet of very coarse gravel and small boulders overlying 3 feet of cross-bedded sand with thin lenses of gravel. A trench in the Massachusetts National Guard Military Reservation (Camp Edwards) north of Snake Pond and many shallow road cuts in the northern part of the reservation exhibit similarly coarse gravel. In contrast, on the outer, seaward edge of the plain, the many low wave-cut cliffs almost exclusively expose only sand. Pits excavated for sand, used in surfacing roads and bogs, are scattered along the main highway leading northeastward from Falmouth and around each cranberry bog in this vicinity. Very little coarse gravel is available in these pits.

There are exceptions, however, to this gradation in coarseness. Most notable is the Lawrence Company pit, just west of Sols Pond in Falmouth. Although this locality is only 2 miles from the outer edge of the plain, it shows a 6-foot layer of very coarse gravel containing rounded boulders, some of which are as much as 18 inches in diameter (Pl. 2, fig. 1). On the floor of the pit are a few boulders as much as 30 inches long. Similar coarse gravel with rounded cobbles 8 inches in diameter occurs in a near-by pit just southwest of Teaticket. The eastward dip of the layers and slope of the surface between pits in this vicinity suggests that this abnormally coarse material was deposited by meltwater issuing from the adjacent Buzzards Bay lobe while the Buzzards Bay moraine was under construction. Apparently streams from that source built a small local fan on top of the more extensive and older deposits of the Mashpee pitted plain.

The bedding of the sand and gravel composing the fan is essentially parallel to the surface of the plain. Although cross-bedding and channelling are common, no foreset delta structure was observed. The "scour and fill" nature of the deposits is evident at many places where lenticular gravel layers lie between cross-bedded and laminated sands.

The features described above—fan shape, definite concave slope, general decrease of grain size from apex to distal edge, and beds parallel to the surface but vertically variable in texture—are characteristic of deposits of present-day anastomosing streams. The lack of foreset delta structure, even in pits at the distal edge of the exposed plain, indicates that this is a subaerial fan and was not built to accord with any lake surface or sea level.

GLACIAL ORIGIN

The very large natural pits in the plain are most convincing evidence that this is glacial outwash. They occur chiefly in groups near Falmouth, Coonamessett, the rifle ranges in the Massachusetts National Guard

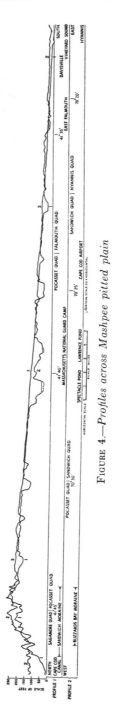

FIGURE 4.—*Profiles across Mashpee pitted plain*

Military Reservation (Camp Edwards), Farmersville, Newtown, the vicinity of Johns Pond, Waquoit, and Osterville. There is no system or regularity to the grouping or to the positions of groups of pits in the plain. More than 30 pits exceed half a mile in diameter. They range in depth from 50 to 120 feet, and several are enclosed completely by gravel walls. The most prominent examples are the depression which contains Snake Pond, west of Forestdale, and the hollows near the center of the Pocasset quadrangle. Some 500 smaller pits are shown as depression contours on the new topographic maps of the United States Geological Survey, and many others observed in the area under discussion are too small to show on maps of the scale used. The side slopes of the pits are inclined 5°, 10° and even as much as 20°. Where closely grouped, the slopes around adjacent depressions may coalesce to form saddle-shaped ridges. The pits vary greatly in shape; some are nearly circular, others elongate oval, and some quite irregular. Within an area of compound pits, the surface may resemble the hummocks and hollows of small "kame and kettle" deposits so common in valleys throughout New England.

These features—the irregular grouping, the great depth, steep slopes, and irregular shape—are prime characteristics of glacial kettle holes. Furthermore, several of these pits, such as those occupied by Long Pond and Oyster Pond near Falmouth, or the elongate depression southwest of Telegraph Hill in Sandwich, extend from the plain into the bordering moraines. As these hollows interrupt the continuous front of the moraine, they must have formed after the moraine was built on top of the outwash plain. Clearly, all these depressions must have resulted from the melting of ice that had been buried during the deposition of the plain.

Two other conditions point to the fact that this is glacial outwash: (1) the plain heads at an altitude of more than 200 feet; river deposits at this altitude could have formed here only during the existence of glacial ice to support the loaded channel of a stream at this high level; (2) fragments of rock of the types known to be in place many miles to the north and west are abundant among the pebbles in the gravel of the plain. Pebbles of the Squantum tillite member of the Roxbury conglomerate, for example, were found at Falmouth Heights and in a pit west of Coonamessett River, northeast of Round Pond. Carboniferous conglomerate pebbles from the Boston or Narragansett Basins were found at a dozen or more exposures. Gray-green melaphyr pebbles like those of the Boston region were observed at a number of localities. The scarcity of these types suggests that they are the few pieces that travelled 25 to 60 miles from their sources and became mixed with far more abundant stones from nearer sources. Varieties of rock like the

TABLE 3.—*Stone-count data*

Composition (in per cent) of pebbles in the Mashpee pitted plain on Cape Cod

Locality*	T-105	G-95	G-59	G-47	G-12	G-40	T-35	G-61	G-81	G-65	T-79	G-35	Averages			
													Mashpee pitted plain	Buzzards Bay moraine	Sandwich moraine	Area of overlap of moraines
Granite (including gneissoid granite, aplite, pegmatite)	61	59	71	62	67	70	71	76	75	47	68	72	67	66	45	60
Quartz-feldspar gneiss	5	2	0	0	0	0	0	0	0	1	0	2	1	1	3	0
Diorite and gabbro (including biotite gneiss)	3	3	2	6	6	2	0	0	1	5	1	2	2.5	3	3.5	2
Basalt (including basalt porphyry)	7	4	1	5	0	6	2	2	1	15	3	2	4	3	9.5	5
Other volcanics (mostly rhyolite and pyroclastics)	9	17	6	9	8	4	10	7	11	11	5	6	8.5	4	13.5	11
Schist (including greenstone)	5	2	3	2	7	2	0	1	0	3	7	2	3	3	5	4.5
Slate (including argillite)	2	2	2	1	0	1	0	0	1	0	0	0	1	t	4	2
Quartzite	5	5	7	8	2	6	7	7	5	7	9	7	6	11	7.5	9
Vein quartz	3	7	7	6	9	9	9	6	5	9	6	6	7	8	7	6.5
Arkose and sandstone	0	1	0	0	0	0	0	0	0	1	0	0	t	0	1	t
Conglomerate	0	0	0	0	0	0	0	1	1	0	0	0	t	t	0	0
Miscellaneous	1	0	1	2	1	1	1	0	1	1	1	1	1	1	1	1
	East to West							North	to	South						

* Localities designated T represent collections by Thiesmeyer; G, Goldthwait.

Dedham granodiorite, which make up a large part of every collection of pebbles (*See* Table 3), probably came from buried ledges of bedrock only a few miles away, although no outcrops are visible within 15 miles of the plain. Regardless of the distance they may have travelled, the only plausible sources for these types of rock are north or west of Cape Cod. Southward or eastward transportation by moving ice and its associated melt-water is implied by their presence in the deposits described here.

RELATIONS TO ICE FRONTS

As outlined in preceding paragraphs the Buzzards Bay and Sandwich moraines were deposited by readvance of the ice after construction of the Mashpee pitted plain and overlap the stream-washed sand and gravel at its western and northern margins. The gravel underlying the till of those moraines is in all respects like that of the pitted plain, but its surface is 50 to 100 feet lower. As may be seen along the western side of the Buzzards Bay moraine, however, the surface of the bedded deposits beneath the till is not smooth like much of the plain, and in places large angular blocks are embedded. Therefore, it is believed that these portions of the plain, now buried under the moraine, were originally at the margin of the great fan. There the outwash was deposited on and around irregular masses of ice at the ragged edge of the ice sheet. Shifting currents, such as would result from movements of water over this rough and changing surface, deposited layers sloping at unsystematic angles. The irregularity of the surface increased as the buried ice melted. From the ice itself angular blocks fell and slid into the gravel at some places.

The water that built the pitted plain carried fragments of rock like those found in both the Sandwich moraine on the north and the Buzzards Bay moraine on the west. Counts of the number of different kinds of rocks among 400 pebbles selected at random at each of the 12 locations bear out the more casual observations made at innumerable pits. (*See* Table 3.) Only one stone count, from west of Peters Pond, is closely similar to counts in the Sandwich moraine, and only two stone counts—one west of Massachusetts National Guard Military Reservation (Camp Edwards) on Turpentine Road, the other east of Shallow Pond on Mill Road—are much like counts in the Buzzards Bay moraine. The average of all pebble counts from the plain shows a proportion of granites, diorites, and basalts like that of the Buzzards Bay moraine, whereas the proportion of other volcanic and quartzite pebbles is like that of the Sandwich moraine. As the types of rock contained in the two moraines are significantly different, it would appear that the water that bore gravel to the pitted plain gathered material from sources within both lobes of the ice sheet. During transportation these pebbles were fairly well commingled.

The uniform slope of the plain toward the south and southeast, away from the apex but parallel to each moraine front, shows that the water entered the plain somewhere near or at the present apex, rather than here and there along the line of the moraines. At only two localities (just southeast of Signal Hill in Bourne and just northwest of Falmouth) does the plain show a marked slope away from the bordering moraine. In Bourne this deposit is made of sand and is shaped like a small fan. In Falmouth it is made of coarse gravel already described in the discussion of the Lawrence pit. Presumably, water flowing through breaks in the ridge on the east side of the Buzzards Bay moraine or directly from the face of the Buzzards Bay glacial lobe superimposed fans on top of the main plain at these two localities.

At two other places constructional gravel deposits rise above the general surface of the pitted plain. The more conspicuous one—Falmouth Heights, on the shore of Vineyard Sound—is elliptical, and its top is 35 feet above the adjacent plain. On the seaward side a 40-foot wave-cut cliff consists entirely of cross-bedded sand and gravel. In the upper layers of the gravel are many ventifacts. This gravel must have been deposited as outwash from the ice prior to the building of the pitted plain, for its altitude implies ice or other gravel immediately surrounding the present site to enable streams to deposit at such a high level. This deposition occurred before any other event registered by the surface deposits on this end of Cape Cod. Presumably it occurred while the ice front was retreating from Martha's Vineyard toward the line of the Buzzards Bay and Sandwich moraines.

Originally the deposit may have been far more extensive than it is now. If so, most of it must have been removed by the braided streams that built the outwash fan around it and by ocean waves now active at present sea level. It is more likely, however, that this body of gravel never extended very much farther toward the north than it does today and that it represents the apex of an outwash fan built southward from the ice front when the latter was temporarily standing close by the Heights. Rather rapid recession of the ice then removed the support from the north side of the fan, and subsequent outwash deposits were never built to so great an altitude.

The other outwash deposit above the pitted plain is a discontinuous, broad, low ridge trending southeast from the edge of the Sandwich moraine, north of the Cape Cod Airport, out into the plain as far as the Barnstable-Falmouth road. At its northwest end, this ridge blends into the moraine front; at its southeast end, it is half a mile south of the moraine. Its undulating crest rises 10 to 25 feet above the adjacent plain, and it consists of well-bedded and cross-bedded sandy gravel without boulders. The gentle south slope and slightly steeper north side of the ridge suggest that it, too, was deposited by glacially fed streams when the ice edge stood close to or along its northern margin.

TILL CLUMPS

Masses of unmodified till are not ordinarily found in glacial outwash at any great distance from the moraines with which the outwash is associated. Irregular clumps of unmistakable till were observed, however, at six localities in the Mashpee pitted plain, far out on the glaciofluvial fan. Five of these occurrences were in pits excavated for sand or gravel: (1) immediately west and (2) a mile north of camp headquarters in the Massachusetts National Guard Military Reservation (Camp Edwards); (3) in the northern part of Bear Hollow Farm south of Snake Pond; (4) a mile west of Bear Hollow Farm; and (5) south of the cranberry bogs in Teaticket. The sixth was in a road cut north of Pimlico Pond. Similar till clumps were later noted at a few localities in the pitted plains of the Sagamore and Wareham quadrangles, west of the Cape Cod canal.

The individual masses of till range from a foot or two in length and thickness to more than 10 feet in longest dimension. They are generally more or less rounded or oval in shape although some have irregular or angular bounding surfaces. Commonly there are several clumps within a few yards of each other at any one locality, each entirely surrounded by sand and gravel (Pl. 2, fig. 2). Each locality is at the rim of a kettle hole. No till was found in the scores of gravel pits and road cuts on the smooth parts of the plain or in the many, small, sand pits near the cranberry bogs that have been developed in the furrows at a distance from the kettle holes. Evidently the till was formerly associated with the ice blocks responsible for the kettle holes near which it is now found.

The rock types represented by the pebbles and angular fragments embedded in the sand and clay of the till clumps are like those of the till in the Buzzards Bay moraine, both with regard to variety and relative abundance of the different components. Ordinarily the compactness and coherence of the till are such that the till clumps stand out on the faces of the artificial cuts after exposure to weathering for only a few days or weeks, during which time the surrounding sand and gravel is loosened and blows away or slumps toward the foot of the cut. The surface of each till clump is commonly stained to a depth of 2 to 8 inches with bright red iron oxides, and streaks of the same color penetrate nearly or quite to the center of some of the clumps. Moreover, the sand and gravel immediately adjacent to the till are ordinarily oxidized similarly so that the same red color appears as an aureole surrounding the till at several localities.

All the clumps that have been observed are in the uppermost 6 to 10 feet of the gravel forming the plain, and some extend upward practically to the surface. Where overlain by sand and gravel the covering is ordinarily so thin that in contrast to the subjacent material the bedding has been obscured or destroyed by frost action, the penetration of roots, and the activities of animals. The filling between clumps is relatively loose, non-bedded, or irregularly bedded sand and fine gravel. The uppermost beds of gravel beneath the clumps are wrinkled into wavelike crests in the spaces 1 to 4 feet wide between the bases of adjacent clumps. At the Teaticket locality, the long till clumps stand on end. Near Bear Hollow Farm the long axes of several clumps are all inclined in the same direction. At the National Guard locality, where the till masses lie more or less horizontally, their ends are twisted into a scroll-like form. Where the cuts were deep enough to expose several feet of thickness of gravel below the till, the gravel layers appeared quite undisturbed at depths greater than 2 or 3 feet beneath the base of the clumps.

Evidently the till masses came into place very late in the building of the plain, and considerable pressure was exerted upon at least some of them as they assumed their present positions. Such pressure may have resulted from contraction and expansion of the ice near the top of a stranded ice block exposed to rapidly fluctuating air temperatures. Till dropped directly on the gravel floor beneath the edge of a melting block of ice may have thus been squeezed into place. Moreover, the streams of meltwater on the surrounding plain would occasionally undercut the sides of the dwindling remnants of the glacial ice and cause chunks of it to fall or slide downward and outward, thus shoving the till vigorously into the gravel to produce the wrinkles in the subjacent beds.

SUBAERIAL FURROWS

The surface of the Mashpee pitted plain is not only pock-marked by kettle holes, but it is also scored by long furrows that head about midway up the plain and extend to its outer edge. The longest measures about 10 miles. The lower parts of the larger furrows are now drained by small streams, such as Coonamessett River and Childs River, but these are not actively enlarging the furrows today. The lower end of each is flooded by the sea and forms a long, narrow bay, such as Great Pond, Green Pond, Bowens Pond, and Eel Pond (Fig. 3).

Erosion rather than deposition is responsible for these furrows, since they are actually cut into the sand and gravel layers of the plain, as indicated by the truncated edges of gravel beds seen in exposures on the walls of furrows. Two of the best exposures available in 1939 were in road cuts half a mile southeast of Jenkins Pond (Pl. 2, fig. 3) and a quarter of a mile southeast of Round Pond. Furthermore, the furrows have the form of stream-cut valleys. Each heads in a narrow, shallow, V-shaped cut. Down valley, most of them increase to approximately 1000 feet in width and 25 feet in depth. Their floors are broad and smooth. Short, V-shaped tributaries enter the main valleys and give each system a stunted dendritic pattern like the veins of a narrow leaf.

The close spacing of the furrows and subparallel orientation of all their main courses are remarkable. Some adjacent furrows are so close where they enter Nantucket and Vineyard Sounds that the intervening flat-topped remnants are rarely as much as half a mile broad. All are oriented within 20° of south. Those near the western edge of the plain are nearly parallel to the Buzzards Bay moraine. Such an arrangement of valleys is not like the dendritic "crows-foot" pattern formed in the normal processes of stream erosion, and it, therefore, requires some special explanation.

As many furrows are not now occupied by streams, present-day erosion is totally inadequate to account for these forms. As the headwaters are not directly connected with areas of moraine, and as the pattern of the furrows is not braided and therefore is not indicative of overloaded streams, the furrows were not cut by the glacial melt water that constructed the plain. It is possible that, after the plain was essentially complete, the extra supply of water from ice melting west and north of the moraines saturated the loose gravels. This ground water might have reached the surface in springs part way down the fan and flowed on the surface from there to the sea. A second possibility is that there was a period of barren, treeless climate, caused either by drouth or cold, during which ephemeral rains cut shallow arroyos in the gravels. It is also possible that the condition of the ground, frozen while the ice sheet still covered the interior of New England, made all save surface gravels impermeable to water and induced shallow cutting in closely spaced channels.

PERSISTENCE OF BURIED ICE

The duration of all these events which contributed to the shaping of the present Mashpee pitted plain is very impressive. Following the isolation of large masses of ice from the wasting ice sheet, braided streams of melt water deposited sand and gravel around and over most of the outlying ice masses. The period required for the construction of such a large plain, which now contains more than 2 cubic miles of gravel, must have been many tens and perhaps hundreds of years.

The surface materials of the plain include many thousands of stones which are plainly cut, grooved, and faceted by the wind. These may be found singly or in quantity in nearly half of the pits now excavated. The deep etching of these ventifacts and their abundance impress all who see them with the vigor and duration of wind action that must have taken place. No comparable wind action is effective on the Cape today. Many of these ancient ventifacts are actually in the beds of gravel near the surface of the plain, but they could not have rolled far in the streams that built that portion of the plain or the delicate polish and edges made by the wind would have been obliterated. It is necessary, then, to assume that the deposition of the upper layers of the plain alone required many tens and perhaps hundreds of years during which wind action was very intense.

Buried ice persisted under the sand and gravel of the plain all the time that the wind was thus active, for none of the sand or silt that must have supplied abrasive tools for the sandblast fills the kettle holes. Furthermore, the furrows were cut before the ice entirely disappeared. Furrows are found to enter and exit from groups of kettle holes with no change of direction. There is no sign of drainage from one kettle hole to another. Dozens of small pits occur on the floors of furrows, and large pits may interrupt and separate a whole headward section of furrow from its lower flooded end at the sea. Had these low kettle-hole depressions existed when the furrows were cut, they would certainly have been filled with stream gravel to the level of the channel floor. As this is not true, the ice must have remained after the plain was built until the furrows had been cut.

Buried ice remained while the adjacent moraines were superimposed upon the edges of the plain. Thin active ice moved forward over this gravel and deposited the till of the Buzzards Bay and Sandwich moraines without destroying the layers of gravel beneath. Masses of stagnant ice caught under the first outwash gravel remained until the active ice overrode them, inasmuch as certain large kettle holes already described extend from the exposed surface of the fan into the moraine-covered area. The ice finally melted away after both the gravel of the fan and the till of the moraine had been deposited.

DROWNED SHORE LINE

The final event affecting the present aspect of the Mashpee pitted plain was drowning by the sea. At no time since completion of the plain has sea level been higher on this portion of Cape Cod than it is today, for no sand bars were found above the modern shore line. These would certainly be present had wave action once been higher on such a sandy shore. Moreover, none of the kettle holes at the lower edge of the plain are filled by wave wash. On the other hand, submerged cedar stumps have been dredged from below low-tide level at Witchmere Harbor in Harwichport, 15 miles east of this pitted plain (Chute, 1939, p. 25), and at Centerville (Sawyer, 1932) near its eastern edge. The lower end of each furrow is obviously drowned; no process of shore erosion or construction could produce such a deeply indented shore line. Following the rise of the sea to its present stand with respect to this part of Cape Cod, sand bars were built across many of the bays and estuaries.

Low wave-cut cliffs facing narrow sea beaches have been eroded along the low margin of the plain. If the original gentle surface of the plain be projected beyond the present cliffs until it touches sea level, it appears that the plain once extended at least a quarter and perhaps half a mile farther seaward. Waves and shore currents have beaten the shore line back by this amount. At distances greater than half a mile off the Falmouth shore, or one mile off the shore at Cotuit, the Coast and Geodetic Survey charts (no. 1209 especially) show a submarine slope far too steep to be the original surface of the pitted plain. If the pitted plain formerly extended beyond that line, recent excavation by currents, or wave cutting at some lower stand of the sea, has modified the bottom contour too much to permit reconstruction of the original outer margin of the plain.

* * * * * * *

SUMMARY OF LAST SECTION

Study of more than 2000 ventifacts embedded in undisturbed deposits and in disturbed surficial zones of western Cape Cod yields corroborative evidence that the deposits were formed during late Wisconsin time. The climate during recession of the ice and sculpturing of the ventifacts was cold and semi-arid. The present shapes of the wind-cut stones were determined by the shapes of the original fragments, textural and structural features of the individual rock, amount of dislodgment and rotation during sandblasting, duration of abrasion while a stone lay in one position, and degree of protection from direct impact afforded by adjacent stones. Because such a large number of lithologic types is represented among these ventifacts a variety of erosional effects which may be used as criteria of sandblasting were observed. Modification of wind-scoured surfaces during later glacial and glacio-fluvial transportation shows that aeolian abrasion occurred during construction of the various landforms made of outwash and prior to the development of each successive recessional moraine. Lack of an appreciable veneer of wind-drifted sand over the surface, even in the bottoms of kettles, indicates that cutting of the ventifacts was completed before all of the residual buried ice blocks had disappeared.

REFERENCES CITED

Bryan, Kirk (1931) *Wind-worn stones or ventifacts—a discussion and bibliography,* Nat. Res. Council, Rept. Comm. Sedimentation, 1929-1930, p. 29-50.

—— (1932) *New criteria applied to the glacial geology of southeastern Massachusetts* (abstract), Geol. Soc. Am., Bull., vol. 43, p. 176.

Chamberlin, T. C. (1883) *Preliminary paper on the terminal moraine of the second glacial epoch,* U. S. Geol. Survey, 3d Ann. Rept., p. 377-379, pl. 33.

Chute, N. E. (1939) *Geology of the coastline between Point Gammon and Monomoy Point, Cape Cod, Massachusetts,* Commonwealth Mass., Dept. Public Works, Spec. Paper 1, p. 1-26.

Davis, W. M. (1895) *Faceted pebbles on Cape Cod, Massachusetts,* Boston Soc. Nat. Hist., Pr., vol. 26, p. 166-175.

Heim, Arnold (1922) *Känguruh-Polituren in Australien,* Zürich, Société Géologique Suisse, vol. 17, p. 349-351.

Hitchcock, Edward (1833) *Report on the geology, mineralogy, botany, and zoology of Massachusetts,* p. 143-144.

Howe, O. H. (1936) *The Hingham red felsite boulder train,* Science, vol. 84, p. 394-396.

Johnson, Douglas (1925) *The New England-Acadian shoreline,* New York, p. 105-117, 295.

MacClintock, Paul (1940) *Weathering of the Jerseyan Till,* Geol. Soc. Am., Bull., vol. 51, p. 103-116.

Mather, K. F., Goldthwait, R. P., and Thiesmeyer, L. R. (1940) *Preliminary report on the geology of western Cape Cod, Massachusetts,* Mass. Dept. Public Works, U. S. Geol. Survey, Coop. Project, Bull. 2, p. 40-43, 46-52.

Sawyer, G. C. (1932) *A drowned forest of white cedar at Centerville, Cape Cod, Massachusetts,* mss. report to Dept. Geology, Dartmouth College, Hanover, N. H.

Sayles, R. W. (1939) *Upper till, two boulder clays, and interglacial flora on Cape Cod* (abstract), Geol. Soc. Am., Bull., vol. 50, p. 1931-2.

Schoewe, W. H. (1932) *Experiments on the formation of wind-faceted pebbles,* Am. Jour. Sci., 5th ser., vol. 24, p. 111-134.

Shaler, N. S. (1889) *The geology of Nantucket,* U. S. Geol. Survey, Bull. 53, p. 21-26; pl. X; p. 46.

—— (1898) *Geology of the Cape Cod district,* U. S. Geol. Survey, 18 Ann. Rept., pt. 2, p. 297-593.

—— and Davis, W. M. (1881) *Illustrations of the earth's surface, Glaciers,* Boston, p. 57-58.

Upham, Warren (1879) *Terminal moraines of North American ice sheet,* Am. Jour. Sci., 3d ser., vol. 18, p. 81-92, 197-209.

Wilson, J. H. (1906) *The glacial history of Nantucket and Cape Cod,* Columbia Univ. Press, p. 4, 7, 48-50.

Woodworth, J. B. (1894) *Post-glacial eolian action in southern New England,* Am. Jour. Sci., 3d ser., vol. 47, p. 63-71.

—— and Wigglesworth, Edward (1934) *Geography and geology of the region including Cape Cod, the Elizabeth Islands, Nantucket, Martha's Vineyard, No Mans Land and Block Island,* Harvard Coll., Mus. Comp. Zool., Mem., Cambridge, Mass., vol. 52, p. 1-322.

22

Reprinted from *U.S. Geol. Survey Prof. Paper 422-A,* pp. 1–5, 6–27, 29–45, 49–51, 56–61, 66–67
(1963)

PHYSIOGRAPHIC AND HYDRAULIC STUDIES OF RIVERS

MORPHOLOGY AND HYDROLOGY OF A GLACIAL STREAM—WHITE RIVER, MOUNT RAINIER, WASHINGTON

By Robert K. Fahnestock

ABSTRACT

This is a study of the processes by which a valley train is formed by a proglacial stream. The area investigated is the White River valley on the northeast flank of Mount Rainier, between the present terminus of Emmons Glacier and the moraine marking the terminal position in 1913. Five square miles of the 7.5-square-mile drainage basin above this moraine are presently covered by active ice.

Measurements of channel characteristics were made in 112 channels developed in noncohesive materials. Channel widths ranged from 0.7 to 60 feet, mean depths from about 0.03 foot to 2.08 feet, and mean velocities from 0.3 to 9 feet per second for discharges of about 0.01 to 430 cfs. The relations between these variables can be expressed by the equations: $w=aQ^b$, $d=cQ^f$, and $v=kQ^m$. The exponents for White River channels were found to be similar to those of streams in the Southwestern States. In contrast, Brandywine Creek, Pa., with cohesive bank materials, had higher velocity exponents and extremely low width exponents. Width and depth of channels in noncohesive materials may change by scour and deposition as well as by flow at different depths in predetermined channels. White River channels, with steep slopes in coarse noncohesive materials, were narrower, slightly shallower, and had much higher flow velocities than the channels of Brandywine Creek in cohesive materials.

Slope of the valley train was related to particle size and discharge. Pebble counting demonstrated a systematic decrease of 60 mm in median diameter of the valley train deposits in a distance of 4,200 feet downstream from the source areas. Discharge was essentially constant through this reach, the stream received no major additions. Discharges of 200 to 500 cubic feet per second were capable of transporting almost all sizes of materials present and thus modified the form of the valley train.

Data on the velocities required to transport coarse materials in White River showed that a curve in which diameter is proportional to velocity to the 2.6 power approximates the relation better than the traditional sixth power law in which diameter is proportional to velocity to the 2.0 power. The few samples contained suspended-load concentrations up to 17,000 ppm.

The most graphic evidence of the large amount of material transported by the White River was the amount eroded and deposited on the valley train itself. Measurements indicated an average net increase in elevation of 1.2 feet during 1958 and a net decrease of 0.12 foot in 1959.

Description and analysis of the change in pattern of the White River were difficult at high flows because of the rapidity of the change. However, a marked change from a meandering pattern to a braided pattern took place with the onset of the high summer flows and the pattern returned to meanders with the low flows of fall.

Explanations offered in the literature for the cause of braided patterns include erodible banks, rapid and large variation in discharge, slope, and abundant load. The common element in all explanations seems to be a movement of bed load in such quantity or of such coarseness that there is deposition within the channel, causing the diversion of flow from one channel into other channels in a valley wide enough to provide freedom to braid. White River braiding took place most actively at large loads and discharges.

Although examples of braiding by an aggrading stream are common, the White River and the Sunwapta River, Alberta, have reaches in which degradation took place while the stream had a braided pattern. The conclusion is reached that both braided and meandering reaches can occur along the same stream, which may be aggrading, poised, or degrading. The pattern alone does not conclusively define the regimen of the stream.

The regimen of the glacier has long-term effects in providing debris to the stream; short-term effects of weather and runoff determine the current hydraulic characteristics, rate of deposition and erosion, and channel pattern.

INTRODUCTION

This is a study of the processes by which a valley train is formed by a proglacial stream. The term valley train, as used here, is an outwash plain laterally constricted by valley walls. Studies of such distinctly different geologic environments improve the understanding of past geologic events and knowledge of the interrelations of hydraulic and morphologic variables. Comparisons of data from distinctly different environments make possible the evaluation of the sensitivity of hydraulic parameters to environment.

It is hard to imagine a more radical departure from normal stream regimen than a glacierized drainage basin (fig. 1). Diurnal fluctuations in discharge bring bankfull or overbank flow for brief periods to many short reaches of the stream. As a large part of the precipitation falls as snow, it may be years before heavy accumulations at high elevations are reflected in changes of position of the glacier and in runoff.

A1

Editor's Note: A row of asterisks indicates material deleted owing to lack of space. Certain figures and tables have also been omitted for reasons of brevity.

Most of the runoff occurs during the months of June, July, and August, which in many environments would be the period of extreme low water. The presence of the Emmons Glacier makes it possible to study the relation of stream and glacial regimen. The stream pattern changes rapidly in response to changes in discharge which cause much shifting of debris on the valley train. The coarseness of the materials being transported, the high stream gradient, and the rapid flow provide situations in which the influence of these factors can be measured under extreme conditions.

Hjulström (1935) and Sundborg (1954) in their detailed studies of river systems summarized contemporary knowledge of sediment transport and open-channel hydraulics and applied this knowledge to natural channels, which had low velocities, tranquil or subcritical flow, relatively fine bed materials, and low slopes. Leopold and Maddock (1953) presented a method of quantitative analysis of channel characteristics of natural streams. They limited their discussion to a number of rivers in the Great Plains and the Southwest. Wolman (1955) applied the method to a stream in a more humid region. In a recent investigation where these methods were used, Miller (1958) studied high mountain streams in regions which had at one time supported glaciers. All of these channels differed from laboratory channels and from the rapid streams which issue from glaciers to flow with steep gradients and constantly changing channel patterns to the sea. Hjulström led an expedition during the summers of 1951 and 1952 to study the alluvial outwash plains (sandurs) of Iceland and the mechanics of braided streams. The stream studied by Hjulström is much larger and has more gentle gradients than the White River and pattern changes appear to take place much less rapidly than on the White River valley train.

The first section of this report describes the regional setting and the recent history of the Emmons Glacier. The second and third sections cover the hydrology and hydraulic characteristics of White River channels and the transportation, erosion, and deposition of materials of the valley train. The fourth section is a description and analysis of the channel pattern change on the valley train. The fifth section is a discussion of the related problems, the application of the concept of equilibrium to valley train formation and the influence of the glacier regime on valley train formation.

The present report is a modification of a thesis [1] submitted to Cornell University. Work in the field was performed by the author as a member of the Geological Survey under the general supervision of C. C. McDonald, Chief, General Hydrology Branch. The

[1] Fahnestock, Robert, K., 1960, Morphology and Hydrology of a glacial stream: Ithaca, N.Y., Cornell University, doctoral thesis.

field assistants were T. G. Bond in 1958 and P. V. D. Gott in 1959. Cornell University professors Marvin Bogema, R. A. Christman, P. G. Mayer, and C. M. Nevin, graduate committee members; E. H. Muller and L. L. Ray, committee chairmen; and M. G. Wolman, U.S. Geological Survey, made many helpful suggestions in regard to the fieldwork and the manuscript.

John Savini, U.S. Geological Survey, was extremely helpful in the field through his aid with problems of stream gaging and in preparation of streamflow records. Ann M. Fahnestock, the author's wife, provided material support in the fieldwork, maintenance of camp, and preparation of the original manuscript.

DESCRIPTION OF STUDY AREA

PHYSICAL FEATURES

The White River study area lies within Mount Rainier National Park on the northeast flank of Mount Rainier (figs. 1 and 2), west of the crest of the Cascade Mountains, 80 miles south southeast of Seattle, Wash. The area of most intensive study was the 1-mile reach of the stream between the present terminus of the Emmons Glacier and the valley loop moraine, which marks the 1913 terminus. The study area is reached from the White River Campground by a truck trail leading up the left bank of the Inter Fork and by a trail across the moraines. This area was selected because it is accessible and includes both a site for a gage where the stream is confined to one channel and a reach where at times the channel pattern changes rapidly. Measurements to determine the relation of size of material and slope were made in several other areas.

An area of 7.5 square miles (M. F. Meier, written communication, 1958) is tributary to the gage at the moraine (fig. 2). Of this area, 5 square miles was covered by active glacier ice, 4.4 square miles by the Emmons Glacier, and 0.6 square mile by that part of the Frying Pan Glacier thought to contribute to the White River above the gage. Flow from the Frying Pan Glacier, which does not pass through the study area, enters the White River about 3 miles below the gage by way of Frying Pan Creek.

In the period 1957–59 the Emmons Glacier (fig. 2) extended from its source area on the summit of Mount Rainier at an elevation of more than 14,000 feet to an elevation of about 5,300 feet through a valley carved into the northeast flank of the mountain. The depth of this erosion is suggested by the following data: The surface of the glacier is about 2,000 feet lower than the summit of Little Tahoma Peak (fig. 1) on a line transverse to the valley; and the terminus in 1910 was at an elevation of about 4,700 feet near the present stream-

FIGURE 1.—Emmons Glacier and the White River study area, Mount Rainier. Photograph by M. F. Meier, October 1, 1958.

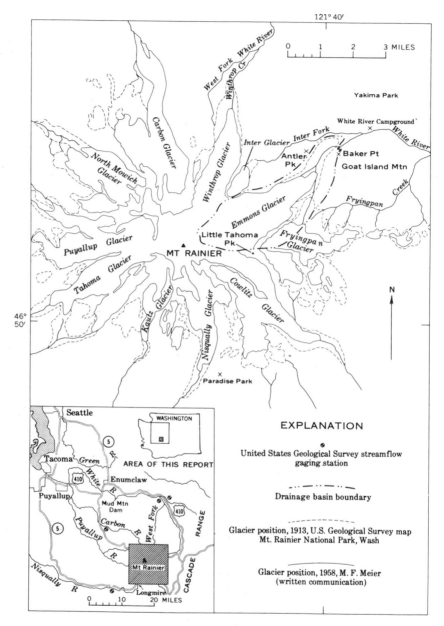

FIGURE 2.—Study area and distribution of active glaciers on Mount Rainier, 1913 and 1958.

gaging station, which is some 1,700 feet lower than the adjacent Yakima Park, Baker Point, and Antler Peak (fig. 2). The White River is tributary to Puget Sound through the Stuck and Puyallup Rivers.

A stream-gaging station equipped with a water-stage recorder is 1 mile below the glacier at the moraine; another station, White River at Greenwater, Wash., is 28 miles downstream, just above the junction with the Greenwater River and 2 miles below the junction with the West Fork White River. The source of the West Fork White River is the Winthrop Glacier, which emanates from the same icefield as the Emmons Glacier.

* * * * * *

EMMONS GLACIER

The study area has been uncovered by the stagnation and melting of the Emmons Glacier since its position was recorded by some of the earliest observers. One of the first authors to record his observations on the "White River Glacier" as it was then called, was S. F. Emmons, who with A. D. Wilson made the second successful ascent of the mountain in 1870. Emmons in a letter to his chief (King, 1871) seems to have been overenthusiastic and exaggerated the glacier's dimensions, stating:

The main White River glacier, the greatest of the whole, pours straight down from the rim of the crater in a northeasterly direction, and pushes its extremity farther out into the valley than any of the others. Its greatest width on the steep slope of the mountain must be four or five miles, narrowing towards its extremity to about a mile and a half; its length can be scarcely less than ten miles.

The map of Mount Rainier and its glaciers by Sarvent and Evans dated 1896, (plate 68, *in* Russell, 1898, p. 363) shows the Emmons Glacier terminus in approximately the same position as shown by the 1913 U.S. Geological Survey map (fig. 2). The 1896 map, however, shows two streams issuing from the glacier—one as shown on the 1913 map and the second flowing into the Inter Fork at the left side of the glacier terminus. F. E. Matthes (1914) stated:

The youngest moraine, fresh looking as if deposited only yesterday, lies but 50 feet above the glaciers' surface and a scant 100 feet distant from its edge; the older ridges subdued in outline and already tinged with verdure lie several hundred feet higher on the slope.

Matthes gave the length of the Emmons Glacier as about 5½ miles and its width as 1¾ miles in its upper half. The area of active ice, measured from the 1913 map as approximately 5.3 square miles, had decreased to 4.4 square miles in 1958 (M. F. Meier, written communication). Most of the 0.9 square mile decrease represents areas of ablation moraine and valley train underlain by slowly melting stagnant ice. Matthes' (1914) estimate of 8.5 square miles could not be verified by rechecking with the 1913 map.

Periodically since 1930 the U.S. National Park Service has studied the position of the glacier terminus; the data show the rate of recession of the point at which the stream emerged from under the glacier to be about 75 feet per year. The positions of the glacier (fig. 3) are sketched from Park Service data and photographs provided by V. R. Bender, park naturalist. In 1943 an ice tunnel was discovered to have caved in upstream from the ice face being measured; the caving produced a second point from which the ice faces receded both upstream and downstream. This second point was near the present junction of East and West Emmons Creeks; thus, an ice mass was left bridging the valley until at least 1953.

Rigsby (1951) spent 6 weeks in 1950 studying ice petrofabrics and glacier motion on the Emmons Glacier. He determined the rate of motion to be as much as 0.75 foot per day in the center of the glacier, about half a mile above the position of the terminus in 1958 (fig. 2).

The general advance of glaciers of the Cascade Mountains (Hubley, 1956, and Harrison, 1956a and 1956b) had also started on the Emmons Glacier by 1953. The National Park Service has measured this advance periodically since 1953 and has found that it averaged about 165 feet per year from 1953 through 1957 and continued in 1962. In 1957 Arthur Johnson, of the U.S. Geological Survey, began taking a yearly series of phototheodolite pictures from which topographic maps of the glacier might be made by terrestrial photogrammetry. Volumetric computations as well as measurements of the change in position of the glacier terminus will be possible when these maps are available.

The present location of the valley-train is in the approximate position of the clear ice shown by the 1913 U.S. Geological Survey map. The debris-covered ice along the valley train appears to have changed little in elevation since the 1913 map, emphasizing the role played by the debris in insulating the ice along the sides of the glacier tongue. The approximate elevation change of the clear ice is shown in figure 4. Downwasting is almost overshadowed by the combination of caving and melting, and by erosion and melting from below by streams of water and air that flowed through ice tunnels in this central part of the glacier. In 1959,

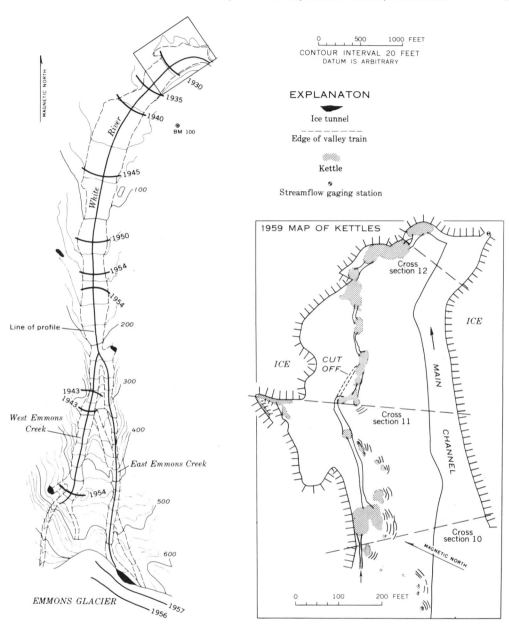

0   500   1000 FEET

CONTOUR INTERVAL 20 FEET
DATUM IS ARBITRARY

EXPLANATON

Ice tunnel

Edge of valley train

Kettle

Streamflow gaging station

1959 MAP OF KETTLES

Cross
section 12

ICE

CUT
OFF

ICE

MAIN

Cross
section 11

CHANNEL

West Emmons
Creek

East Emmons Creek

Cross
section 10

MAGNETIC NORTH

EMMONS GLACIER

0   100   200 FEET

FIGURE 3.—Map showing approximate positions of Emmons Glacier ice front, 1930–58, from National Park Service glacier surveys, line of profile (fig. 4), and inset map showing kettles in 1959.

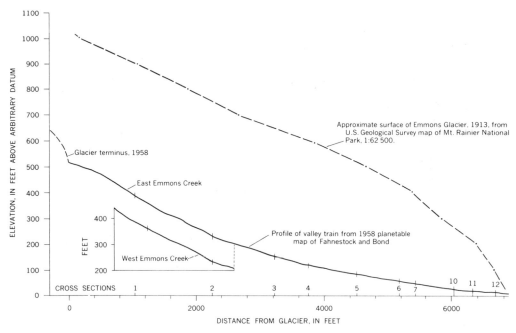

FIGURE 4.—Profiles of valley train and former glacier surface. Vertical exaggeration ×4.

an ice tunnel about 10 feet wide was observed to double in width and height throughout its length by melting due to the flow of warm air. There was no apparent change in elevation of the debris-covered ice above the tunnel. Melting from below soon leads to collapse of tunnels, producing ice blocks that rapidly waste away. Such ice blocks are the source of many well-rounded boulders of ice deposited on bars downstream.

The presence of the numerous kettles shown in figures 3, 5, and 6 is evidence that at least some part of the valley train is deposited on ice. These kettles appeared during August 1958 when the stream with its heavy load of debris had shifted for a few days from the areas of valley train underlain by ice. Several new kettles were seen in June 1959; they developed rapidly in size, occasionally coalescing to form larger kettles. By June 25, 1959, the development of kettles demanded attention, and on this day they were mapped by planetable methods. (Inset map fig. 3.) The kettles were mapped again on July 31, after they had been filled with debris by the stream. All were filled in during 5 hours on July 20, 1959, with the exception of one which the stream did not reach. This kettle was filled in a few hours on July 23. Cracks in the ground and incipient

kettles formed within 24 hours after the most actively enlarging kettles had been filled. It was evident that material was disappearing from the bottom of the kettle because they became no shallower by the caving of their banks. The kettles that could be waded had cross sections similar to those shown in figures 5 and 6. Noticeable sifting out of the fine materials occurred as the kettles formed, the coarser materials being concentrated toward the bottom of the funnel-shaped depression. It is probable that this cross section was characteristic of all kettles not receiving materials from through-flowing streams.

CHARACTERISTICS OF WHITE RIVER CHANNELS

Streams may vary in discharge of water, size and amount of sediment load, width, depth, velocity, slope of water surface, hydraulic roughness and pattern of their channels. It was possible to measure each of these variables for the White River with the exception of the amount of sediment in transport. Leopold and Maddock (1953, p. 33) suggested that discharge and sediment load are the result of the climatic and geologic environment within the basin and that the other variables adjust to these imposed conditions.

FIGURE 5.—Small kettle. Note the absence of finer material and concentration of the coarser material toward the center.

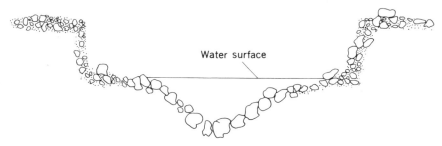

Water surface

FIGURE 6.—Diagrammatic cross section through kettle, figure 5.

They also presented a method for the analysis of channel characteristics that permits comparisons with variation in discharge at the same channel cross section and successive cross sections downstream. Wolman (1955), Leopold and Miller (1956), Miller (1958), Brush (1961), and Wolman and Brush (1961) have applied this method to the analysis of a variety of natural streams and flume channels.

The shape of a channel is reported to be a function of the type of materials that make up its bed and banks (Blench, 1956; Schumm, 1960; Wolman and Brush, 1961), and of the quantity of water and sediment transported by the stream, including possibly their distribution in time. Where a stream flows on its own deposits, the earlier water and sediment discharges have determined the size and disposition of the material composing the bed and banks. Streams will adjust in different ways to increases or decreases in discharge of sediment and water, in accordance with the character of bank materials. It is thought that two streams with the same mean annual discharge will develop somewhat different channel characteristics if the flow distribution for one is uniform throughout the year and for the other is the product of periodic flooding.

The bed and bank materials of the White River, to be described in detail later, are deficient in silt- and clay-size materials and are therefore noncohesive and relatively easily eroded. They provide abundant bed and suspended loads. Observations indicate that a change in the channel shape by erosion or deposition, or both, accompanies any change in discharge. As this was not true for those streams previously described by Leopold and Maddock (1953) and Wolman (1955), some of their data may not be strictly comparable with those of the White River. In spite of this, comparisons can be made that reflect and help define the differences in characteristics of channels in readily erodible materials and channels in more resistant materials.

DISCHARGE

Measurements of channel characteristics were made in 112 channels that were selected as representative of a wide variety of channel sizes and shapes. It was necessary to make discharge measurements with reasonable accuracy—a criterion that eliminated channels in which bed-load movement was extremely heavy and which were rapidly changing their shape. Of these representative channels, 81 were included in the 48 measurements of total discharge (table 7) made for the primary purpose of rating the gaging station. This station was established on June 19, 1958, at the point where the stream flows through the moraine (figs. 2 and 3). Although this position did not provide the most desirable hydraulic characteristics for the stage-discharge relation because of the shifting control, it was the only position where the stream was confined to one channel and could be expected to flow past the gage at all times.

The frequent changes in the stage-discharge relation were caused by scour and fill in the vicinity of the gage as a result of the large bed load and mean velocities up to 9 feet per second. Frequent measurements of discharge were required to compute these changes.

Without doubt the variation in the stage-discharge relation introduced inaccuracies in discharges determined from the recorded stage, but the frequency of discharge measurements and adjustment of the stage-discharge relation probably kept these errors within reasonable limits. The adjustments were made by John Savini according to standard Geological Survey methods for gaging streams with shifting control. He utilized numerous discharge measurements, temperature records, and records from the nearest gaging station, White River at Greenwater, Wash., as well as records from the station recording runoff at South

Cascade Glacier, in the northern Cascade Mountains, to compute the adjustments.

Discharge measurements were made in reaches of the stream where 1 to 4 channels were found. At the time of highest discharge, measurements could not be made in multiple channels because streamflow shifted rapidly from one channel to another during the period of measurement. The channels in which rating measurements were made, although typical of the White River in general, are not representative of some of the widest, shallowest channels, in which much of the flow was around boulders, preventing accurate measurements. These wide, shallow channels were extremely unstable and often divided into several narrower channels.

A number of measurements were made to obtain data for channels with a wide range in discharge and shape. Measurements were made also in very small channels with a discharge of 1 cfs or less, for comparison with the larger channels of the White River and with data obtained for flume channels by Wolman and Brush (1961).

Most measurements of discharge were made by an expedient method, adopted to reduce the wear and tear on equipment and personnel. Several current meters were broken, and feet and shins were bruised by the coarse material being transported by the stream. The velocity and depth were determined at 10 to 15 stations rather than the 20 or more stations customarily used to average out variations in flow across the channel. The Price type-A current meter was used at 0.6 depth for 20 to 35 seconds rather than at 0.2 and 0.8 depth for the usual 40 to 70 seconds. Whatever inaccuracies were introduced by these expedients were in part offset by the fact that the shorter period of measurement gave less opportunity for changes in stage to affect the accuracy. A pigmy current meter was used to measure the velocity in small channels. When the pigmy meter could be used at only a few stations because of extremely shallow water, an accurate cross section was measured with a steel tape, and this area was used with adjusted float velocities to compute the discharge. The mean velocity in a vertical was taken as 0.6 of the float velocity, on the basis of 20 discharge measurements where both current meter and float velocity data were available for the same channel.

The estimated error in current-meter measurements of discharge ranged from 5 percent at moderate discharges to 15 percent at high and low discharges. The accuracy of discharges estimated from gage heights for use in the study of stream pattern depends both on the accuracy of the rating measurements and on the length of time elapsed between such rating measurements and the event for which discharge was estimated. The possible error is as much as 25 percent of the estimated discharge.

The hydrograph (fig. 50) computed by Savini shows the fluctuations in discharge and differences in runoff for the summers of 1958 and 1959. The year 1958 was unusual for the Pacific Northwest because snowfall was light during the winter of 1957–58 and the summer was warm with long spells of clear weather, which produced high ice melt. By midsummer, areas of the mountain at high elevations, which had not been uncovered before in the memory of local inhabitants, were free of snow. Winds blowing over such areas raised a column of dust above the southwestern side of the mountain during July and August. The snow and ice on the mountain were brown from fallen dust, which served to increase melting by lowering the albedo of the snow and ice.

In 1959, snow at all elevations lasted much longer; many areas that were uncovered the previous summer remained · under snow. Melt water produced during the summer was considerably less than in 1958 (hydrograph, fig. 50). The average flow for July 1959 was about 20 percent less, for August about 35 percent less, than for the same months of 1958. The variations from the long-term mean for average temperature and total precipitation for the periods October 1957 to September 1958 and October 1958 to September 1959 at Longmire on the southwestern side of the mountain, and at Stampede Pass 20 miles to the north, might be considered representative of the difference in weather in 1958 and 1959. These two stations were the nearest for which records were available. In 1958, the temperature at Longmire averaged 2.7°F, and at Stampede Pass, 2.6°F, above the long-term mean. In 1959, the temperature averaged 0.2°F above the mean at Longmire and 1.0°F below the long-term mean at Stampede Pass. Precipitation records for Longmire show a deficiency of 5.41 inches over the long-term mean in 1958 and a surplus of 24.44 inches in 1959. At Stampede Pass a precipitation deficiency of 19.25 inches in 1958 and a surplus of 29.15 inches in 1959 were recorded. At both locations the 1959 total was increased in an extraordinarily rainy September by 9.72 inches of precipitation at Longmire and 11.20 inches at Stampede Pass.

The order of magnitude of the winter flow given by discharge measurements 21a, 49, and 50 (table 7), made in November 1958 and 1959, is 20 to 40 cfs. The gagewell froze in late October and the gaging station remained inoperative for low flows until May, so that these measurements are the only flow data available for this period.

The reasons that the melting snow of April, May, and June do not produce flows at the moraine gage (figs. 2, 3) as high as those produced by the melting of July and August may be the following: (a) Unlike nonglacierized drainage basins, which usually have lost their snow cover by late June, more than 70 percent of the White River drainage basin above the moraine has a year-round snow and ice cover. The warm weather of the summer months produces rapid melting. (b) Flow continues throughout the winter. It is thought to result from ground-water discharge as well as from subglacial melting, as the daily mean temperature often falls below freezing for periods of several days. Thus, in the spring, ground-water recharge must take place before snowmelt can produce much direct runoff. (c) The albedo of snow is much higher than that of ice (Sauberer and Dirmhirn in Hubley, 1957, p. 77, table 5), so that the same amount of radiation produces more melt water per unit area from exposed ice than from snow and the area of exposed ice enlarges throughout the melt season. (d) Cloud cover is a factor that should be considered, although records are not available for analysis. The shielding from radiation of the heavier cloud cover of spring may well be offset by heavier precipitation. Summer precipitation was slight in both 1958 and 1959.

No records or field evidence of flooding prior to 1957 were found for the study area. As the surface of the valley train is constantly changing owing to lateral shifting of channels and deposition, all evidence might well be obscured a few years after a flood. That floods do occur in the downstream part of the basin is suggested to even a casual observer by the size of Mud Mountain Dam on White River, 47 miles downstream. It is the highest earth-cored rock-fill dam in the world, with a flood storage capacity of 106,000 acre-feet.

The highest discharges from June 1958 to October 1959 were recorded during the summers and were about 20 times the discharges measured in winter. These discharges, although capable of radically altering the appearance of the valley train, can hardly be considered floods because of the relatively small discharges and the limited reaches in which overbank flow occurred.

In the fall and winter of 1959, there were two severe storms at Mount Rainier. On October 22 a rainfall of 5.5 inches in 24 hours was recorded at Paradise on the southwestern side of the mountain. It produced a rise of about 4 feet at the staff gage at Longmire (fig. 2) and a rise of about 1 foot at the station on the White River at the moraine. A second storm on November 22, 1959, produced a rise of about 1.2 feet in the White River; the peak flow was estimated by Savini to be about 1,000 cfs. Where the flow was concentrated in one channel at the gage, boulders estimated to be as large as $8 \times 8 \times 5$ feet were undermined and moved; however, photographs taken by Savini 1 week after this storm show no major changes in the valley train deposits, although some scour is evident in that part of the valley train within 100 yards above the gage. The effects of flooding were much more evident on the Inter Fork, the West Fork, and other tributaries where roads and bridges were washed out.

Surveys run in June 1960 showed that there was several feet of deposition over parts of the valley train. This storm may have produced a greater proportion of overbank flow than usual, but in most of the area the flow was probably contained in definite channels. The frequency of recurrence of such a flow cannot be established with such a short period of record; but on the White River at Greenwater, Wash., 28 miles downstream (fig. 2), with 30 years of record, such a storm was estimated to have a recurrence interval of 45 years. There was a third major storm during December, but its effects were not separable from those of the other storms when the area was revisited in June 1960.

Diurnal fluctuations in discharge for the White River range from an estimated 10 cfs on a winter day to more than 700 cfs after several consecutive warm summer days. Figure 7 is a reproduction of the water-stage recorder chart for July 21-23, 1958, which shows three types of fluctuations recorded by the gage: (1) diurnal, (2) intermediate, (3) minor. The largest is the diurnal fluctuation, which reaches its peak in the late afternoon and its low point in the early morning. Its magnitude, on July 22, was about 0.65 foot. This represents a change in discharge from 140 cfs to approximately 600 cfs. The fluctuations of intermediate size, such as the one at 2 p.m. on July 22, with a magnitude of 0.3 foot and a duration of about 1 hour, could represent either bar formation and removal that temporarily

altered the stage-discharge relation, or periodic storage and release of water from the crevasse system of the glacier. Other more probable explanations for such variations are the storage and release of water by the formation and destruction of antidunes, or a change in bed roughness in the vicinity of the gage similar to that described by Dawdy (1961). The minor fluctuations of 0.1 foot and 5-minute duration, superimposed upon the other fluctuations, are thought to be related to the movement of small amounts of material or to changes in position of standing waves in the control section of the gaging station, although it is possible that some of these minor fluctuations reflect changes in discharge.

WIDTH, DEPTH, VELOCITY, AND AREA

In the process of measuring discharge, the width, depth, velocity, and cross-sectional area of flow were determined. Widths ranged from about 0.7 foot at 0.01 cfs to a maximum of 60 feet at 330 cfs (channels 38 and 25, table 7). Mean depths of the White River ranged from 0.03 foot to 2.08 feet for discharges of 0.01 to 430 cfs (channels 38 and 88, table 7). Mean velocities of the White River ranged from 0.3 to 9 feet per second for discharges of 0.01 to 300 cfs (channels 38 and 83, table 7).

These values, plotted against their respective discharges on logarithmic paper, using the method of Leopold and Maddock (1953), showed a straight-line relation for the channels of the White River (fig. 8). For such a straight line, the relation between the variables can be expressed in terms of the following equations:

$$w = aQ^b$$
$$d = cQ^f$$
$$v = kQ^m$$

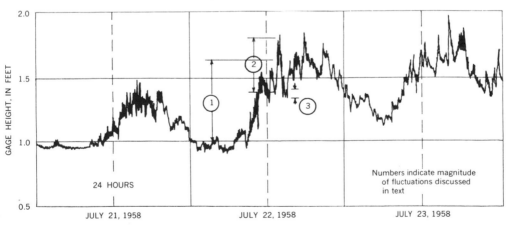

FIGURE 7.—Gage height record of the White River at moraine gage for July 21-23, 1958.

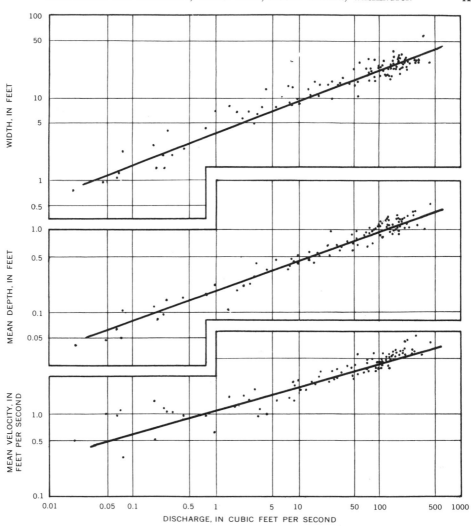

FIGURE 8.—Relation of mean velocity, mean depth, and width to discharge of the White River channels.

The exponents b, f, and m, which represent the slopes of the respective lines, and the coefficients a, c, and k, which represent the intercepts at a discharge point equal to 1 cfs, are parameters characteristic of a given situation and remain constant for that situation.

Values of the parameters determined for the White River, and those for other rivers for which values could be determined from the literature, are summarized in table 1. Changes in width, depth, and velocity in relation to discharge were compared using the values for changes "at-a-station," because the scatter caused by measuring in a number of different channels is thought to be analogous to that shown by Wolman (1955) to be caused by measuring at different points in the vicinity of a gaging station. The wide range of values in table 1, and in some cases lack of information about the parameters, is due to the great diversity of streams considered. Only one paper (Wolman,

681–370 O—63——3

TABLE 1.—Channel parameters of the White River and other streams

Stream	At a station						Values at 100 cfs			Downstream [1]						Data based on—
	Intercepts			Exponents			width	depth	velocity	Intercepts			Exponents			
	a	c	k	b	f	m				a	c	k	b	f	m	
White River [2]	4.0	0.22	1.1	0.38	0.33	0.27	24	1.0	4.2	4.0	0.22	1.1	0.38	0.33	0.27	112 channels.
Southwest and Great Plains: [3]																
Median	26	.15	.37	.26	.40	.34				4.0	.5	.9	.5	.4	.1	20 stations.
Range				(.03–.59)	(.06–.63)	(.07–.55)				(1.1–7)	(.1–.9)	(.6–1.2)				
Brandywine Creek: [4]																
Median	54	.23	.10	.04	.41	.55	67	1.5	1.3	5.7	.19	1.04	.57	.4	.03	7 stations.
Range	(37–80)	(.12–.52)	(.022–.16)	(.0–.08)	(.32–.46)	(.48–.69)	(45–90)	(1.1–2.0)	(.7–2.0)							
Ephemeral streams: [5]																
Median	10	.1	1	[6].26	[6].33	[6].32	45	.5	3.3	5.2	.11	1.5	.50	.28	.22	9 stations.
Range	(3–26)	(.03–.2)	(.7–1.5)				(20–80)	(.25–.9)	(2.5–5.0)							
Flume channels: [7]																
D_{50} 0.0022 foot	4.2	.18	1.3	.50	.39	.16				4.2	.18	1.3	.50	.39	.16	29 channels.
D_{50} 0.0066 foot	2.0	.27	1.8	.33	.52	.16				2.0	.27	1.8	.33	.52	.16	13 channels.

[1] Values based on mean annual discharge except for White River and flume channels.
[2] White River: sandy gravel bed and banks, no vegetation, adjust at all stages, Q=0.01–400 cfs.
[3] Leopold and Maddock (1953): streams of Southwestern States and Great Plains, bed and bank material not noted.
[4] Wolman (1955): Brandywine Creek, cohesive banks, gravel and sand bed, little scour or fill. Q=5–2,000 cfs.

[5] Leopold and Miller (1956): ephemeral streams, sandy bed, silty clay banks, adjust at low flows. Q=0.15–800 cfs.
[6] Unadjusted median values (Leopold and Miller, 1956 p. 15).
[7] Wolman and Brush (1961): flume channels, uniform sand in bed and banks, adjust readily to equilibrium.

1955) limits the discussion of the variables and parameters to a single stream and its tributaries. The study of flume channels by Wolman and Brush (1961) supplies data considered comparable to data from the White River. It must be noted that some of the variation of White River channels is due to the slopes, which are higher than those for any of the other streams reported.

Values of the exponents for White River channels were found to be similar to the average of those found for streams of the Southwestern States (table 1) by Leopold and Maddock (1953). The value for b, the width exponent, was 0.38 for the White River whereas the average for Southwestern streams was 0.26. The value of 0.38 lies well within the range of values observed for Southwestern streams. In contrast, the Brandywine Creek stations (table 1) have the extremely low values for b of 0.04 to 0.08, which are thought to reflect the cohesive nature of bank materials.

The value of f, the depth exponent, was 0.33 for the White River, which is close to the 0.40 average for Southwestern streams and the 0.41 for Brandywine Creek, although the manner in which depths change is dissimilar. Width and depth of streams flowing in noncohesive materials may increase or decrease by scour and deposition, or by changes in effective boundary roughness as well as by changes in discharge. Streams in cohesive bank materials change primarily by flowing at different depths in a channel determined by previous discharges.

The value of m, velocity exponent, for the White River was 0.27, which was not materially different from the 0.34 average of the Southwestern streams data but differed considerably from the value of 0.55 for Brandywine Creek.

According to Leopold and Maddock (1953, page 9),

Width and depth for a given discharge vary widely from one cross section to another, and therefore, the intercept values of a, c, and k, [the values at a discharge of 1 cfs] will vary. Further work is necessary to determine the factors which govern these variations and to determine the extreme limits.

A comparison of these intercepts, computed from the studies listed in table 1, may allow insight into their significance. The value of the parameter a for White River channels is 4.0. That is, a width of about 4 feet was measured for channels with a discharge of 1 cfs. A similar value can be determined by projection of the line of the graph (fig. 8) for White River channels of much higher discharges. Other intercepts for White River channels are mean depth c, 0.22 foot, and mean velocity k, 1.1 feet per second.

The corresponding values for discharges of 1 cfs for Southwestern streams are: width, 26 feet; depth, 0.15 foot; and velocity, 0.26 foot per second. Intercepts for Brandywine Creek, averaged from values projected from discharges of 30 cfs or less, are width 54 feet, depth 0.23 foot, and velocity 0.1 foot per second.

It is apparent that a width of 26 feet, and especially one of 54 feet, is unrealistic for a discharge of 1 cfs. The extremely low values of velocity, 0.1 to 0.26 foot per second, are those of a stream incapable of moving any of its bed or bank materials. The Brandywine Creek data would seem to indicate that the extreme values of width and velocity cannot be explained entirely by error introduced through extrapolation, since the extrapolation is not large, but reflect instead the failure of the channel to adjust to small discharges.

It is interesting to note that the intercept values for ephemeral streams (Leopold and Miller, 1958) 9 feet wide, 0.1 foot deep, with a velocity of 1 foot per second, show neither extreme widths nor extremely small velocities, but present a picture of a stream able to transport material and determine its channel shape at discharges near 1 cfs.

If stream channels adjust readily to changes in discharge as do White River channels, then there is no reason to expect a difference between at-a-station and downstream relations. Wolman and Brush (1961, table 5) found that there was no difference for channels in noncohesive materials where the materials were at or above the point of incipient motion. A comparison based on this conclusion is made in table 1 between White River data, flume data, and data from other stream systems.

The values (table 1) of the intercepts a, c, and k for White River (at-a-station) were 4.0, 0.22, and 1.1; the exponents, b, f, and m, were 0.38, 0.33 and 0.27. These values are similar to the ranges in downstream values of a, 1.1–7; c, 0.1–0.9; and k, 0.6–1.8 and the ranges of b, 0.33–0.57; f, 0.28–0.52; and m, 0.03–0.22 for the other streams discussed. This much greater similarity for all types of channels represents adjustment in all cases. In White River, ephemeral, and flume channels it represents adjustment to changing discharge in short reaches of the stream, and in the other streams it represents adjustment to the mean annual or larger discharges, which increase downstream.

A more realistic picture, at least in the case of streams in cohesive materials, is given by the evaluation of width, depth, and velocity at a discharge of frequent occurrence. Table 1 contains a summary of the values obtained for the intercepts and the average or median values of width, depth, and velocity at a discharge of 100 cfs. Brandywine Creek, with a discharge of 100 cfs, has average values of 67 feet for width, 1.5 feet for depth, and 1.3 feet per second for velocity. The White River has an average width of 24 feet, a depth of 1.0 foot, and a velocity of 4.2 feet per second at a discharge of 100 cfs. White River channels, in coarse noncohesive materials, have been shown to be narrower and slightly shallower and to have much higher velocities than the channels of Brandywine Creek, with cohesive bank materials. The higher velocities in White River channels, which are due to the high slope, permit smaller cross-sectional areas to carry the same flows.

SHAPE

Individual elements of the cross-sectional shape of White River channels have been described in the preceding section in relation to the flow that produced that shape. The shape of the channel cross sections now will be illustrated and the elements of shape compared by considering the width-depth ratio and a shape factor.

Width-depth ratio.—Width-depth ratio—that is, the ratio of top width to mean depth—may be a more useful and valid criterion for comparison of individual channels, as well as comparison of the relation of channel shape to bed and bank material, than the consideration of regression and intercepts alone. In no instance have several channels been lumped for computational purposes, as is sometimes the practice, for it is felt that only if channels are treated individually can width-depth ratios be used to characterize channel shape. White River channels have a mean width-depth ratio of 23.1 (table 7). The range of width-depth ratios of the 112 channels is from 10 to 71, with a standard deviation of 9.9.

It has been suggested previously that White River channels were readily able to adjust their width, depth, and velocity at all the discharges measured. Figure 9 shows that the mean width-depth ratio for White River channels changed less with discharge than did the width-depth ratio of any of the other streams with which it has been compared. The values of width and depth for the White River were obtained from figure 8; those for the other streams were extrapolated, interpolated, or read directly from data provided in the respective studies. White River channels differ from those of other rivers not only in the constancy of width-depth ratios but also in their relatively small value. The high width-depth ratio for the other types of streams at discharges of 10 cfs or less is to be expected because these flows are unusually low for these streams and the flows occupy a channel developed at much higher discharges.

Shape factor.—In addition to the use of width-depth ratio for comparison of channels, one might use as a shape factor the ratio of maximum depth to mean depth. Considerations of geometry show that a triangular channel would have a shape factor of 2; a parabolic channel, 1.5; and a rectangular channel, 1.0. The mean shape factor for White River channels is 1.62, and the standard deviation is 0.26. That is, although most cross sections are nearly parabolic, some tend to be triangular. This relation seems to hold for channels of all sizes. As maximum depths have not been reported in the other publications listed in table 1, a comparison with other types of streams cannot be made. Only with further use can the value of this shape factor as an indicator of channel shape be estimated. It must be recognized, however, that a shape factor of 1.5 is a necessary, but not sufficient, condition to prove that a channel is parabolic in cross section. Although it is possible that a variety of weird cross sections could have a shape factor of 1.5, it is probable that a channel in alluvial materials that has such a shape factor is nearly parabolic.

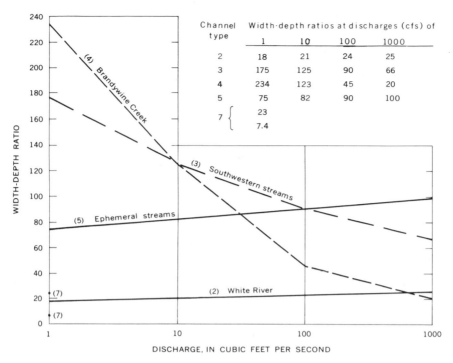

Channel type	Width-depth ratios at discharges (cfs) of			
	1	10	100	1000
2	18	21	24	25
3	175	125	90	66
4	234	123	45	20
5	75	82	90	100
7 {	23			
	7.4			

FIGURE 9.—Relation of average width-depth ratios to discharge. Channel type numbers refer to table 1 footnotes, where channels are described.

Figure 10 illustrates the manner in which shape factor and width-depth ratio represent the shape of some White River channels. Neither width-depth ratio nor shape factor indicates the irregularity of the cross section.

MODE OF CHANGE

The rapid adjustment to increasing flow and therefore increasing bed load in non-cohesive materials appears to take place by increasing width and depth in the following manner. The initial channel widens and may even grow shallower, until at some point the critical tractive force is less than that necessary for transport of the bed load. The material deposited causes further widening and a decrease in the tractive force. Deposition or cutting of adjacent channels eventually brings the bar above water, resulting in the development of two channels similar in shape to the original channel but adjusted to the higher flow condition. Additional flow repeats the sequence of events in either or both of the new channels.

Adjustment of the White River channel shape at low discharges takes place by both erosion and deposition.

With decreasing discharge, the adjustment might take place as follows:

1. The bed load decreases sharply.
2. The velocities, especially near the bank, are lower due to presence of shallower water at the sides of the channels and occasional boulders in the channel near the bank. These boulders remain after being undermined and dislodged from the bank when the surrounding finer materials were removed by velocities incapable of moving the boulders. As a result of these lower velocities, deposition of the coarser fraction of the suspended load takes place at the sides of the channel.
3. At the same time, velocities remain high enough to permit scour in the center part of the channel as the depth decreases with the decreasing discharge. The result is that similar channel shapes are possible for widely different discharges, whether increasing or decreasing.

Wolman and Brush (1961) noted that in flume channels, at a constant slope and with a constant introduction of load at the same rate that material was being

391

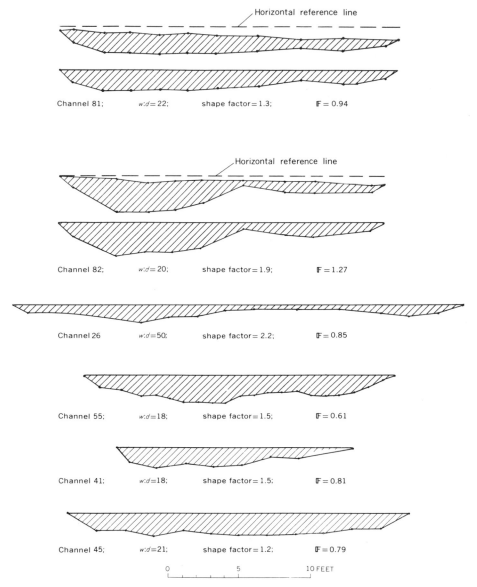

FIGURE 10.—Cross sections of several White River channels showing width-depth ratio, shape factor, and Froude number. Water surface assumed level except in upper cross sections of channels 81 and 82; maximum difference in water surface elevation: channel 81, 0.6 foot; channel 82, 0.6 foot.

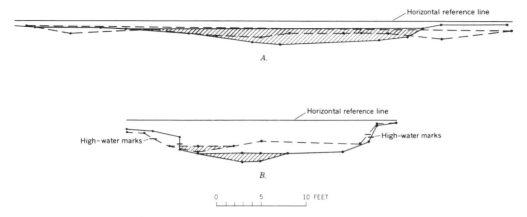

FIGURE 11.—Cross sections of White River channels showing adjustment to load and discharge. *A*, Main channel of White River at cross section 7, July 7, 1958; dashed lines show cross section 1 hour later. *B*, Cross section of a channel to the right of the main channel between cross sections 6 and 7, July 16, 1959; dashed line indicates the same cross section on July 25, 1959.

transported out of the reach, the adjustment to a discharge larger than the initial channel could accommodate was primarily by widening, with little or no change in depth. Figure 11 shows several cross sections where a similar process appears to have operated in White River channels. In the channel at cross section 7, where a large bed load was noted, the rate of widening at one point was so great that the rodman asked the instrument man to hurry his shots, as the bank for a period was receding faster than the rodman was approaching.

WATER-SURFACE SLOPE, HYDRAULIC ROUGHNESS, AND FLOW CHARACTERISTICS

The water-surface slopes, in White River channels with discharges greater than 10 cfs, range from about 0.01 to 0.08 foot per foot. The greater variation in the slopes of smaller channels was due, at least in part, to the difficulty of measuring small differences in elevation over the short reaches in which these small channels were relatively uniform in character.

Figure 10, showing some of the channel cross sections, illustrates one of the difficulties encountered in making accurate measurements of slope. The water surface, because of the type of flow in most White River channels, is neither flat nor horizontal normal to the direction of flow, making it difficult to estimate the average water-surface elevation at any point along the channel. Two estimates of elevation were included in every measurement of water-surface slope.

The uneven water surfaces shown in figure 12 and channels 81 and 82, figure 10, are characteristic of channels in which the Froude number,

$$\mathbb{F} = \frac{v}{\sqrt{gd}},$$

is approximately 1 or larger. At a Froude number of 1.0, the transition to shooting flow takes place (Robertson and Rouse, 1941; Chow, 1959, page 13). Many of the White River channels have Froude numbers which approach 1.0 (table 7). It must be noted that the other channels are shown in figure 10 with an even water surface only because the cross sections were drawn assuming a horizontal water surface.

The Froude number for the main anabranches of braided reaches ranged from about 0.8 to 1.5, most being greater than 0.9. Froude numbers for single channels containing the entire flow of the river, and with few exceptions for smaller anabranches with discharges larger than 10 cfs, ranged from 0.5 to 0.9. The

FIGURE 12.—Standing waves in a White River channel.

high Froude numbers associated with the main channels of braided reaches are thought to reflect their relative instability.

Other flow phenomena, such as antidunes and extremely rough water, are discussed in the section on transport and may be seen in figures 27 and 28.

TRANSPORTATION, EROSION, AND DEPOSITION ON THE VALLEY TRAIN

CHARACTER OF THE SOURCE MATERIALS

The original source of the materials now being transported and deposited by the White River was the highly diverse bedrock of the area, described briefly in the section on geology. From 1957 to 1959, however, the stream did not flow on bedrock at any point from its source at the glacier terminus to a point several miles below the gage. The materials transported by the river consisted of supraglacial, englacial, and subglacial debris. Although the stream had some load before it issued from the glacier, most of its load was derived by erosion of morainic debris, mudflow deposits, and earlier valley-train deposits in the reach immediately below the glacier. These deposits range widely in form, erodibility, and size distribution. Their areal distribution can be seen in figures 1 and 13.

MORAINIC DEBRIS

Dense till and loose debris, the result of the two modes of deposition, compose the morainic debris. The dense till, which stands in vertical walls up to 15 feet high, has been compacted by the weight of overriding ice. The author saw no ice beneath this till, which may be the lodgment till described by Flint (1957, p. 120). In most places the dense till is overlain by loose debris deposited from melting ice. Where still underlain by melting ice this loose debris shifts continually and rolls easily underfoot (fig. 14). The ablation debris mantles both the active snout and the stagnant ice of the glacier and forms the medial and lateral moraines. Both the dense till and the loose debris are composed of materials from clay size to occasional boulders as large as 10 feet in intermediate diameter, but material of the dense till appears to be more rounded than the angular ablation debris. The supraglacial ablation deposits are characterized by irregular topography. Locally they appear to have been sorted. Occasionally material of one size predominates, as in some pockets of coarse boulders that contain no fine materials; apparently continual shifting due to the melting of underlying ice has "sifted out" the fines. Walking in such areas is treacherous, owing to the delicate balance of some of the large boulders and the large spaces between them. Within this loose morainic

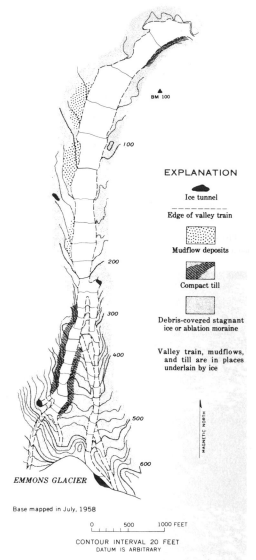

EXPLANATION

Ice tunnel

Edge of valley train

Mudflow deposits

Compact till

Debris-covered stagnant ice or ablation moraine

Valley train, mudflows, and till are in places underlain by ice

MAGNETIC NORTH

EMMONS GLACIER

Base mapped in July, 1958

0   500   1000 FEET

CONTOUR INTERVAL 20 FEET
DATUM IS ARBITRARY

FIGURE 13.—Distribution of deposits on and near the White River valley train.

debris there are occasional small patches of bedded deposits. Ablation debris stands at what must be its normal angle of repose in banks cut by the stream. In

FIGURE 14.—Tills. Ablation moraine underlain by till compacted by overriding ice.

contrast to the dense till, the ablation debris offers very little resistance to stream erosion.

MUDFLOW DEPOSITS

The extent and character of mudflow deposits in the area are indicative of rainfall of high intensity or long duration, during which the steep faces of stagnant ice were washed clean and large amounts of morainic debris fell, slumped, or flowed from the ice. This movement was evidently concentrated in amphitheatrelike basins, from which the debris-laden mud flowed onto moraine and valley-train deposits, forming fans that were often truncated by stream erosion. Minor mudflows also occurred during periods of intense melting.

Mudflow deposits (figs. 15–19) are dense and unsorted and usually occur in fans composed of a series of flows. The top part of each flow is sorted material usually of sand size. Levees and boulder fronts (figs. 18, 19, and 38) are surficial features similar to those described by Sharp (1942), and Sharp and Nobles, (1953). The levees are depositional features thought to have formed when coarse debris at the sides of the flow lost mobility while material in the center of the flow continued. The height of the levees is usually emphasized by subsequent erosion by small streams. The boulder fronts are similar to the levees; they were apparently formed when coarse materials lost mobility and formed a dam

FIGURE 15.—Cutbank in mudflow sequence. Note stream deposits at the top of each mudflow layer. The noncohesive materials in the stream deposits have fallen from between the more cohesive clayey layers.

395

in the path of flow, stopping the flow or diverting part of it.

Most sizes of debris apparently were present in these mudflow deposits. Boulders 2 to 3 feet in maximum diameter were moved by mudflows about 2 feet deep. In the course of the fieldwork, a bank of mudflow material 3 feet high, undercut as much as 3 feet by the stream, was observed to support the weight of a person jumping on it, probably because of the presence of a large percent of clay-size material (fig. 17), whereas the moraine and valley-train deposits were observed not to withstand more than 6 inches to a foot of undercutting without collapsing under their own weight.

In places the wind had moved the finer fraction of stabilized mudflow and morainic deposits and produced deflation pavements and small drifts or dunes. Where there was shifting of the supraglacial materials as a result of melting, a new supply of fine debris was continually brought to the surface. Gusty winds

FIGURE 16.—Detail of cutbank in mudflow sequence.

FIGURE 18.—Mudflow truncated in foreground by stream. Note ice faces in wall of amphitheater in background, coarse material in levees, boulder fronts on surface of mudflow to the right rear of man.

FIGURE 17.—Mudflow deposit. This material stands although undercut as much as 3 feet.

681–370 O—63——4

FIGURE 19.—Boulder front on surface of mudflow shown in figures 15 and 16.

reworked these materials and produced numerous "dust devils" and occasional dust clouds.

VALLEY-TRAIN DEPOSITS

Valley-train materials provide a source of load to the stream throughout its length. Much of the material deposited downstream from the junction of East and West Emmons Creeks had previously been eroded from valley-train deposits upstream. Because such deposits are also a source of material, they are discussed here in relation to the other source materials, although the size of the materials in the valley train is discussed in detail in the section on pebble counts.

The valley-train materials are as large as 3 feet in diameter and are angular to subrounded. The fine gravel and smaller sizes show little rounding. Bedding can usually not be traced for more than a few feet (figs. 20, 21). Sand carried by the stream fills most interstices between the coarser particles as soon as the coarse particles come to rest. In places coarse particles appear to be widely scattered at the surface because the most recent streams on the surface were small ones which deposited fine materials, burying the coarser deposits of earlier streams (fig. 22). Even with such burial, coarse materials invariably occur close to the surface if not on the surface at any point.

Hjulström (1935, p. 326–327) suggested that empirical data show that a 10-cm-diameter particle can be transported by a stream without erosion over a bed of silt and clay or over a bed of material 3 to 4 cm in diameter and larger, but not over material ranging from 0.0001 to 4 cm. A 5-cm particle could be transported over silt and clay, and materials 2 cm and larger, but not over material from 0.0001 to 2 cm. He concluded that

FIGURE 21.—Cutbank in valley train. Note poised boulders, a possible lag concentration at surface. Mixing of sizes is apparent in the top of the sandy layer halfway up the bank. Bedding is limited in its horizontal extent.

it would be unusual to find such coarser materials on top of the finer ones noted and suggested:

Should this be the case, two alternatives appear:
1. The coarser material was not brought there by running water. In this case there may be a clearly defined contact surface.
2. The coarser material was brought there by a stream which eroded away the finer material, but had no time to erode it away entirely before the velocity of the water diminished and deposition occurred. There is a border zone where the two kinds of material are mixed.

Evidence of the correctness of Hjulström's conclusions is offered by figures 15–22. Figures 15–19 show mudflow deposits from the stagnant ice adjacent to the valley train. Figure 15 shows large boulders deposited immediately above undisturbed thinly bedded sands, which could not have been done by running water. Figure 20, a typical valley-train deposit, shows concentration of fine materials only at the top, although fines are mixed throughout the rest of the deposit. Figure 21 shows the mixed border zones mentioned in alternative 2 above.

The nearly vertical cutbanks of valley-train deposits, to 10 feet in height, demonstrate an apparent cohesion yet may give way when a person steps close to the edge. After drying, the banks lose this ability to stand vertically and assume an angle of repose. Drying rarely occurs, however; the gravel is usually damp within a few inches of the surface. After a few humid cloudy days without precipitation, the surface of the valley train becomes darkened by moisture thought to be the result of decreased evaporation at the surface.

FIGURE 20.—Cutbank in valley train. Note concentration of fine debris at surface. Vertical bank is due to stream erosion of materials with apparent cohesion.

FIGURE 22.—Small stream depositing fines which bury coarser deposits. Note antidunes.

At times, drops of water are actually seen on the surface of the material. It is thought that this moisture is either brought to the surface by capillary action or results from condensation. Deposits of a white substance which tastes somewhat like sodium bicarbonate form on rocks lying on the wettest areas of the valley train, suggesting that capillary action is present.

Cohesion due to the moisture in the deposits appears to be weakened by the addition of more water. Bank erosion is thus due, at least in part, to the saturation of a "cohesive" bank by the wash of a wave; material is left overhanging until it caves. The rate of bank erosion appears to depend on the saturation of vertical banks as well as the vigor of attack by the current.

ANALYSES OF PARTICLE SIZE

Size analyses of the three types of materials help to explain the differences in appearance, erodibility, and stability. The results, while based on a few samples, suggest that analysis of the finer sizes could be quite helpful in determining the mode of deposition of materials of questionable origin.

SIEVE ANALYSES

With the exception of the valley-train materials, two sieve analyses of the finer fraction (less than 53 mm or 2.1 inches) of each of the three types of source material showed no consistent differences. The valley-train deposits were deficient in the smaller sizes. Analyses by Richard Arnold (written communication) of the fraction less than 2 mm in size of one sample of the valley train and one of the mudflow, using a combined sieve analysis and bottom-withdrawal tube method, led him to conclude that the stream deposit showed a washing out of silt and clay size materials. He found that 40 percent of the mudflow material was silt and clay (less than 0.105 mm) as opposed to 10 percent of the valley-train material.

These results help to explain why dry moraine and mudflow materials when disturbed are much dustier than the valley-train materials. In addition, small clear streams emerge at various points on the valley train—some from kettles developed in the valley train and others from springs in abandoned channels—

whereas small streams from ice faces and mudflow areas are usually very muddy.

PEBBLE COUNTS

Size changes with distance from the glacier were observed in the valley-train materials. To measure these changes by sieve analysis was impractical because of the large samples that would have been required because of the coarseness of the materials. Pebble counting was adopted as the most rapid and accurate method of measurement. By means of pebble counts size classifications were made of randomly selected particles from the surface of the deposits, in order to develop size-distribution curves. For a detailed discussion of the technique used, see Wolman (1954).

Pebble counting demonstrated a systematic decrease in median diameter of the valley-train materials with distance from the source. Sampling at each cross sec-

tion was limited to the area that appeared to have the coarsest materials at the surface, because very small streams, present almost everywhere on the valley train at some time during the summer, tended to bury the coarser deposits thought to be more characteristic of the load of the main channel (fig. 22). The sample sites were selected near the center of the valley train in order to minimize the effects of local bank erosion of coarser materials. Sizes less than 4 mm in diameter were not differentiated, and this size class is shown in table 2 only to indicate the percentage of area covered by finer material. Comparisons are, thus, based on the fraction greater than 4 mm in diameter.

The range in median diameter of samples was from 180 mm to 120 mm (0.59 to 0.39 foot) in table 2. A median diameter of 180 mm was found for material sampled near cross section 2 on a valley slope of 0.11 foot per foot, 2,350 feet below the glacier, whereas a

TABLE 2.—Size distribution of valley-train materials

[Percentages and D_{50}, D_{25}, and D_{75} are based on samples more than 4 mm in size]

Miles from source	Valley slope (ft/ft)	From.. / To....	4–8	8–16	16–32	32–64	64–128	128–256	256–512	512–1024	1024+	<4	D_{50} (mm)	D_{25} (mm)	D_{75} (mm)	Location
White River between Emmons Glacier and moraine																
0	0.20	--------	0	0	8	16	16	25	29	6	0	4	180	71	300	Upper West Emmons Creek
.36	.20	--------	0	2	0	13	20	36	22	4	2	8	177	95	270	Lower West Emmons Creek.
.43	.11	--------	1	0	3	7	21	39	24	5	0	8	180	106	270	Cross section 2.
.64	.075	--------	1	2	6	12	20	29	28	2	0	11	175	77	270	Cross section 3.
.83	.05	--------	2	3	8	10	15	37	25	0	0	10	165	71	255	Cross section 5.
.95	.04	--------	1	3	4	7	25	47	13	0	0	5	145	94	200	Cross section 6.
1.12	.03	--------	1	4	14	18	14	38	11	0	0	43	125	40	188	Cross section 10.
1.23	.03	--------	0	1	3	16	36	39	5	0	0	3	120	73	160	Cross section 12.
White River below moraine																
1.7	0.066	--------	0	0	1	4	12	47	33	3	0	3	210	150	320	Just below moraine.
2.2	.055	--------	1	3	5	5	27	38	19	2	0	10	200	104	230	Cross section 13.
4.5	.018	--------	0	0	2	9	37	36	16	0	0	4	130	84	215	Near park dump.
14.8	.018	--------	0	1	3	15	55	26	0	0	0	21	105	73	130	Cross section 14.
24.8	.027	--------	0	0	0	5	52	43	0	0	0	3	115	83	155	Near lumber company gate.
28.4	.0085	--------	0	1	2	10	29	55	3	0	0	25	143	94	185	Below West Fork River junction.
West Fork White River below Winthrop Glacier																
7.5	0.024	--------	1	0	5	20	31	41	2	0	0	4	110	62	165	Below end of road.
14.0	.020	--------	0	0	1	23	54	21	1	0	0	2	89	64	125	At logging bridge.
Carbon River below Carbon Glacier																
0.9	0.058	--------	1	0	1	8	11	34	41	4	0	2	230	145	350	½ mile above end of road.
3.5	.034	--------	0	6	8	5	18	39	23	1	0	10	160	91	250	Ipsut Creek campground.
9.0	.015	--------	0	1	8	17	42	27	5	0	0	4	94	62	145	Just outside park.
Nisqually River below Nisqually Glacier																
0	0.128	--------	2	3	5	5	8	20	25	24	8	17	310	140	630	Below stagnant ice.
.5	.094	--------	3	3	3	3	8	16	38	22	4	11	325	170	520	Above Glacier Bridge.
2.8	.057	--------	7	4	4	11	19	20	21	13	1	17	153	61	360	Above Power Plant Bridge.
4.3	.040	--------	2	4	4	4	9	30	39	8	0	27	240	130	380	Longmire campground.

median diameter of 120 mm was recorded for materials at cross section 12 on a slope of 0.03 foot per foot, 6,600 feet from the glacier. The change of 60 mm in median diameter for the valley-train materials over a distance of some 4,200 feet can hardly be explained by diminution of the materials by wear. Stream materials batter each other, but a one-third reduction in median diameter by this means would seem quite unlikely, and this change would strongly suggest that selective transport must have produced radical changes in size.

In general, the median diameter of the valley-train materials decreases with distance below the glacier (fig. 23). Increases in median diameters such as those at pebble count sites 1.7, 25, and 28 miles downstream from the glacier call for a logical explanation. The major increase at the 1.7-mile site can be readily accounted for by the presence of a new source of extremely coarse material in the massive valley-loop moraine through which the stream flows 1.2 miles below the glacier. An explanation for the size increase at the 25-mile locality developed when D. R. Crandell (oral communication) reported a previously unknown moraine about 17 miles downstream from the Emmons Glacier. This suggests that size analysis of valley-train materials may offer clues to the presence of moraines. Time did not permit investigation of the cause of the

even greater increase in size between the 25- and 28-mile sampling sites on the White River or the cause of the increase in median diameter at the 4.3-mile locality on the Nisqually River.

The relation between median diameter of material and valley slope shown in figure 24 demonstrates that slope decreases very rapidly with decreasing median diameter.

Both the Carbon and the Nisqually Rivers have histories of frequent and destructive floods. On the Nisqually River the average intermediate diameter of the 10 largest boulders deposited on and near the Glacier Bridge parking lot by the 1955 flood was 9 feet; one maximum diameter was 24 feet. Such extremely coarse deposits were avoided in sampling, but it is apparent from the scatter of points for the Nisqually River that in selecting pebble-count sites some of these deposits were included. A much more extensive and systematic sampling would be necessary to characterize these streams.

The relation between transportation, declivity, and discharge discussed at length by Gilbert (1877) has been summarized by Lane (1955b, p. 6) in the form of the proportionality.

$$Q_s d \propto Q_w{}^s$$

where Q_s and Q_w are quantity of sediment and water,

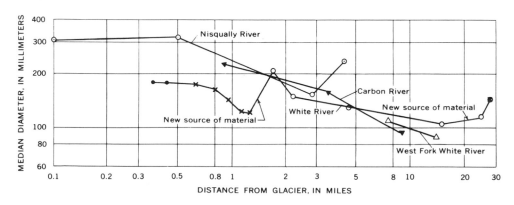

FIGURE 23.—Relation of size of material to distance below glacier.

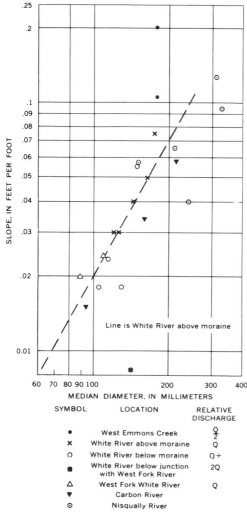

FIGURE 24.—Relation of size of material to slope.

SYMBOL	LOCATION	RELATIVE DISCHARGE
●	West Emmons Creek	$\frac{Q}{2}$
×	White River above moraine	Q
○	White River below moraine	$Q+$
◉	White River below junction with West Fork River	$2Q$
△	West Fork White River	Q
▼	Carbon River	
⊙	Nisqually River	

respectively, d is particle diameter or size of sediment, and s is the slope of the stream.

At any given time, at all points in the reach of the White River between the junction of East and West Emmons Creeks and the gage, it can be assumed that the discharge (Q) is approximately the same, because the contribution from the stagnant ice is small. From the proportionality above, a change in slope should bring about a corresponding change in size of sediment or quantity of sediment or both. Measurements of erosion and deposition (figs. 36, 37, tables 4 and 5) revealed no location or slope at which there was always erosion or always deposition; thus, it is necessary to look for a change in particle size to keep the proportionality balanced. Figure 24 shows that for the White River in the reach above the moraine the slope varied approximately as the square of the particle size.

The slope of the White River (with its greater discharge, $2Q$) below the mouth of the West Fork White River is far less than would be predicted from the relation between slope and intermediate diameter for the smaller discharge of the White River above the moraine. The values of slope for West Emmons Creek $\left(\frac{Q}{2}\right)$ above its confluence with East Emmons Creek are higher than for an area with similar-sized material and higher discharge below the junction.

If one accepts the hypothesis of Wolman and Miller (1960) that the dominant process is one that occurs with sufficient frequency and magnitude to cause most of the changes observed and is neither the frequent event, which is less than the threshold value, nor the infrequent catastrophic event, the slope-forming discharge lies between 200 and 500 cfs. These discharges are clearly capable of modifying not only the channels in which they flow but of changing the elevation of the valley train and transporting all sizes of materials present, and they occur with sufficient frequency (hydrograph fig. 50) to determine the form and slope of the valley train.

TRANSPORTATION OF BED LOAD

METHODS OF MEASUREMENT AND ANALYSES OF DATA

Because of its high velocities and turbulence, White River carries a large volume of debris in the form of bed and suspended load.

The attempts to measure both the amount and the size of the materials in the bed load proved to be rugged sport. The device used to trap samples as they moved along the bed (fig. 25) consisted of a screen or sieve with a wooden frame. Legs were added later (fig. 29) so that the screen and its load could be rocked back out of the water. In fast deep water with large quantities of materials in motion, a coarse screen with 0.175-foot (53-mm [2]) openings was used because it was easier to handle in the current and did not immediately become clogged with the smaller size fraction of the bed load. With lower velocities and depths and smaller quantities of material, a finer screen with 0.05-foot (15-mm) openings was used. Where the bottom

[2] This size was an expedient determined by the coarsest chicken wire that could be purchased in the nearest town.

FIGURE 25.—Sampling bed load with screen. Sample is from the antidune phase of transportation.

FIGURE 26.—Large load moving over rough bed in shallow channel. Note flow over boulders.

was rocky it was difficult to be sure that material was not being swept under the screen. This was a greater problem with the finer material trapped by the 0.05-foot screen. Velocities at sample sites were measured with a current meter or with floats. Often, when a sample was trapped, part would be lost in the struggle to remove it from the current and bring it to the bank, making it necessary to discard the sample. The small number of successful measurements of the amount of bed load (table 8) indicates the difficulties encountered. The method showed promise, and with better equipment and concentrated effort, much additional information on bed-load transportation might be obtained.

Figure 26 shows the extremely rough water associated with quantities of coarse bed load moving in a shallow channel with rough bottom. This is thought to be shooting or supercritical flow, for it shoots into the air over the top of obstacles instead of flowing around them. The boulders battering each other in such a reach make a terrific din and shake the ground for as much as 15 feet back from the bank. This shaking occasionally dislodges material poised on the banks. If one overcomes his awe of the noise, these rough,

shallow reaches can be waded more easily than those that are quieter, but deeper and swifter.

During the measurements of amount of bed load, its movement was observed to be discontinuous both in position in the channel and in time. Only at the highest discharges were large quantities of material in motion. Occasionally the materials could be traced to nearby bank erosion, but usually the source of the material was not readily apparent. With the exception of the measurements made on the troughs and crests of one set of antidune ripples, no conclusions as to the amount of bed-load movement were drawn because of the scarcity of data.

* * * * * * *

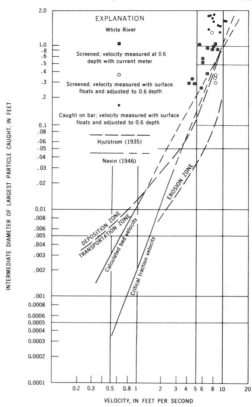

FIGURE 30.—Relation of particle size to velocity.

ANALYSIS OF COMPETENCE OF THE WHITE RIVER

Attempts to measure the largest materials in transport were more successful than attempts to measure bed load using the same screens. Some of the largest boulders were caught by hand, and if they could not be lifted from the water, they were rolled onto bars for measurement. Measurements were made of the largest boulders deposited on bars and the velocity in the channel immediately upstream from these bars. The wide range of materials available to White River provided an excellent opportunity to measure competence as all sizes at or near the range of competence of the stream were readily available. For a summary of the size ranges see table 2.

Figure 30 is a graph of the maximum size particle moving at a given velocity. Several factors that must be kept in mind when using such graphs are: the method of velocity measurement, the condition of the bed over which the particle was transported, and the conditions under which the data were gathered.

681–370 O—63——5

Velocities must not only be accurately measured, but must be comparable, preferably the velocities actually impinging on the particle.

Methods used to measure velocity ranged from the use of surface floats and current meters to mean velocities based on measurements of cross-sectional area of flow and discharge. Most of the values for White River point velocities were obtained by measurements with a Price type-A current meter set at 0.6 depth (normal setting for average velocity in a vertical section of a shallow turbulent stream). These current-meter velocities for White River were taken over a period of 20 seconds or a little more, about the practical maximum for the shifting channels being measured. As the greatest depth for any velocity observation was 2.1 feet, the maximum setting of the current meter was 0.8 foot from the bottom. As the particles in transit were up to 1.8 feet in intermediate diameter, the velocities measured were those in the immediate vicinity of the moving particle in question. When current-meter measurements of point velocity could not be made, surface floats (inflated paper bags ballasted with sand) were used. Their velocities were adjusted to 0.6-depth velocities using the formula:

0.6 depth velocity=0.87 × float velocity

The coefficient of 0.87 was based on 40 measurements where both float and current-meter velocities were available. The value of 0.87 falls within the usual range of 0.85 to 0.95 (Hoyt and Grover, 1916).

The relation of particle size to velocity (figure 30) indicates that the boulders from White River were moving at lower stream velocities than would be predicted from the other data cited. White River data show that boulders up to 1.8 feet in intermediate diameter moved in currents of about 7 fps. For comparison, data that Nevin (1946) selected from Gilbert (1914), the U.S. Waterways Experiment Station (1935), Rubey (1937), and his own traction-tube experiments have been used to define bed velocity and critical traction velocity lines shown in figure 30. Projections of these lines intersect at a velocity of about 9 fps and a particle diameter of about 0.8 feet. It appears from White River data that for coarse materials the critical traction velocity (velocity measured near the bed or mean velocity in shallow flow, Nevin, p. 665) gives lower velocities and therefore is closer to the effective velocity than the bed velocity computed with Rubey's formula. The projection of the calculated bed-velocity curve is based on the assumption that the size of particles in motion is a function of the square of the effective velocity (sixth power law).

Parts of Hjulström's curves (1935, p. 298) which separate the zones of erosion, transportation, and deposition are also shown in figure 30. Most data for White River boulders in motion plot in the zone of deposition that is defined by projections of these curves; however, these curves were based on average velocity and uniform bed materials.

Hjulström's curves, based on Schaffernak's (1922, p. 14) data for uniform materials from 8 to 70 mm, indicate that size is proportional to the 2.6 power of velocity, the power found by Nevin (1946), who used average velocity. Such a line (slope of 2.6) fits the White River data better than one with a slope of 2.0 (the "sixth power law" expressed in terms of linear dimensions); thus, it is suggested that for materials having intermediate diameters larger than 0.1 foot, streams may have greater competence than some applications of the "sixth power law" would predict.

The White River has larger particles in motion than would be predicted by either of the curves shown in figure 30 because the measured velocities were transporting and not necessarily eroding velocities. Most experiments related to erosion and traction velocities have been made in laboratory flumes with uniform materials less than 0.1 foot in diameter, usually sand. The few experiments that have been made with mixed grain sizes show that the mobility of such materials is quite different from that of uniform materials. Gilbert (1914, p. 178) stated:

* * * when such hollows are partly filled by the smaller grains its [the coarser particle's] position is higher and it can withstand less force of current. In other words, the larger particles are moved more readily on the smoother bed * * *. The promotion of mobility applies not only to the starting of the grain but to its continuance in motion.

Ippen and Verma (1953) noted a similar effect of bed roughness. Thus because of the large range of particle sizes in the bed and banks of White River channels, one can expect the erosion of particles larger than those predicted by formulas based on data for uniform materials.

Another factor that must be considered when attempting to determine the significance of the largest boulders within a deposit are the methods by which such boulders may be set in motion by velocities far less than those that would disturb them from a bed composed solely of boulders of the same size. Some of these methods have not been considered in laboratory experiments. From the pictures of the White River source materials (figs. 14–22) it is obvious that coarse materials are not limited to the bed of the stream. Boulders of all sizes are poised in and on the cutbanks, ready for launching by any current that can erode the finer supporting materials. The relative instability of bank materials when compared to bed materials is due to the large gravity component which tends to make the material roll or slide down the bank. This gravity force component acting with force of the flow produces a resultant force much larger than the force acting on similar materials on the bed. Lane (1955a) suggests that the force on the bank materials be taken as the limiting force for stable canal design. Boulders on the bed may be set in motion by the blows from other boulders loosened from the banks. Finer materials may be scoured from under boulders setting them in motion. Boulders that are not moved from the bed on one day may tumble into the stream from the cutbank of the next day's channel as a result of the rapid cutting and filling that takes place in the White River. Illustrations of the rapidity of this change will be given in the section on pattern change (fig. 39).

The relation of size (fig. 31) to tractive force is similar to the relation of size to velocity as the sampled materials were coarser than would be predicted from experimental data on uniform materials. Lane's (1955a) data on the D_{75} (the diameter for which 75 percent of the material is smaller) of the material through which stable canals were constructed falls in a

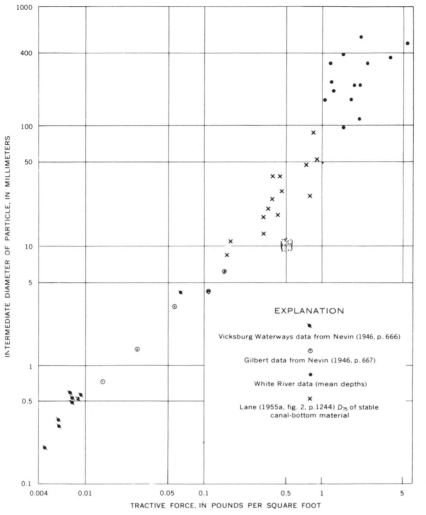

FIGURE 31.—Relation of particle size to tractive force.

line between the tractive forces calculated by Nevin (1946) and the White River data, suggesting that the total increase in mobility due to presence of mixed sizes must be limited.

Studies by White (1940) and by Kalinske (1942) emphasized the importance of turbulence in the initiation of motion of particles on the bed of natural streams. White (p. 332) stated:

If the turbulence extends up to and into the walls * * * then a speed variation of 2 to 1 implies a stress variation of 4 to 1, and the maximum drag is four times the mean.

The so called effective velocity and the tractive force are only indicators of the degree of turbulence and of the probable maximum values of the velocity which may impinge upon a particle. Turbulence may well play the dominant role in the entrainment of materials so small that channel boundary layer effects are important. Turbulence is probably also quite important in the entrainment and transport of coarser materials, for the momentum, acquired from threads of current with higher velocities, helps the particle to keep moving in rough or soft parts of the bed or when the particle is under the influence of the slower threads of current. It must be noted that because the force exerted is proportional to the square of the velocity, the effects of the higher-than-average velocities are proportionally greater than the effects of those lower than average. The higher the turbulence, the more effective a given mean velocity will be. The White River's high turbulence favors high competence.

TRANSPORTATION OF SUSPENDED LOAD

A few suspended-load samples (table 3) were taken by using a US DH–48 depth-integrating sampler to determine the relative amount and size of the material being carried in suspension. Most of the silt and clay in suspension is probably carried out of the area studied and some of it undoubtedly reaches Puget Sound as the "milk" of the White River.

The appearance of the river water varies greatly. Rarely during the summer is the river white as there is usually a brownish color to the water and at highest flows the river has the appearance of fluid mud. The color differences seem to be due to the nature of the materials being attacked by the stream at a given time combined with the contributions of countless muddy rills that flow from the melting stagnant ice. After a period of cloudy cool weather the stream becomes milky in appearance. On close inspection, grains of sand can be seen carried upward in rising threads of current, indicative of the sand being moved along the bottom. Such a condition will change with time because the amount of sand readily available for traction load decreases rapidly as it is removed from around and between the coarser particles, armoring the stream bottom.

SIZE OF MATERIAL IN SUSPENSION

When the suspended load is derived primarily from valley-train materials, the proportion of finer size is less than when moraines or mudflows are the dominant source materials. This might well explain the relative coarseness of the two samples of fairly low concentration, Nos. 6 and 11 (table 3), which have median diameters of 0.13 and 0.19 mm for concentrations of 3,600 and 4,000 ppm respectively. The relatively fine median diameter of 0.082 for the sample with the highest concentration, No. 18, would seem to indicate the large contribution of the tiny muddy rills from the melting ice faces. These rills pick up silt, clay, and sand from the slumping and flowage of oversaturated ablation moraine on the melting stagnant ice. One or 2 days after being washed clean by rain the ice faces are covered again with a thin film of debris and with dust that is airborne in all but the wettest weather. All material picked up with the US DH–48 sampler was less than 2 mm in diameter, but in the most turbulent reaches coarser materials are carried in suspension.

CONCENTRATION AND DISCHARGE

Figure 33 shows the relation of sediment concentration to discharge for all the samples taken. Figure 32 shows the relation of suspended sediment concentration to the diurnal fluctuations in discharge for a day of moderate flow. The change in sediment concentration where the stream was divided into two channels of unequal discharges in shown in figure 34. Approximately one-fifth of the total flow is in the smaller channel. The samples were taken about 50 feet downstream from the point of division. It must be noted that the differences in sediment concentrations of samples taken at about the same time in large and small channels (fig. 34) are about the same magnitude as the variations across the channel (fig. 32); thus, the variation could be the result of the change in discharge due to the division, or the

TABLE 3.—*Suspended load of White River, August 1958*

Sample No.	Date	Time	Channel [1]	Station [2]	Water Temperature (° F)	Depth (feet)	Velocity (fps)	Discharge (cfs) [3]	Concentration (ppm)
1	8/19/58	0720			36.5			120	520
2	8/19/58	0810	18		36.5			110	687
3	8/19/58	0950	18	54	36.5			180	2,610
4	8/19 58	1005	17		36.5			180	1,650
5	8/19/58	1135	17	14	38.0			220	1,860
6 [4]	8/19/58	1145	18		38.0			240	3,630
7 [5]	8/19/58	1330	20	55	38.0			240	2,985
8	8/19/58	1330	19	14	38.0			240	2,360
9	8/19/58	1545	19	14	37.5			270	3,070
10	8/19/58	1547	20	55	37.5			270	3,480
11 [6]	8/19/58	1730	20	55	36.0			240	4,049
12	8/19/58	1730	19	14	36.0			240	1,830
13	8/19 58	1945	23	55				200	2,580
14	8/19/58	1945	22	14				200	1,030
15	8/22/58	1650						320	14,600
16	8/22/58	1650						320	15,500
17	8/22/58	1700						320	14,100
18 [7]	8/25/58	1445			36.5			700	17,220
19	8/25/58	0830			35.0			200	989
20	8/28/58	1220	46	15	36.0	1.1	6.02	165	2,060
21	8/28/58	1225	46	15	36.0	1.1	6.02	165	2,180
22	8/28/58	1220	46	21.5	36.0	1.35	4.46	165	1,410
23	8/28/58	1225	46	21.5	36.0	1.35	4.46	165	1,540
24	8/28/58	1215	46	27	36.0	1.7	4.51	165	1,060
25	8/28/58	1225	46	27	36.0	1.7	4.51	165	1,260
26	8/28/58	1520	47	17	37.0	1.4	5.96	190	3,190
27	8/28/58	1520	47	17	37.0	1.4	5.86	190	4,550
28	8/28/58	1520	47	22	37.0	1.4	5.86	190	2,680
29	8/28/58	1520	47	22	37.0	1.4	5.86	190	2,930
30	8/28/58	1520	47	28	37.0	1.6	6.29	190	2,490
31	8/28/58	1520	47	28	37.0	1.6	6.29	190	2,720
32	8/28/58	1715	48	15	36.0	1.2	6.97	165	2,800
33	8/28/58	1715	48	20.5	36.0	1.3	5.52	165	2,190
34	8/28/58	1715	48	26	36.0	1.6	4.50	165	1,910
35	8/28/58	1715	48	30	36.0	1.5	5.63	165	1,560

[1] Sampling done before or after discharge measurement in channel described in table 7.
[2] Station number refers to a point on the channel cross section.
[3] Discharge estimated from gage height record and is given for sum of all channels.
[4] Median diameter of 0.13 mm.
[5] Median diameter of 0.066 mm.
[6] Median diameter of 0.19 mm.
[7] Median diameter of 0.082 mm.

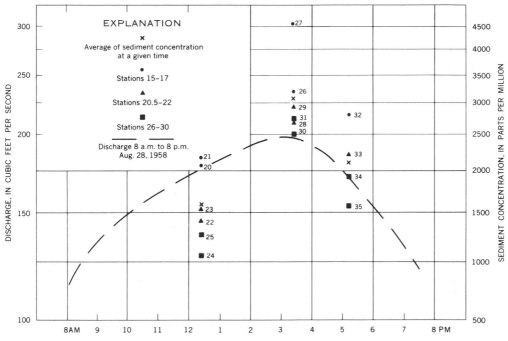

FIGURE 32.—Diurnal variation in suspended sediment concentration with discharge.   Stations and sample numbers refer to table 3.

result of sampling water-sediment mixtures that had already been differentiated upstream from the division point.   Only a much more extensive program of suspended-sediment sampling could define the cause of variation in sediment concentration observed in the White River.   The data and tentative analysis are presented here only to indicate the order of magnitude of suspended load carried in suspension and some possible causes of the variations.

VALLEY-TRAIN ELEVATION CHANGE

The most graphic evidence of the great amount of material transported by the White River is the amount of erosion and deposition on the valley train itself. Measurements of the elevation changes due to erosion and deposition can give only minimum figures for transport because much of the material once set in motion is carried from the area.   Measurements of elevation change are used later, in the study of the significance of channel pattern and in the study of the conditions of equilibrium.

METHODS OF MEASUREMENT

During late August, 1957, cross sections were surveyed across the White River valley train between the

terminus of the Emmons Glacier and its enclosing moraine some 6,600 feet downstream (fig. 35 and table 4). These cross sections were surveyed using a transit as a level to establish elevation and to determine horizontal distance.   The maintenance of horizontal and vertical control was a serious problem as both the valley train and the moraine were, at least in part, resting on stagnant ice.   Primary control was set up by surveying level circuits from bench mark 100 (fig. 35), established on bedrock at the base of Baker Point, both downstream to the gagehouse and upstream to bench mark "C."   Bench mark "C" (fig. 35) was established on bedrock near the channel of West Emmons Creek by the National Park Service for use in their studies of Emmons Glacier.   During the surveying of the line of levels, the elevations of a series of reference points near each cross section were established.

On subsequent surveys of cross sections, the height of instrument, referred to as HI, was established by computing an HI from each "known" reference point in the vicinity and taking the average of those that agreed within a tenth of a foot.   Differential settling was assumed when the movement was more than 0.1 foot as the reference points were usually separated by

407

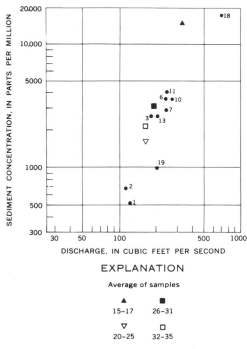

EXPLANATION

Average of samples

▲   ■
15–17   26–31

▽   □
20–25   32–35

FIGURE 33.—Relation of suspended sediment concentration to discharge. Stations and sample numbers refer to table 3.

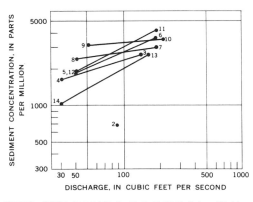

FIGURE 34.—Relation of suspended sediment concentration to discharge 50 feet downstream from the division into two channels. Lines connect samples taken at same time. Stations and sample numbers refer to table 3.

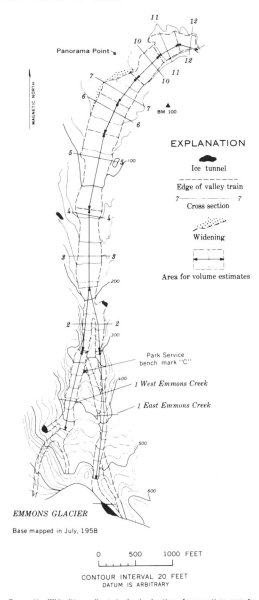

EXPLANATION

Ice tunnel

Edge of valley train

Cross section

Widening

Area for volume estimates

EMMONS GLACIER

Base mapped in July, 1958

0   500   1000 FEET

CONTOUR INTERVAL 20 FEET
DATUM IS ARBITRARY

FIGURE 35.—White River valley train showing locations of cross sections, areas for volume estimates, and widening.

20 feet or more. Points showing movement were not used. Horizontal control was established by: (a) driving center stakes; (b) locating cross sections so that reference points could be painted on large boulders along the line, and beyond the ends of the line, (c) measuring magnetic azimuths; (d) painting marks on

bedrock high along the right valley wall and building cairns along the lateral moraine on the left.

COMPUTATION OF NET ELEVATION AND VOLUME CHANGE

The net elevation change at each cross section during a period was determined as follows. Cross-sectional areas of erosion and deposition were measured from the cross section surveys at the beginning and end of the period (table 4; fig. 37) and a mean depth was determined by dividing that area by the length of cross section over which the areas had been computed. The net elevation changes (table 5) are shown graphically in figure 36. These changes do not include the erosion that took place as a result of the widening of the valley train. The area involved in this widening could not be computed because the height of the terrace or bank that had been cut away was not known. The width of the valley train was measured during each survey but could not be considered in calculating the net area of change. Attempts were made to extend the cross sections beyond the edge of the valley train in 1959, after the plotting of the cross sections surveyed in 1957 and 1958 showed the importance of this widening. During the 1959 season, however, there was not as much widening as in 1958; so the computations for all periods were made using the original lengths of the cross sections. This simplification should not have seriously affected the computation of net elevation change, as similar elevation changes took place on the newly added part of a cross section. The changes in length of cross sections are given in table 4.

The method used in computing volume of erosion and deposition is similar to the end-area method used by surveyors in computing cut and fill. As shown in figure 35, the area of valley train measured halfway to adjacent cross sections was determined from the plane-table map and multiplied by the net elevation change. This method allowed computations for the cross sections at the ends of the valley train. The total net volume change was divided by the sum of the map areas to determine a weighted mean elevation change for the whole valley train for each period (table 5).

Checks on the net errors in elevation are provided by resurveys of parts of cross sections that had not changed and by comparing the net change over several periods between surveys with the sum of the net changes for each period. The net change for five parts of different cross sections (total length of 670 feet), which showed no evidence of change, averaged 0.03 foot and ranged from +0.12 to −0.13 foot for individual parts. These parts where no change had taken place had not been reached by the stream during the period; footprints, and marks made by the base of the rod, were still present after a month or more.

The net elevation change over several periods should equal the sum of the changes for the periods. For example, for cross section 3, the elevation changes are +0.9, +1.9 and −1.0 for periods 1, 2, and 3 (table 5). The sum is +1.8, or 0.1 foot less than the +1.9 feet calculated for period 4, the sum of periods 1, 2, and 3. Because periods 4, 8, 9, and 10 are composed of periods of shorter duration, similar checks are provided for all periods at each cross section and for the sum of the

TABLE 4.—*Cross section survey of the White River valley train, 1957–60*

Cross section No. (fig. 35)	Duration of periods and dates of surveys											Valley train					
												Dimensions				Widening (feet)	
	Period 10																
	Period 4			Period 9													
				Period 8													
	Period 1	Period 2	Period 3	Period 5	Period 6	Period 7	Period 11					Width (feet)	Slope (feet per foot)	Surface area (square feet)	Distance from glacier in 1958 (feet)	Period 4	Period 9
0					7/ 1/59		8/19/59					115	0.16		[1] 3,000		0
1W					7/ 3/59		8/19/59	6/ 8/60				162	.14		[1] 1,000		10
1E					7/ 1/59		8/19/59	6/10/60				182	.16		1,250		0
2					6/21/59	7/26/59	8/25/59	6/ 8/60				240	.11	180,000	2,350	50	[2] −52
3 [3]	8/ 1/57	7/15/58	8/ 5/58	9/ 2/58	6/21/59	7/26/59	8/25/59	6/10/60				290	.075	320,000	3,500	65	14
	8/28/57	7/15/58	8/ 5/58	9/ 2/58	6/21/59	7/26/59						300	.06		3,700		0
4								6/10/60				425	.05	340,000	4,500	0	0
5	8/22/57	7/15/58	8/ 6/58	8/29/58	6/20/59	7/25/59	8/26/59	6/10/60				435	.04	200,000	5,150	34	[2] −14
6	8/21/57	7/ 2/58	8/ 6/58	8/28/58	6/20/59	7/25/59	8/26/59	6/10/60				440	.04	150,000	5,400	74	[4] 70
7	8/19/57	7/ 7/58	8/ 6/58	8/28/58	6/20/59	7/25/59	8/26/59	6/ 9/60				290	.03	110,000	6,000		6
10	8/20/57	7/ 8/58	8/ 7/58	8/25/58	6/19/59	7/24/59	8/25/59	6/ 9/60				360	.03	110,000	6,250	0	10
11	8/20/57	7/ 1/58	8/ 8/58	8/28/58	6/19/59	7/24/59	8/25/59	6/ 9/60				170	.03	50,000	6,600	0	6
12	8/20/57	7/ 9/58	8/ 8/58	8/28/58	6/19/59	7/24/59	8/19/59	6/ 1/60				151	.055		11,600		
13								6/ 3/60				400	.018		78,500		0
14				9/ 4/58													

[1] Approximate distance upstream from cross section 2 is given for 0 and 1W.
[2] Narrowing caused by mud flows from adjacent stagnant ice.
[3] Cross section 3 also surveyed 7/25/58, periods 2a and 2b, figures 36 and 37.   2a=7/15–25/58 and 2b=7/25–8/5/58.
[4] Estimated.

TABLE 5.—*Changes in elevation and volume of the valley train of the White River, 1957–60*

[See table 3 for dates of periods and figure 36 for plot of data for cross sections 2-12. Italic figures show discrepancy as compared to the sum of the individual periods.]

Period No.	1	2	3	4 (1+2+3)	5	6	7	8 (6+7)	9 (5+8)	10 (4+9)	11
Duration (months)	10	1	1	12	10	1	1	2	12	24	10
Elevation change (feet)											
Cross section											
0								−1.2			
1W									−.5		
1E									−.2		+.4
2	−0.6	−0.1	−0.2	−0.9	+0.5	−0.2	−0.3	−.4	*+0.2*	−0.7	+3.3
3	+.9	¹+1.9	−1.0	+1.9	−.2	−.1	−.2	−.3	−.5	+1.4	+2.4
4						+.4	0	+.3			+1.4
5	+1.1	+.9	0	+2.0	+.4	−.2	−.3	−.4	+2.1		+.6
6	+.9	+.1	+.9	+1.9	+.1	−.4	−.1	−.5	−.4	+1.5	+.4
7	+1.0	+.8	−.3	+1.4	+.9	−.3	+.1	−.1	+.6	+2.0	+.9
10	+.3	−.3	0	0	−.3	+.7	+.1	+.7	+.3	*+.4*	+.6
11	0	+.3	0	+.3	−.3	+.2	+.2	+.3	+.1	*+.3*	−.5
12	+.5	+.4	+.2	+1.1	−.2	−.5	+.7	+.1	0	+1.1	−.4
13									0		−1.0
14									−.2		
Total volume change (10³ cubic feet)²											
All cross sections	+900	+1,100	−200	+1,800	+300	−200	−200	−400	−200	+1,700	+1,800
Average elevation change (feet)²											
All cross sections	+0.6	+0.7	−0.2	+1.2	+0.2	−0.1	−0.1	−0.3	−0.1	+1.2	+1.2

¹ Survey in midperiod showed deposition of 3.6; subsequent erosion of 1.7 gave net for period of 1.0.
² Total volume change and average elevation change are based on cross sections 2, 3, 5-7, 10-12.

net changes of all cross sections. The italicized values in table 5 for periods 4, 8, 9, and 10 are those which differed by 0.1 foot from the sum of the included periods. An error of 0.1 foot can be produced by rounding the calculations to tenths. The frequency with which an error of 0.1 occurs, as well as the limit in accuracy imposed by the ruggedness of the terrain surveyed, suggests that 0.1 foot is about the limit of accuracy for procedures outlined. The fact that larger errors are infrequent (bold type in table 5) attests to the general reliability of the procedure used to determine elevation change.

In estimating volume change there is also the possibility of error in the determination of area. No check is available for this. In addition, the assumption has to be made that the elevation change for the cross section applies to all points within the area of valley train (fig. 35) in which the cross section is centered. The main purpose of volume determination is the calculation of an average elevation change for the entire area weighted to allow for differences in spacing of the cross sections. The discrepancies that appear when the volume changes for cumulative periods are compared with the sum of those for the component periods are the same discrepancies noted in the section on elevation change.

ELEVATION CHANGE ON CROSS SECTIONS OF THE VALLEY TRAIN

Eleven cross sections (table 4) were surveyed in 1957 (figs. 35, 36, 37). When resurveyed in early July 1958, it was discovered that two-thirds of cross section 0 had been overridden by a slight advance of the glacier. The controls for Nos. 4 and 8 had been eroded or buried and these cross sections could not be resurveyed. Although other controls were lost, enough points remained or were reestablished to allow all other cross sections to be resurveyed with reasonable certainty that the elevation changes were not due to faulty location or faulty HI.

These 8 cross sections (Nos. 2, 3, 5–7, 10–12) were resurveyed 3 times in 1958—between July 1 and 15, between August 5 and 8, and between August 28 and September 2—and 3 times in 1959—between June 19 and 21, July 24 and 26, and August 25 and 26. No. 3 was also surveyed on July 25, 1958, by plane-table methods because striking deposition had taken place (fig. 37). In addition, 4 new cross sections were established, 3 of them upstream from No. 2 and the 4th in the vicinity of No. 4, which had been lost in 1958 (fig. 35). Nos. 13 and 14, established on September 4, 1958, were resurveyed on August 19 and 20, 1959. Dates on which each cross section was surveyed are

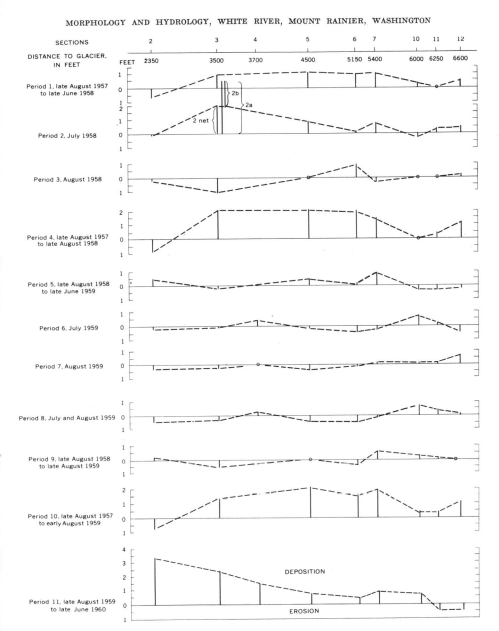

FIGURE 36.—Erosion and deposition on cross sections.   For exact dates of surveys for each section, see table 4.

indicated in table 4. Most cross sections were resurveyed in early June 1960 to measure the changes brought about by the winter storms of 1959–60.

In period 1 deposition took place at all cross sections with the exception of No. 11, which showed no net change, and No. 2, where there was a net degradation of 0.6 foot (tables 4, 5). If the East and West Emmons Creeks portions of No. 2 are analyzed separately, the picture is different. East Emmons Creek, its flow primarily derived from the active snout of the glacier, showed sufficient net erosion to offset the net deposition of West Emmons Creek. Most of the change for this period is thus thought to have taken place during late June and early July 1958. When the valley train was first visited in mid-June 1958, all the stakes in the centers of the cross sections were in place, and the area appeared as it had in late August 1957. Within 10 days, before surveys could be made, the channel pattern had changed, and half the center points were lost. The average elevation change for all cross sections was an increase of about 0.6 foot, at Nos. 3, 5, 6, and 7, averaging about 1 foot.

In period 2, the last 2 to 3 weeks in July, deposition was predominant at most cross sections, with the exception of minor reductions in surface elevation at Nos. 2 and 10. The greatest erosion and deposition for any cross section in any period took place during the first part of period 2 between July 15 and 20, 1958, at No. 3 (fig. 37). Deposition raised the elevation of the valley train an average of 3.6 feet. Activity noted on July 17 was accompanied by rapid pattern changes, rough water similar to that in figure 26, and the roar of clashing boulders. On July 20 East and West Emmons Creeks, which join just below cross section 2, ceased flowing over the valley train near cross section 2 and entered a tunnel beneath stagnant ice on the right side of the valley train, for 6 or 7 days (July 20 to July 26 or 27); the stream emerged near cross section 5. Within 10 days after returning to the valley train, the stream had removed an estimated 450,000 cubic feet of its earlier deposits, most of it in the first few days. This volume estimate was based on three cross sections surveyed after the erosion had taken place. The 1.7 feet of erosion at No. 3 was computed from the difference between the surveys made on July 25, when the stream was in the ice tunnel, and on August 5. Two additional cross sections were surveyed on August 5 in areas of obvious erosion between No. 3 and the constriction below No. 2. The lower line of the cross section was surveyed on the ground surface and the upper line was inferred from erosion remnants. On July 28, during the removal of this material, one-third of the valley train in this reach was covered with water flowing in at least 4 channels in the vicinity of cross

section 3 (fig. 52). An average increase in elevation for all cross sections of 0.7 foot took place during period 2.

During period 3, the last 3 weeks in August, there were only minor changes at most cross sections, with the exception of 1 foot of erosion at No. 3 (fig. 37), which was a continuation of the erosion that had begun during the last week of period 2, and deposition of 0.9 foot at No. 6. The net for the period was about 0.2 foot, and the erosion and deposition served to bring Nos. 3 and 6 back into relative agreement with adjacent cross sections.

In period 4, the aggregate of changes during periods 1, 2, and 3 showed net deposition at all cross sections with the exception of No. 2, which lost 0.9 foot, and No. 10, which showed no change. Nos. 3 (fig. 37) and 7 received about 1.9 feet of material and No. 6, 1.4 feet. No. 5 gained 2 feet, and two small remnants of moraine that partly blocked the valley 200 feet downstream from the cross section were almost completely buried by the deposition that took place between Nos. 5 and 6. The average net gain in elevation was about 1.2 feet over an area of about 1,500,000 square feet. Approximately 1,800,000 cubic feet of material was deposited during 1958.

In period 5, early September 1958 through mid-June 1959, there was minor deposition and erosion at most cross sections. The exceptions were Nos. 2 and 7, which received 0.5 and 0.9 foot of deposition respectively. The net for the period was about 0.2 foot of deposition.

Period 6, mid-June through August 1959, showed net erosion for all cross sections (Nos. 2–7) above the constriction at Panorama Point (fig. 35) with the exception of No. 4, which gained 0.4 foot. No. 10 gained 0.7 foot and No. 12 lost 0.5 foot. The net change for the period, excluding changes on No. 4, was a loss of about 0.1 foot. No. 4 was not used in computations of volume and elevation change because data were not available for all periods.

Period 7, the month of August 1959, showed minor changes on all cross sections with the exception of No. 12, which gained 0.7 foot. The net for the period was a degradation of about 0.1 foot.

Period 8, mid-June to late August, 1959, showed erosion on most cross sections: 1.2 feet on cross section 0, 0.5 foot for No. 1 West Emmons Creek, and 0.2 foot for No. 1 East Emmons Creek. The last two sections were established to evaluate these reaches as possible sources for materials deposited in the valley train. Exceptions to the erosion were Nos. 4, 11, and 12, where minor deposition had taken place, and No. 10 which had received 0.7 foot of material. The net for

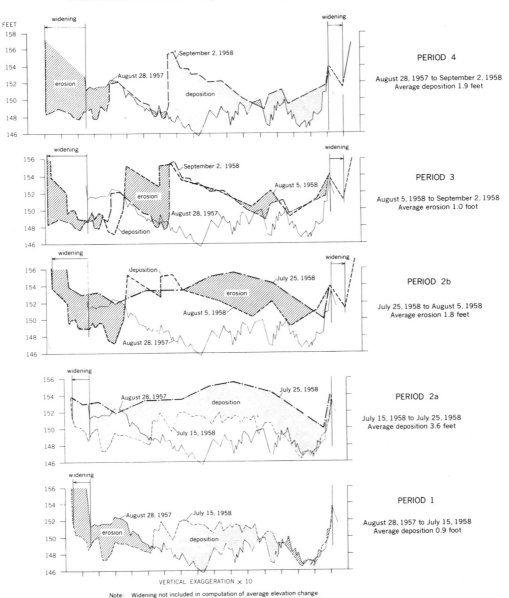

FIGURE 37.—Erosion and deposition on cross section 3 from August 28, 1957, thru September 2, 1958.

the period, excluding the cross sections that had not been previously surveyed, was about −0.3 foot.

In period 9, the aggregate of changes for the year ending in late August 1959 showed minor changes on most cross sections. The exceptions were No. 3, which lost 0.5 foot, and No. 7, which gained 0.6 foot. The net elevation change for this year was about −0.1 foot, an estimated loss of about 200,000 cubic feet of material, or roughly, 10 percent of the volume change for the previous year.

In period 10, the aggregate of changes over 2 years of study ending in late August 1959 showed a gain in elevation at all cross sections except No. 2, which lost 0.7 foot. The greatest gains, made in the first year of the study, were retained through the second year. Gains on cross sections 3, 5, 6, and 7 were 1.4, 2.1, 1.5, and 2.0 feet respectively. The net gain for the 2 years was about +1.2 feet or about 1,700,000 cubic feet. The two years are similar in that the greatest rate of change occurred from late June through late July, the time of greatest discharge (fig. 50). During the periods from late August to June of both 1958 and 1959, there was net deposition, although in 1958 most of, if not all, the deposition during this period took place during late June.

Surveys were made in June 1960 to check on the erosion and deposition that took place during three major storms of October, November, and December 1959. One of these storms produced the second highest flow of record at White River near Greenwater, Wash. Major deposition was recorded on cross sections Nos. 2 and 3 with 3.3 and 2.4 feet, respectively. No. 4 received 1.4 feet of material and all other cross sections lesser amounts. The average net elevation change for the period was 1.2 feet. This large average elevation change may be more apparent than real as Nos. 2 and 3 are more heavily weighted in this average because their elevation changes were multiplied by large areas in computing the volume and average elevation change. Most of the deposition on No. 2 was on the right end of the section and it is thought that mudflows from stagnant ice along East Emmons Creek were the major source of the deposits. Since these deposits were localized at the cross section, the calculated average elevation change for the period was larger than the true elevation change.

Additional evidence of general alluviation of the valley train was observed in the fate of kettle pools. Morning Glory Pool, in the adjacent moraine, so-named for its beautiful color in 1957, still deserved this name in early June 1958. At this time it was brimming with potable clear water, the surface of which was several feet above the adjacent valley train. Periodic invasion by debris-laden streams built a birds-foot delta into the pond, gradually filling it (fig. 40). By the end of August 1958 the same thing was occurring along the left edge of the valley train, in two other ponds in the next 300 yards upstream. The rise in the level of the valley train threatened ponds that had been considered completely out of reach in 1957.

SOURCE OF DEPOSITS

A search was made to determine the source of the 1,650,000 cubic feet of material deposited over the 2 years. The streams, as they issue from beneath the glacier, usually appeared to carry less debris than further down downstream and were often potable. East Emmons Creek, which flowed from the active snout of the Emmons Glacier after receiving a tributary from Frying Pan Glacier, did not deafen the observer with a din of clashing boulders. West Emmons Creek before reaching the area shown on the map (fig. 35) flowed for about 1 mile from the left lobe of the glacier (fig. 1) over, under, and through debris covered stagnant ice, valley train, and mudflow deposits (fig. 38) on a very steep gradient (as steep as 0.20 ft per ft). It was usually the muddier of the two creeks and the one which was thought to be the main supplier of material. Evidence supporting this conclusion was provided by the erosion recorded during the summer of 1959 on cross sections 0 and 1 of West Emmons Creek which on slopes of 0.16 and 0.14 foot per foot lost 1.2 feet and 0.5 foot of material respectively. During the same period cross section 1 on East Emmons Creek lost only 0.2 foot. During period 1 a comparison of the two creeks at cross section 2 showed that the net erosion for the cross section was due to a large amount of erosion by East Emmons Creek while West Emmons had been depositing material, which might be expected of a stream carrying an abundant load derived farther upstream.

During the summer of 1958, conditions appear to have been especially favorable for the deposition of large quantities of material in a short period of time. An unusually high melt of ice and snow occurred during the summer and conditions on upper West Emmons Creek provided a large load. This area is thought to be the major source of the material deposited on the valley train.

In June 1958, West Emmons Creek began flowing over the ridge of debris-covered stagnant ice which connected the left side of the advancing glacier with stagnant ice along the left side of the valley train. Prior to this diversion over the ice ridge, the stream had flowed relatively quietly under the ridge in one or more ice tunnels. These tunnels must have been blocked in early June by debris or collapse. They may have been blocked in part by materials from a mud-

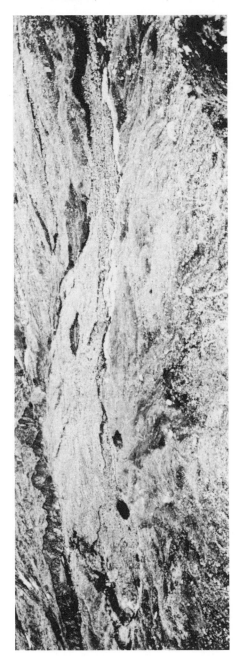

FIGURE 38.—Valley train near head of West Emmons Creek, July 3, 1958. See also figure 1. Stream plunges into ice tunnel in center of picture. Two weeks earlier it flowed across the field of view. Note mudflow levees (parallel strings of boulders in right center). Valley train is almost completely enclosed and underlain by stagnant ice. As an indication of scale, the boulder in right foreground is 10 to 15 feet in diameter.

flow that had descended on top of snow, probably during the spring, and flowed as far as cross section 5, where remnants of the snow it had buried and protected were still melting in late June. When the stream began flowing over the ridge, it reached its former bed on the downstream side of the ice ridge after plunging in a spectacular cascade, called during field studies "chocolate falls" after the color of the water. This cascade was about 70 feet high at first but within a few days the stream with its load of debris had cut deeply into the ice. The rate of cutting was estimated at 5–10 feet per day. Within a few weeks the stream had cut down leaving a vertical-walled ice gorge shown in figure 1 and, in part on the map, figure 35. High on the sides of this gorge, parts of abandoned meanders could be seen at various levels. They appeared as debris-covered terraces or, when undercut, as shelves. The roar of the stream within the gorge made it impossible to be heard at its margin.

The high melt of late June, acting with the changes in gradient brought about by this stream diversion over the ice and the rapid erosion, delivered large quantities of debris to the upper reaches of the valley train. (Note the changes in cross sections 2 and 3 during the second period, table 5.) This material was subsequently removed and distributed over most of the valley train. The change in gradient (knickpoint) at the site of the falls was almost entirely removed by early August, and the lower discharges of August 1958 and of the summer of 1959, on the lowered gradient, brought far less material to the valley train.

Above the ice ridge, West Emmons Creek is very unpredictable, frequently disappearing under the valley train into previously unknown tunnels; this disappearance indicates that the valley train in this area is underlain by stagnant ice. One day it is a formidable stream to cross, and the next day it may not even be present on the surface (figure 38).

An additional source of material was provided by widening of the valley train by erosion of the deposits along its margins. The changes in width which were recorded for the 2 years are shown in table 4, and the area of maximum widening has been sketched on the map, figure 35. Here the channel widening reached a maximum of about 100 feet when the stream removed a bank 10 feet high and found access to an area of kettles behind.

Negative changes in length of cross section (table 4)—that is, narrowing—took place by slumping of banks and by the addition of material from the areas of stagnant ice. Widening of the valley train took place by lateral erosion and filling of depressions that had been separated from the valley train by moraine or mudflow deposits. The pond, well back from the river, shown in the foreground of figure 39 A (photographed July 18, 1958), is the pond (elevation 56.0) at the edge of the valley train in figure 46, mapped September 3, 1958. The pond had become part of the valley train by the time the widening was mapped the next summer. The materials added to the valley train by this lateral erosion are estimated to make up about 10 percent of the total deposition. Since widening showed much less continuity than erosion or deposition, extrapolations to the areas between cross sections in order to calculate volume were not thought justified.

It is interesting to note that the maximum widening (fig. 35) took place not at the narrowest point of the valley train but at the outside of the valley curve toward the east, at the junction of the White River valley with the valley of Inter Fork (fig. 2). The coarser fraction of the large quantity of material furnished by stream cutting of the high banks in this area was deposited first as a bar within 100 yards downstream from the cut banks. This bar may well have served to protect the constriction downstream from erosion during the highest flows of the summer.

In summary, the abundant discharge and high gradients of both East and West Emmons Creeks, with assistance from unique conditions on West Emmons Creek and widening of the valley train, provided an abundant load during 1958. The high gradients on the Emmons Creeks with the flows of 1959 continued to supply materials to the valley train but in greatly reduced quantities, and erosion exceeded deposition for most of the valley train. There was a slight flattening of the valley profile, owing to the predominance of erosion above the junction of the two creeks and depositions over most of the downstream part of the valley train.

CHANNEL PATTERN

The description and analysis of the pattern change of White River channels presents a problem owing to the rapidity with which the changes take place. Slight changes in water-surface elevation which caused the stream to occupy or abandon old streambeds provided as much change in appearance of the channel pattern and valley train as major erosion and deposition (fig. 39). In order to analyze the mechanism of pattern change panoramic photographs (fig. 40) and time-lapse motion pictures were taken (Fahnestock, 1959), planetable maps and cross sections surveyed, and measurements of channel characteristics and elevation change were made.

DESCRIPTION OF PATTERN
CHANGES DURING THE PERIOD OF STUDY

To record the change in pattern of White River channels during the summers, panoramas like figure 40 were taken as frequently as several times per day when channels were changing rapidly and every 3 to 4 days when there was little or no change. Table 6 summarizes the number of channels in the vicinity of each cross section and the estimated total discharge for White River at the time each panorama was photographed. In addition, figure 50 shows the number of channels at the odd numbered cross sections in relation to changes in discharge.

Both tranquil and shooting turbulent flow commonly occur in natural channels. Wolman and Brush (1961) have shown that meandering began with the onset of shooting flow in some of their flume channels in non-cohesive sand materials. They suggested that because these meanders occur at much higher Froude numbers than those of major rivers they are not dynamically similar in all respects and, therefore, termed the former "pseudomeanders."

Photographs of White River taken after extended periods of low flow (less than 200 cfs) show what appears to be a meandering pattern from the junction of West and East Emmons Creeks to the gage. As the flows in these meanders have relatively high Froude numbers (table 7), they are not dynamically similar to common meanderings with much lower Froude numbers. The meanders of the White River shown in figures 41 to 43 may approach the condition of pseudo-meanders in a natural channel.

* * * * * * *

FEATURES OF THE WHITE RIVER CHANNEL PATTERNS

Features of the White River patterns are shown in figures 41 and 46 to 48. Figure 46 shows the braided pattern and water-surface profile near cross section 7 Inspection of the panoramas suggests the flows of August 20–25 were probably responsible. For a detailed map of the feature called topographic nose, see figure 47 (August 26). On that date the nose, which had already developed, was gradually being modified. The depths of water are shown by means of bottom and water-surface contours. Such features appear to be a type of small alluvial fan and are quite common on the valley train. Other examples are shown in figures 48 and 41. In each example there is a relatively deep and swift reach upstream which is confined by either cutbanks or "levees." At the end of this reach, the water spreads and flows in several directions in sheets or in a number of poorly defined channels. When the feature first develops, boulders rolled through the swift reach are deposited with a great amount of noise and splashing on the bars at the end of the chute. Often the deposition of these bars blocks other channels (right foreground, fig. 39), forming pools. A delta is then built into the pool from the main stream and from the blocked stream with fine material settling out in the deeper and quieter portions of the pool. These deposits are the only concentrations of fine materials on the valley train.

The subsequent history of the nose depends on the amount of water and sediment delivered to it. Often a decrease in load occurs at some time after deposition of the nose with a consequent channeling of the deposits and concentration of flow. If sufficient material is available as a result of this scour, the nose may then be left in the form of cut banks and levees and a second nose deposited downstream.

The term "nose" has been used here, as these features actually project above the general slope of the valley train. It is possible to sit on the streambank near one of these features and watch the water rolling boulders at eye level only a few feet away across the stream. These noses or alluvial fans may be similar in some respects to the horseshoe bars described by Hjulström (1952, p. 340).

Contrary to what might be considered normal behavior for a stream, White River channels are often not in depressions but along the top of a ridge. This ridge is formed by the natural levees like those shown in figures 45 and 48. Striking changes in alinement and pattern are caused by deposition of levees that shift the position of the channel, confine it, or block off a channel entrance, thus diverting the flow to another channel as if a valve had been closed. These levees may be similar to the features described by F. E. Matthes (1930, p. 109). Levees may be formed also by the coalescing of bars in a bed of the river, confining the flow to a narrower and deeper channel. These bars appear to be similar to those described by Hjulström (1935, p. 340). Hjulström pointed out that the thalweg of a glacial stream is not necessarily on the outside of these bends, as in a normal channel, but may occur in the middle or on the inside of the bend. Bars are formed quite rapidly under favorable conditions (fig. 39). Time-lapse photography is quite useful in recording and analyzing the changes which take place.

Figures 41 to 43 show reaches of meandering channel. Water-surface slopes are shown for both figures 41 and 43, and the depth of water is shown along the thalweg in figure 41. It is interesting to note the common riffles or crossovers and pools are missing but there are periodic changes in slope which correspond to bends in the plan view. The water surface shows

A. 1:00 p. m.

B. 1:15 p. m.

FIGURE 39.—Pattern changes, 1:00 to 2:00 p.m., July 18, 1958. *A*, 1:00 p.m.; *B*, 1:15 p.m.; *C*, 1:30 p.m.; *D*, 1:45 p.m.; *E*, 2:00 p.m. Discharge is about 350 cfs. Arrows indicate new channels or bars which have appeared since the previous photograph.

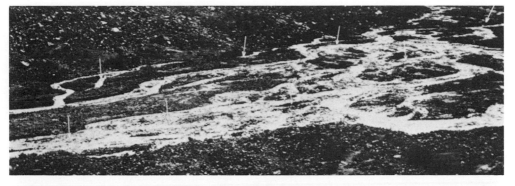

C. 1:30 p. m.

D. 1:45 p. m.

E. 2:00 p. m.

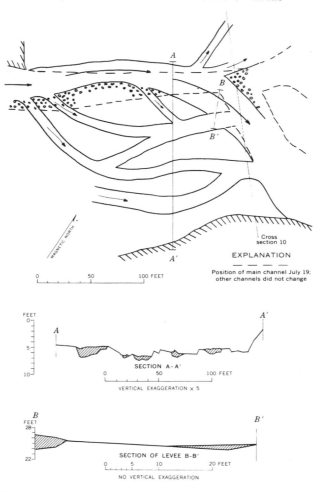

FIGURE 45.—White River near cross section 10, July 21, 1959. Planetable map by Fahnestock and Gott.

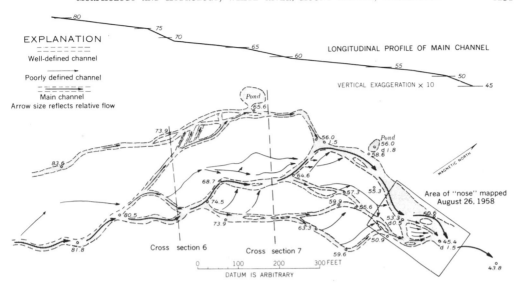

FIGURE 46.—Relict braided pattern of the White River, September 3, 1958.   Planetable map by Fahnestock and Bond.

alternations of steep and gentle slopes as does the channel bottom. The gentle slopes are characterized by swift, relatively smooth flow, whereas that in the steeper part appears to be much more turbulent and the bottom much rougher. The fact that along the thalweg there appears to be little if any shallowing in the steeper reaches may well be misleading, as the mean depth may be less because of the greater irregularity of the bed in these sections.

The "meanders" have the peculiar appearance of a series of bends connected by short straight reaches with swift quiet flow. The steeper parts are usually associated with bends. From the sketch maps (figures 41, 43, 45, 46 and 48), it is apparent that no greater sinuosity is associated with the meanders than with the anastomosing channels. As Leopold and Wolman (1957) observed in several rivers, individual anabranches may meander, and meandering and braiding reaches may alternate along the stream channel.

* * * * * * *

RELATION OF PATTERN TO ELEVATION CHANGE

The concept of a braiding stream as an instrument of aggradation is generally accepted. A braided stream as an agent of degradation may not be quite as familiar.

Mackin (1956) mentioned braiding in a "stable or slowly degrading reach." Leopold and Wolman (1957, p. 53) stated:

Braiding is not necessarily an indication of excessive total load. A braided pattern, once established, may be maintained with only slow modifications. The stability of the features in the braided reaches of Horse Creek suggests that rivers with braided patterns may be as close to quasi-equilibrium as are rivers possessing meandering or other patterns.

At two places during the White River study degradation and braiding were closely associated.

Degradation took place at cross section 3 during period 2b, July 25 to August 5, 1958 (table 5), when there was a net erosion of 1.9 feet. Figure 52 shows that at least half of the valley train at cross section 3 is covered by water in 3 or 4 channels. This photograph, the only record of the channel pattern for this period, shows that the stream which removed the material was braided.

Cross sections Nos. 5, 6, and 7 showed a net elevation loss for period 6, June 20–July 25, 1959, of 0.2, 0.4, and 0.3 foot, respectively (table 5). As this degradation took place over less than half the length of the cross sections and as these calculated values are based on the entire length, the loss in elevation in the vicinity of the stream was at least twice as great. During this same period, there were from 1 to 6 channels at No. 5, 1 to 5 at No. 6, and 2 to 6 at No. 7 (table 6 and fig. 50). The greatest number of channels at each cross section

FIGURE 52.— Braided pattern during the degradation of the valley train between cross sections 2 and 4.   Arrows indicate the ends of cross section 3.   4:00 p.m., July 28, 1958.

occurred during the highest discharges of the summer, when one would expect the greatest elevation change to take place. Active braiding was again associated with net degradation.

A similar situation appears to exist on the Sunwapta River below Athabaska Glacier, Alberta, Canada. There the recession of the glacier from its terminal moraine has created a small lake which serves as a settling basin to clarify the water of the river as it leaves the glacier. This river, as a result of lack of load, has cut 3 to 4 feet below the level of its former valley train, which now forms a terrace. This mechanism of terrace formation was described by Ray (1935). The terrace along the Sunwapta River is now beginning to develop a sparse vegetation of grasses and small plants. The stream, observed in September 1959 at a low stage, had numerous islands and reaches with multiple channels. It appeared as though it had braided actively with the higher discharges of the summer.

These three examples illustrate braiding of a degrading stream. It is concluded that both braided and meandering reaches can occur along the same stream, which may be aggrading, poised, or degrading. Braiding is an indication of channel instability and does not conclusively define the regimen of the stream.

CAUSES OF A BRAIDED PATTERN

Several explanations have been offered for the phenomenon of braided channels by numerous authors who have dealt with the subject. Explanations are quoted for examination in light of the findings on the White River. At the risk of oversimplification it appears that they can be summarized under the headings: erodible banks, rapid and large variation in discharge, slope, abundant load, and local incompetence.

ERODIBLE BANKS

Fisk (1943, p. 46) suggested that the character of the Mississippi River reflects that material through which it flows. He stated that the Mississippi has—

1. A tendency to braid where bank caving is active. Bank caving is active where sediments are easliy erodible. Sands are the most easily erodible sediments, less permeable sediments are tougher.
2. Tendency to braid where slope is steep and sediments easily erodible, and where slope is excessively low and load great.

Friedkin (1945, p. 16) in his model studies of meandering, produced braiding in one test. He reported:

This stream at first developed a series of bends, but erosion of banks was so rapid that the channel in the bendway became as shallow as the [point] bars. The flow overran the bars and dispersed through the channel . . . No sand was fed at the entrance of this stream; braiding resulted solely from excessive bank erosion. As the channel shoaled and the bed of the river raised, the slope became steeper.

Mackin (1956) in describing the braiding of the Wood River of Idaho stated that the river

. . . meanders in a forest for many miles, braids in a 3-mile segment where the valley floor is prairie, and returns to a meandering habit where the river re-enters a forest. The river is stable or slowly degrading in all three segments. The essential cause of the drastic difference in channel characteristics in adjoining segments is a difference in bank resistance due to presence or absence of bank vegetation.

Much of the material in transport by the White River is probably derived from bank erosion; but this factor alone, although it may occasionally develop a braided pattern, is insufficient to cause the active braiding of White River channels. Frequently at the same time and discharge, both braided and meandering reaches were present on the valley train. At times the number of channels was increasing in one area while decreasing in another. Braiding and meandering were not limited to specific parts of the valley train. All such changes on the White River cannot be explained in terms of changing erodibility of banks but must reflect other factors that are more easily changed. It is possible that winnowing by the high flows of 1958 may have slightly decreased the erodibility of the reach of the White River between cross sections 2 and 4, giving it a stability in 1959 relative to 1958. A more probable explanation for this relative "stability" is that in 1959 less bed load was transported through this reach, affording less opportunity for deposition within the channel.

RAPID VARIATION IN DISCHARGE

Doeglas (1951) stated that none of the frequently mentioned causes of braiding ("greater slope, loose debris, or greater discharge") accounted for the pattern of some rivers during the Pleistocene. He suggested (p. 297):

Large and sudden variations in the runoff seem to form braided rivers. A more regular runoff throughout the year gives a meandering river. The gradient and the available sedimentary material seem to be of little importance.

The White River does show a tendency toward adjustment to higher discharges where they are continued for a period of several days. In both 1958 and 1959, a discharge sufficient to produce radical changes in pattern at some profiles during June and early July produced relatively minor changes in August; for example, the number of channels at cross sections 3 and 5 during August 1958 varied from zero to 7 (fig. 50). The change in discharge and previous flow history appear to be almost as important as quantity of water

in producing changes in pattern. This may be an illustration of Doeglas' idea of discharge variation as a cause of braiding. It is a likely mechanism for deriving more load for the same discharge and might work as follows in a White River channel. During a period of low flow, the fine sediments would be winnowed from the bed of the channel in some places and deposited in others, but coarser sediments would be left on the bed. With a significant increase in discharge, these coarser sediments would start to move and new supplies of finer materials would be uncovered. The increased discharge might also reoccupy abandoned channels and set in motion all the materials deposited in them by waning stages of the former stream. This excess of load would cause local deposition, which in turn would result in scour of adjacent banks and an increase in load. A more gradual increase in flow might allow the stream to develop a channel that could transport the increased load without forming bars and developing a braided pattern.

The frequency with which braided reaches are interspersed with meandering reaches (Leopold and Wolman, 1957, and Mackin, 1956) and laboratory studies (Friedkin, 1945; and Leopold and Wolman, 1957) during which braiding was produced with no variation in discharge would seem to indicate that rapid discharge variation is eliminated as a cause for braiding in most streams.

SLOPE

The suggestion that change in slope alone is sufficient to cause braiding does not explain phenomena observed along the White River. Braiding developed on slopes which range from 0.01 to 0.20, but only coincident with bed-load movement of coarse materials. The river frequently braids in one part of the valley train on slopes both higher and lower than slopes of other parts where it has only one channel. In some cases, slope may serve to aid the stream in setting in motion enough material to form a braided pattern. In others, it may serve only to maintain velocities so high that the deposition of bars does not take place.

ABUNDANT LOAD

Hjulström (1952, p. 310) stated:

A fundamental fact for understanding the braiding of rivers is the great sedimentary load which they carry.

Russell (1939, p. 1200, 1201) stated:

Available load seems to be the factor separating meandering from braided streams. A smaller load, in proportion to carrying capacity, at the moment, makes for meandering, a larger load for braiding * * * Those [streams] flowing through sandy or gravelly flood plains are more readily overloaded and therefore tend to anastomose, or display the effects of braiding.

Rubey (1952) in his report on the Hardin and Brussels quadrangles in Illinois suggested that the form of the Mississippi and Illinois Rivers in that area resembled a braided stream more closely than a meandering one, stating (p. 123)—

* * * they follow somewhat crooked courses, it is true, but the curves are due not to meander growth but to diversion of the channel by large alluvial islands.

He noted (p. 124)—

It is probably a significant fact that the islands are larger and more numerous at the mouths of tributary streams * * * An alternative interpretation explains most of the islands much more satisfactorily. Tributary streams commonly have steeper gradients and carry more debris per unit volume of water than the river; hence they deposit part of their load as deltas and as submerged bars across the tributary mouths. Some of the sand bars and flats built at times of flood stand well above the water level at low and normal stages of the river * * * some of the willow bars grow larger and higher until they become "timber islands" the surface of which is built to the level of the mainland flood plain and covered with dense growths of large hardwood trees. Once an island reaches this stage it becomes an essentially permanent feature.

The pattern changes of the White River in 1958, which were both more frequent and more widespread than those in 1959, illustrate the role that abundant bed load played when it was introduced to the section of valley train under observation. That the material was not derived by bank erosion within the area is evident from the net gain in elevation for the valley train (1.2 ft for the year, table 5). In 1959 the decrease in activity and relatively little braiding in the reach between cross sections 2 and 5 suggest that the braiding that did take place was largely caused by bed load derived within the reaches under observation. The net loss of elevation on cross sections 5 to 7 during the summer of 1959 suggests that material was derived from both bed and banks and that braiding occurred in a degrading reach of the stream.

To the writer's knowledge, it has not been suggested that a braided channel pattern can be developed without an appreciable bed load. The common element in all the above explanations appears to be a movement of bed load exceeding local competence or capacity, and consequent deposition within the channel causing the diversion of flow from one channel into one or more other channels. Observations of braiding by the White River suggest that the rate of change of the pattern in a stream which is not restricted by resistant banks is controlled by the amount of bed load. From observations of White River and the suggestions of authors cited, braiding is favored by the presence of erodible banks and fluctuations in discharge.

RELATED PROBLEMS

LONGITUDINAL PROFILE OF VALLEY AND EQUILIBRIUM IN A REGRADING GLACIAL STREAM

Many similarities exist between the White River as it deposits and reworks the materials of the valley train and the graded streams described by Gilbert (1877), Davis (1902), Kesseli (1941), Mackin (1948) and others. Mackin (1948, p. 471) defined a graded stream as—

* * * one in which, over a period of years, slope is delicately adjusted to provide, with available discharge and with prevailing channel characteristics, just the velocity required for the transportation of the load supplied from the drainage basin. The graded stream is a system in equilibrium; its diagnostic characteristic is that any change in any of the controlling factors will cause a displacement of the equilibrium in a direction that will tend to absorb the effect of the change.

He emphasized the idea of a system in long-term equilibrium while recognizing similar short-term changes in the regimen. Rubey (1952), Leopold and Maddock (1953), and Wolman (1955) have discussed the equilibrium adjustment of the channel cross section. Wolman (1955, p. 41) applied the term "quasi-equilibrium" to adjustments made by streams that were not related to the length of time a set of conditions exists. This term is perhaps unnecessary, as the diagnostic characteristic of a graded stream cited by Mackin above is just this tendency to adjust.

Mackin (1948, p. 477) recognized the similarities between streams that were raising the level of their beds and graded streams and suggested (p. 478) that the term "aggrading" be restricted to upbuilding at grade. The White River, during the period of study, would appear to fit this definition in the reach from the junction of the Emmons Creeks to the gage. As it appears to be cutting down upstream from the junction, it then could be described as a regrading stream showing adjustment between the variables of slope, channel, cross section, discharge, bed material size, and load.

INFLUENCE OF GLACIER REGIMEN ON THE DEVELOPMENT OF A VALLEY TRAIN

Emmons Glacier has retreated about 1 mile since 1910, most of this distance since 1930, and has advanced about 0.1 mile recently as shown in figure 3. The stream in 1958 and 1959 has deposited upon and eroded the valley train, resulting in net deposition in 1958 and net erosion in 1959. This response of the stream is probably related to the difference in weather and availability of material in the two years. It is apparent that the response is not related to the glacial advance of approximately 100 feet during this period. Observations of the stream, at the point at which it issues from the glacier and downstream, suggest that most of its load is derived by erosion downstream from the glacier. The effect of the stream on the valley train would then seem to be a function of the amount of water provided by glacial melt rather than a function of the load introduced directly to the stream from the glacier.

The glacial regimen would seem to be related to the valley-train deposits only through the quantity of water that it makes available to the stream and the debris that it has deposited, subject to erosion by the stream. For example, the melting back of Emmons Glacier has uncovered large quantities of debris which are now available to the stream. Continued advance of the Emmons Glacier may radically alter the availability of this material and thus cause changes downstream. Similarly, changes in position of the termini of continental glaciers of the past must have radically altered the availability of debris to their drainage streams. Such effects are probably restricted to the immediate vicinity of the glacier, as a stream on alluvium will derive a load within a short distance and this load will be determined by its discharge, gradient, and availability of readily erodible material.

If periods of glacial advance and retreat can be considered analogous to winter and summer on the White River, advance (winter) would provide much smaller discharges and less debris; and retreat (summer) larger discharges and more debris. This is obviously a great oversimplification, as there were winters and summers, in the glacial climate. An additional assumption would be a constant rate of glacier flow because fluctuation in the glacier flow rate might change the position of the terminus without changing the discharge. The removal of a continental glacier from a drainage basin could well bring a decrease in the discharge of the stream, bringing to a close the deposition of a valley train.

While the regimen of Emmons Glacier has long-term effects in providing debris to the stream, the short-term effects of weather and runoff determine the rate of deposition and erosion, the hydraulic characteristics, and the pattern of the stream.

SUMMARY

1. Channels of the White River, developed in coarse noncohesive materials, are narrower, slightly shallower, and have higher velocities of flow at higher slopes at discharges of similar magnitude than channels with cohesive bank materials as exemplified by Brandywine Creek.

2. The extreme values of the intercepts (in the equations $w=aQ^b$ and $v=kQ^m$) for width and velocity, and the rate of change of width-depth ratio with dis-

charge extrapolated for data on Brandywine Creek, may reflect the failure of the channel to adjust to small discharges and indicate a relation of intercept to bank material.

3. Although the stream had some load before it issued from the glacier terminus, most of its load was provided by erosion of morainic debris, mudflow, and valley-train deposits in the reach below the glacier.

4. Analysis of a sample of mudflow and valley-train deposits showed that 40 percent of the mudflow material was less than 0.105 mm in comparison to 10 percent of the valley-train material, indicating removal of the finer fraction from the valley-train materials. Measurements demonstrated a systematic decrease in median diameter of the valley-train materials coarser than 4 mm with distance from the source. The change of 60 mm in median diameter for the valley train materials that took place over a distance of some 4,200 feet is attributed to selective transportation. Increases in median diameter such as those 1.7 and 25 miles below the glacier terminus occur at new sources of coarse material.

5. If one accepts the hypothesis of Wolman and Miller (1960) that the dominant process occurs with sufficient frequency and magnitude to cause most of the changes observed, and is neither the frequent event, which is less than the threshold value, nor the infrequent catastrophic event, then it follows that the slope forming discharge for White River channels in this reach lies between 200 and 500 cubic feet per second.

6. When conditions are such that antidunes are developed, much material is carried through the section without deposition. The rate of movement of antidunes, therefore, cannot be used to estimate the total bed load, as is sometimes done with dunes.

7. Measurements of competency in White River channels as well as data from Hjulström (1935) and Nevin (1946) suggest that for coarse materials the "sixth power law" should be the 7.8 power law. Logarithmic plotting shows that a line with a slope of 2.6 fits the White River data better than one with a slope of 2.0 (the "sixth power law" expressed in terms of linear dimensions rather than mass).

8. Evidence of the great amount of material transported by the White River is provided by the amount of erosion and deposition on the valley train. The average net elevation change for all profiles in 1958 was +1.2 feet and in 1959, −0.1 foot. The abundant discharge and high gradients of both East and West Emmons Creeks, together with the unique conditions described for West Emmons Creek, and the widening of the valley train, provided this abundant load during the summer of 1958. The high gradients of the Emmons Creeks combined with the lower flows of 1959

supplied materials to the valley train in greatly reduced quantities, so that erosion exceeded deposition.

9. The channel pattern changed from "meandering" to braided with the onset of high summer discharges and returned to "meandering" with the lower discharges of late summer and fall.

10. Flows in the "meanders" of the White River have relatively high Froude numbers. The "meanders" are a series of bends connected by short straight reaches of swift quiet flow; steeper gradients are usually associated with bends. It is apparent that there is no greater sinuosity associated with the meanders than with the anastomosing channels.

11. The wider the valley train and the farther it is downstream from a constriction, as exists at the junction of the Emmons Creeks, the more numerous the channels when braiding occurs.

12. Leopold and Wolman (1957) described the formation of a braided pattern in a flume channel by the deposition of a center bar which subsequently became in island. Observations of the White River show that a braided pattern may also result from the reoccupation of numerous old channels due to deposition within the main channel as well as to highflows which raise the water surface. This was an important mechanism in the formation and alteration of White River channel patterns. A pattern developed in this manner does not show an appreciable increase in slope for the divided reach as did the channel divided by the deposition of a center bar.

13. Both braided and meandering reaches can occur along White River in reaches where it is aggrading, poised, or degrading. The pattern alone does not conclusively define the regimen of the stream.

14. Braided channel patterns cannot be developed without bed load. The common element in all explanations of braiding appears to be a movement of bed load with local deposition within the channel, causing the diversion of flow from one channel into one or more other channels, or the deposition of channel bars and the development of islands. The rate of pattern change appears to be directly related to the amount of bed load.

15. The White River is a regrading stream showing adjustment between the variables of slope, channel cross section, discharge, load, and size of the bed and bank materials.

16. Observations of the White River show that although the regimen of the glacier has long-term effects in providing debris to the stream, the short-term effects of weather and runoff determine the rate of deposition and erosion, the hydraulic characteristics, and the pattern of the stream.

SELECTED BIBLIOGRAPHY

Arnborg, Lennart, 1955a, Hydrology of the glacial river Austurfljot, chap. 7 *of* The Hoffellssandur—a glacial outwash plain: Geog. Annaler, v. 37, nos. 3–4, p. 185–201.

—— 1955b, Ice marginal lakes at Hoffellsjökull, chap. 8 *of* The Hoffellssandur—a glacial outwash plain: Geog. Annaler, v. 37, nos. 3–4, p. 202–228.

Blench, Thomas, 1956, Scale relations among sand-bed rivers including models: Am. Soc. Civil Engineers Proc., Paper 881, p. 19–25.

Brockman, C. F., 1947, Flora of Mount Rainier National Park: National Park Service, 170 p.

Brush, L. M., 1961, Drainage basins, channels, and flow characteristics of selected streams in central Pennsylvania: U.S. Geol. Survey Prof. Paper 282–F, p. 145–181.

Chow, V. T., 1959, Open channel hydraulics: New York, McGraw Hill, 680 p.

Coombs, H. A., 1936, The geology of Mount Rainier National Park: Washington [State] Univ. Pubs. in Geology, v. 3, no. 2, p. 131–212.

Crandell, D. R., and Waldron, H. H., 1956, A recent volcanic mudflow of exceptional dimensions from Mount Rainier, Washington: Am. Jour. Sci., v. 254, no. 6, p. 349–363.

Davis, W. M., 1902, Base level, grade and peneplain: Jour. Geology, v. 10, no. 1, p. 77–111.

Dawdy, D., 1961, Depth-discharge relation of alluvial streams—discontinuous rating curves: U.S. Geol. Survey Water-Supply Paper 1498–C, 16 p.

Doeglas, D. J., 1951, Meanderende en verwilderde rivieren (Meandering and braided rivers): Geologie en Mijnbouw, v. 13, no. 9, p. 297–299.

Fahnestock, R. K., 1959, Dynamics of stream braiding as shown by means of time lapse photography (abs.): Geol. Soc. America Bull., v. 70, no. 12, pt. 2, p. 1599.

Fisk, H. N., 1943, Summary of the geology of the lower alluvial valley of the Mississippi River: Vicksburg, Miss., U.S. Army Corps of Engineers, U.S. Waterways Exper. Sta., 30 October 1943.

—— 1944, Geological investigation of the alluvial valley of the lower Mississippi River: Vicksburg, Miss., U.S. Army, Corps of Engineers, U.S. Waterways Exper. Sta., December 1944.

Flint, R. F., 1957, Glacial and Pleistocene geology: New York, John Wiley and Sons, 555 p.

Friedkin, J. F., 1945, A laboratory study of meandering of alluvial rivers: Vicksburg, Miss., U.S. Army, Corps of Engineers, U.S. Waterways Exper. Sta., 1 May 1945.

Frödin, Gustaf, 1954, The distribution of late glacial subfossil sandurs in northern Sweden: Geog. Annaler, v. 36, no. 1, p. 112–134.

Gilbert, G. K., 1877, Report on the geology of the Henry Mountains: U.S. Geog. Geol. Survey Rocky Mtn. Region (Powell), 160 p.

—— 1914, The transportation of debris by running water: U.S. Geol. Survey Prof. Paper 86, 263 p.

Harrison, A. E., 1956a, Glacial activity in the western United States: Jour. Glaciology, v. 2, no. 19, p. 666–668.

—— 1956b, Fluctuations of the Nisqually Glacier, Mount Rainier, Washington, since 1750: Jour. Glaciology, v. 2, no. 19, p. 675–683.

Hjulström, Filip, 1935, Studies of the morphological activity of rivers as illustrated by the river Fyris: Univ. Upsala [Sweden] Geol. Inst. Bull., v. 25, p. 221–527.

—— 1952, The geomorphology of the alluvial outwash plains of Iceland and the mechanics of braided rivers: Internat. Geog. Congr., 17th, Washington 1952, Proc., p. 337–342.

—— 1954a, An account of the expedition and its aims; chap. 1 *of* The Hoffellssandur—a glacial outwash plain: Geog. Annaler, v. 36, no. 1, p. 135–145.

—— 1954b, Geomorphology of the area surrounding the Hoffellssandur, chap. 4 *of* The Hoffellssandur—a glacial outwash plain: Geog. Annaler, v. 36, no. 1, p. 169–189.

—— 1955, The ground water, chap. 9 *of* The Hoffellssandur—a glacial outwash plain: Geog. Annaler, v. 37, nos. 3–4, p. 234–245.

Hoppe, Gunnar, and others, 1959, Glacialmorfologi och isrörelser i ett lappländskt fjällområde (Glacial morphology and inland-ice movement in a mountain area of Swedish Lapland): Geog. Annaler, v. 41, no. 1, p. 1–14.

Hoyt, J. C., and Grover, N. C., 1916, River discharge: 4th ed., John Wiley and Sons.

Hubley, R. C., 1956, Glaciers of the Washington Cascade and Olympic Mountains; their present activity and its relation to local climatic trends: Jour. Glaciology, v. 2, no. 19, p. 669–674.

—— 1957, An analysis of surface energy during the ablation season on Lemon Creek Glacier, Alaska: Am. Geophys. Union Trans., v. 38, no. 1, p. 68–85.

Ippen, A. T., and Verma, R. P., 1953, The motion of discrete particles along the bed of a turbulent stream, *in* Proc. Minn. Internat. Hydraulics Convention: Joint meeting of Am. Soc. Civil Engineers and Internat. Assoc. for Hydraulic Research.

Jónsson, Jón, 1954, Outline of the geology of the Hornafjördur region, chap. 2 *of* The Hoffellssandur—a glacial outwash plain: Geog. Annaler, v. 36, no. 1, p. 146–161.

—— 1955a, Tillite in the basalt formation in East Iceland, chap. 5 *of* The Hoffellssandur—a glacial outwash plain: Geog. Annaler, v. 37, nos. 3–4, p. 170–175.

—— 1955b, On the formation of frontal glacial lakes, chap. 8 *of* The Hoffellssandur—a glacial outwash plain: Geog. Annaler, v. 37, nos. 3–4, p. 229–233.

Kalinske, A. A., 1942, The role of turbulence in river hydraulics: Hydraulics Conf., 2d, State Univ. Iowa 1942, Proc., Iowa, State Univ. Bull. 27.

Kesseli, J. E., 1941, Concept of the graded river: Jour. Geology, v. 49, no. 6, p. 561–588.

King, Clarence, 1871, On the discovery of actual glaciers on the mountains of the Pacific slope: Am. Jour. Sci., 3d ser., v. 1, no. 3, p. 157–167.

Lane, E. W., 1937, Stable channels in erodible materials: Am. Soc. Civil Engineers Trans., v. 102, p. 123–194.

—— 1955a, Design of stable channels: Am. Soc. Civil Engineers Trans., v. 120, p. 1234–1260.

—— 1955b, The importance of fluvial morphology in hydraulic engineering: Am. Soc. Civil Engineers Proc., v. 81, paper no. 745, 17 p.

Langbein, W. B., 1942, Hydraulic criteria for sand-waves: Am. Geophys. Union Trans., v. 23, pt. 2, p. 615–618.

Leopold, L. B., and Maddock, Thomas, Jr., 1953, The hydraulic geometry of stream channels and some physiographic implications: U.S. Geol. Survey Prof. Paper 252, 57 p.

Leopold, L. B., and Miller, J. P., 1956, Ephemeral streams—hydraulic factors and their relation to the drainage net: U.S. Geol. Survey Prof. Paper 282–A, 37 p.

Leopold, L. B., and Wolman, M. G., 1957, River channel patterns—braided, meandering and straight: U.S. Geol. Survey Prof. Paper 282–B, p. 39–85.

Mackin, J. H., 1948, Concept of a graded river: Geol. Soc. America Bull., v. 59, no. 5, p. 463–511.

———— 1956, Cause of braiding by a graded river (abs.): Geol. Soc. America Bull., v. 67, no. 12, pt. 2, p. 1717–1718.

Matthes, F. E., 1914, Mount Rainier and its glaciers: National Park Service, 48 p.

———— 1930, Geologic history of the Yosemite valley: U.S. Geol. Survey Prof. Paper 160, 137 p.

Matthes, G. H., 1941, Basic aspects of stream meanders: Am. Geophys. Union Trans. 22d Ann. Mtg., pt. 3, p. 632–636.

Miller, J. P., 1958, High mountain streams—effects of geology on channel characteristics and bed material: Socorro, N. Mex., New Mexico Inst. Mining and Technology, State Bur. Mines and Mineral Resources memo. 4, 53 p.

Nevin, C. M., 1946, Competency of moving water to transport debris: Geol. Soc. America Bull., v. 57, no. 7, p. 651–674.

Pierce, R. C., 1917, The measurement of silt-laden streams: U.S. Geol. Survey Water Supply Paper 400, 42 p.

Ray, L. L., 1935, Some minor features of valley glaciers and valley glaciation: Jour. Geology, v. 43, no. 4, p. 297–322.

Rigsby, G. P., 1951, Crystal fabric studies on Emmons Glacier, Mt. Rainier: Jour. Geology, v. 59, no. 6, p. 590–598.

Robertson, J. M., and Rouse, H., 1941, On the four regimes of open-channel flow: Civil Eng., v. 11, no. 3, p. 169–171.

Rouse, Hunter, 1946, Elementary mechanics of fluids: New York, John Wiley and Sons.

Rubey, W. W., 1937, The force required to move particles on a stream bed: U.S. Geol. Survey Prof. Paper 189–E, p. 121–141.

———— 1952, Geology and mineral resources of the Hardin and Brussels quadrangles (Ill.): U.S. Geol. Survey Prof. Paper 218, 179 p.

Russell, I. C., 1898, Glaciers of Mount Rainier: U.S. Geol. Survey Ann. Rept. 18, pt. 2, p. 349–415.

Russell, R. J., 1939, Louisiana stream patterns: Am. Assoc. Petroleum Geologists Bull., v. 23, no. 8, p. 1199–1227.

Schaffernak, F., 1922, Neue Grundlagen für die Berechnung der Geschiebeführung in Flussläufen: Deuticke, Leipzig und Wien.

Schumm, S. A., 1960, The shape of alluvial channels in relation to sediment type: U.S. Geol. Survey Prof. Paper 352–B, p. 17–30.

Sharp, R. P., 1942, Mudflow levees: Jour. Geomorphology, v. 5, no. 3, p. 222–227.

Sharp, R. P., and Nobles, L. H., 1953, Mudflow of 1941 at Wrightwood, Southern California: Geol. Soc. America Bull., v. 64, no. 5, p. 547–560.

Sigafoos, R. S., and Hendricks, E. L., Botanical evidence of the modern history of Nisqually Glacier, Washington: U.S. Geol. Survey Prof. Paper 387–A, 20 p.

Simons, D. B., Richardson, E. V., and Haushild, W. L., 1961, Studies of flow in alluvial channels—Flume studies using medium sand (0.45 min): U.S. Geol. Survey Water-Supply Paper 1498–A, 76 p.

Sundborg, Åke, 1954, Map of the Hoffellssandur, chap. 3 of The Hoffellssandur—a glacial outwash plain: Geog. Annaler, v. 36, no. 1, p. 162–168.

———— 1955, Meteorological observations, chap. 6 of The Hoffellssandur—a glacial outwash plain: Geog. Annaler, v. 37, nos. 3–4, p. 176–184.

———— 1956, The river Klarälven: Geog. Annaler, v. 38, nos. 2–3, p. 127–316.

Tanner, W. F., 1960, Helicoidal flow, a possible cause of meandering: Jour. Geophys. Research, v. 65, no. 3, p. 993–995.

Tiffany, J. B., Jr. and Nelson, G. A., 1939, Studies of meandering of model streams: Am. Geophys. Union Trans., v. 20, pt. 4, p. 644–649.

U.S. Waterways Experiment Station, 1935, Studies of river bed materials and their movement, with special reference to the lower Mississippi River: Vicksburg, Miss., Paper 17, 161 p.

White, C. M., 1940, Equilibrium of grains on bed of streams: Royal Society (London) Proc., ser. A, v. 174, no. 950, p. 322–334.

Wolman, M. G., 1954, A method of sampling coarse river-bed material: Am. Geophys. Union Trans., v. 35, no. 6, p. 951–956.

———— 1955, The natural channel of Brandywine Creek, Pennsylvania: U.S. Geol. Survey Prof. Paper 271, 56 p.

Wolman, M. G., and Brush, L. M., 1961, Experimental study of factors controlling the size and shape of stream channels in coarse noncohesive sands: U.S. Geol. Survey Prof. Paper 282–G, p. 183–210.

Wolman, M. G., and Miller, J. P., 1960, Magnitude and frequency of forces in geomorphic processes: Jour. Geology, v. 68, no. 1, p. 54–74.

VI
Glacial Lakes and Varves

Editor's Comments on Papers 23 Through 25

In all glaciated continents today, including Antarctica, lakes are abundant along the glacier margins. There are many reasons for extra ponding near glaciers: (1) glaciers block valleys and exits sloping into them; (2) overdeepening by ice abrasion where rock is weak or closely jointed forms enclosed basins; (3) deposits, especially end moraine, dam basins previously scoured by ice; (4) ice, buried by its own dirt load, melts out after centuries, leaving small kettle hole ponds; and (5) the mass of the ice produces isostatic depression of the earth's crust, tilting valleys down toward thick ice centers. Except for some minor moraines (in Part III) glaciolacustrine deposits are not made by the ice itself but as glacier-dependent and glacier–controlled features they are an integral part of glacial deposits.

A separate Benchmark volume is concerned with the chronology worked out from glaciolacustrine varves. As a compromise here, and in order not to duplicate, I have chosen two short classical articles (Papers 23 and 24), plus the abstract and diagram of one longer modern sedimentological study (Paper 25). Here only the bottom sediments (varves) are discussed; shoreline features (beaches, benches, deltas) are presented in a Benchmark volume (Andrews, 1974) and by such classics as Jamieson (1863), Gilbert (1890), Upham (1896), Leverett and Taylor (1915), and Hough (1958). Neither is anything included here about those most ephemeral ice-dammed lakes which drain suddenly and have many shoreline positions (e.g., Thorarinsson, 1939; Stone, 1963).

Again (as in Parts III and IV) it was De Geer who first explored the use of varves and excited members of the 1910 Geological Congress with his findings, so his first English summary of 1912 is reproduced here (Paper 23).

From 1930 to 1950, there was one very active issue: whether varves really are annual, as the name implies, or not. Serious question was raised in correlation studies even within one basin (e.g., Deane, 1950). The concensus from radiocarbon time checks (Wenner, 1968; Tauber, 1970) and detailed varve check counts in specific locations (Järnefors and Fromm, 1960; Strömberg, Paper 13) indicates that varves are mostly annual, with occasional no-deposit summers and double-melt-season years. Kindle's (1930) sediment-source hypothesis and Kuenen's (1951) turbidity hypothesis are certainly benchmarks, but could not be included. I chose the work of a physicist, Fraser (Paper 24), who confirms the annual physical changes affecting glacial-lake sedimentation. Still another early short sedimentation study is by Johnson (1922).

Finally, although many people study glacial shorelines, deltas, and outlets, few have traced each annual sediment layer throughout a glacial lake (Fulton, 1965). An outstanding detailed recent study of the long-known glacial Lake Hitchcock by Ashley (Paper 25) is a modern classic.

References Cited

Andrews, J. T. (1974) *Glacial Isostasy*. Dowden, Hutchinson & Ross, Inc., Stroudsburg, Pa., 491 p.

Deane, R. E. (1950) Deposits of glacio-lacustrine origin: Canadian Geol. Survey, Mem. 256, p. 34–41.

Fulton, R. J. (1965) Silt deposition in late-glacial lakes of southern British Columbia: Amer. Jour. Sci., v. 263, p. 553–570.

Gilbert, G. K. (1890) Lake Bonneville: U.S. Geol. Survey, Monogr. 1, 438 p.

Hough, J. L. (1958) *Geology of the Great Lakes:* University of Illinois Press, Urbana, Ill., 313 p.

Jamieson, T. F. (1863) On the parallel roads of Glen Roy and their place in the history of the glacial period: Geol. Soc. London Quart. Jour., v. 19, p. 235–259.

Järnefors, B., and E. Fromm (1960) Chronology of the ice recession through middle Sweden: 21st Internat. Geol. Congr., Norden, Rept. 21, pt. 4, p. 93–97, 173–174.

Johnston, W. A. (1922) Sedimentation in Lake Louise, Alberta: Amer. Jour. Sci., v. 204, p. 376–386.

Kindle, E. M. (1930) Sedimentation in a glacial lake: Jour. Geol., v. 38, p. 81–87.

Kuenen, P. H. (1951) Mechanics of varve formation and the action of turbidity currents: Geologia Fören. Stockholm Förh., v. 6, p. 149–162.

Leverett, F., and F. B. Taylor (1915) The Pleistocene of Indiana and Michigan, and the history of the Great Lakes: U.S. Geol. Survey Monogr. 53, 529 p.

Stone, K. H. (1963) Alaskan ice-dammed lakes: Assoc. Amer. Geographers, Ann. v. 53, p. 332–349.

Tauber, H. (1970) The Scandinavian varve chronology and C-14 dating: in *Radiocarbon Variations and Absolute Chronology*, ed. I. U. Olsson, Wiley–Interscience, New York, p. 173–196.

Thorarinsson, S. (1939) The ice-dammed lakes of Iceland with particular reference to their values as indicators of glacier oscillations: Geografiska Annaler, v. 21, p. 216–242.

Upham, W. (1896) The glacial lake Agassiz: U.S. Geol. Survey Monogr., v. 25, 658 p.

Wenner, C. G. (1968) Comparison of varve chronology, pollen analysis and radiocarbon dating: Acta Univ. Stockholmiensis, Stockholm Contrib. Geol., v. 18, p. 75–97.

23

Reprinted from *Compt. Rend. 11e Congr. Géol. Internat., Stockholm*, **1,** 241–244, 245–247, 253 (1912)

A Geochronology of the last 12 000 years.

BY

GERARD DE GEER,
Professor at the University of Stockholm.

With two plates.

Geology is the history of the earth, but hitherto it has been a history without years. It is true that many attempts have been made to obtain time-computations for certain parts of that history, but none of them has been capable to stand a closer trial. Thus, the very able authors of one of our lately published textbooks of geology say(1): "The desire to measure the great events of geological history in terms of years increases as events approach our own period and more intimately affect human affairs. The difficulties attending such attempts are, however, formidable, and the results have an uncertain value. At best they do little more than indicate the order of magnitude of the periods involved. Geological processes are very complex, and each of the co-operating factors is subject to variations, and such a combination of uncertain variables introduces a wide range of uncertainty into the results."

Under such circumstances it may be suitable here to place briefly before you a new, exact method of investigation, through which it is possible, by actual counting of annual layers, to establish a real geochronology, for a period reaching from our time backwards some 12 000 years.

As a basis for this chronology have been used certain late glacial and postglacial, periodically laminated sediments in which the deposition for every single year can be discriminated. By actual countings and successive combinations of a great number of sections with regular intervals along a line extending from the southernmost to the central part of Sweden it has been possible not only to sum up the whole

Editor's Note: A row of asterisks indicates material deleted owing to lack of space.

series of centuries it has taken for the ice-border to retire this distance, or some 800 km, but also to estimate the length of the postglacial epoch after the disappearance of the ice and up to our days.

Of the late glacial sediments the most important is a glaci-marine clay, the *"varvig lera"* (hvarfvig lera), so called from its *"varves"*(2) or its periodical laminæ of different colour and grain.

Already at my first field-work as a geologist, in 1878, I was struck by the regularity of these laminæ, much reminding of the annual rings of the trees. The next year, therefore I commenced, and during the following years pursued, detailed investigations and measurements of these laminæ in different parts of Sweden. The laminæ were found to be so regular and so continuous that they could scarcely be due to any less regular period than the annual one. I therefore ventured in 1882 to advance the view that there might be a close connection between the periodical laminæ of the clay and the annual ablation of the land-ice(3). Two years afterwards the investigations had proceeded so far that, being confirmed in my opinion that the laminæ were really annual, and having found out a way for correlating annual layers at different places by means of diagrams, I could, in a lecture read before our Geological Society in Stockholm, indicate the way by which a real chronology for the last part of the Ice-age could be obtained(4). A few months afterwords I also succeeded in finding the first correlation between the clay-layers at three points, though not very far from one another. In 1889 I found — and mapped — in the neighbourhood, NW of Stockholm, a thereto overlooked kind of certainly quite small, but very characteristic terminal moraines, which proved to be periodically arranged in rows with somewhat regular intervals of about 200—300 m. This led me to point out the possibility that these ridges might correspond to the stop in the recession of the ice-border, which was probably caused by each winter, and that this might be ascertained by investigation of the successive annual clay-layers between some neighbouring ridges(5). This kind of moraines has since that time been found to be quite common in the lower parts of the land, and at first I had therefore the intention of pursuing the chronological investigations by means of a careful mapping of the annual moraines.

A series of those characteristic and marked small ridges will be visited during the excursions in the neighbourhood of Stockholm, during which also the *"varve"*-clay and the method of determining the late

glacial ice-recession will be demonstrated. The whole material of measurements, maps, and clay-samples from those different regions of Sweden
upon which the chronology is founded will be accessible during the whole
congress meeting in the Geological Institute of the University of Stockholm.

By detailed studies of some oses, especially at Stockholm and Uppsala, and, later on, also at Dal's Ed it had turned out that also the oses
are of a pronounced periodical structure, marked by centres of coarser
material, on their southern side gradually passing into finer gravel and
sand. This led me to a new explanation of their formation as successive
submarginal delta-deposits, formed in the glacier-arches of the receding
land-ice, and probably corresponding to the annual "*varves*" of the finest,
clayey sediment and to the annual moraines(6).

Finally, in 1904, I happened te get a very good correlation between
two clay-sections 1 km apart from each other, and now I determined
to make an earnest attempt to realize my old plan for a clay-chronology.

By investigating some forty points in the Stockholm region it
was soon found that the clay-correlation offered less difficulty than
thereto suspected and was — the localities of observation being well
chosen — as a rule performable at distances of 1 km. This being ascertained, I secured the assistance of a number of students from the
universities of Stockholm and Uppsala, ten from each, and after some
training they went all out on a summer morning in 1905, each of
them to his special part of a line about 200 km long, running, as seen
from the map Pl. 1, past Stockholm and Uppsala through the Söderman-
land—Uppland peninsula, from the great Fennoskandian moraines at its
southern end to the river Dalälfven to the north, and going as nearly
as possible in the direction of the ice-recession. The main work was
performed in four days; though the filling up of some lacunæ at difficult points could be performed only after several repeated attempts.

Among the different results it may suffice here to mention, that I
now finally got the conclusive proofs for the assumption that the individual "*varves*" had a very wide distribution. Thus it was shown that
it often exceeded some fifty km, and that the cubic-mass of the "varves"
must be measured by millions of m^3. This together with their regular
structure definitively showed, that they could not be due to any local or
accidental cause of smaller importance or less pronounced periodicity than
the climatic period of the year. On the other hand, it seems equally
impossible that every sharply marked *varve* should correspond to any

hypothetical and, in every case, indistinctly limited series of years without showing any registration of the in fact so sharply accentuated period
of the single year. Indeed, it seems to me quite as improbable that
the melting-season of the land-ice should not put its stamp upon the
annual sedimentation, as that this should not be the case with the
annual period of vegetation in relation to the annual rings of the trees.

*   *   *   *   *   *   *

Thus every ose-centre is nothing else than the proximal glacier-
arch portion of an annual layer and, if this be compared to a fan, corresponds to the very handle of it.

Every year, by the melting during the warm season, followed also
a recession of the steep ice-edge with the glacier-arch and its river-
mouth. This retreat, on the whole quite dominating, was during winter-
time somewhat counter-acted by a slight advance, at many places wonder-
fully well registered by the small, but well-marked winter-moraines.

Every following mild season caused a new recession and a formation of a new fan of gravel, sand, and clay. Thus the whole series of
those fans are placed as tiles, one over another, the uppermost always
having their northern, or proximal border extending so much over that
of the underlying as the ice-border had receded and the sea extended
since the last year. As the recession was often very regular, the
handles of the fans were gradually combined to ridges, thereby giving
rise to the oses, the periodical structure of which has been afterwards
often more or less concealed by the smoothening wave-action during
the later land-emergence.

From this cause and owing to the thickness and coarseness of the
material together with the casualities in its deposition the most proximal
parts of the annual layers are as a rule not well adapted for direct
chronological determinations, though, of course, a regular development
of the very ose-deposits is a reliable sign that the ice-recession in such
a region has been of a corresponding regularity.

Yet, it is the fine, extraglacial, clayey sediment that affords the
most valuable means for the chronological investigations. For determining the temporary situation of the receding ice-border during certain
years, the following method was used. As the laminated, glaci-marine
clay originally formed a continuous covering over all the deeper parts
of the ancient sea-bottom and afterwards has been cut away from all
hills and other exposed places, the bottom-layers, the northern limit of
which was to be fixed, ought to be most easily reached near the borders
of the remaining clay-areas. At such points, when possible, railway-

435

sections were examined or new diggings made down through the bottom-layer of the clay where it had the thickness of a man's height, the main point being to determine, at each locality, which of the layers was there lying immediately upon the ground from which the land-ice had just retired that year. As such determinations had to be made at so short intervals as one kilometre, it was not necessary to measure a greater number of *"varves"* above the bottom than was wanted for a reliable correlation with the next section to the north, where only so many of the bottom-layers were missing as corresponded to the number of years that the land-ice had at the last locality been a hinderance to the deposition of clay. Thus, to avoid unnecessary loss of time and money, deeper diggings were not made, except when it was necessary to use the thicker parts of the layers in the neighbourhood of the oses, or near to the ancient river-mouths.

In the diggings the vertical clay-section was carefully cut clean with a transversally sharpened brick-trowel. Then the limits between the annual layers were marked and numbered with a lead-pencil upon a narrow, long strip of paper, and afterwards the thickness of the single layers was marked out at equal distances on a diagram, Pl. 2 B, the tops of those lines of thickness being also combined. In this way it was possible at the same time to compare the whole series of identical layers from two or more different localities, to recognize the corresponding shiftings in the variation-curve, and thus to determine which of the layers at every locality was at the bottom, or, in other words, close to its northern limit.

Of course, it is always necessary to avoid such points where the annual layers have got their original thickness disturbed and falsified by stranding ice-bergs or different kinds of slidings.

By plotting the points of observation on a map and dividing the distances between them with the number of the years which had elapsed during the corresponding ice-recession, the mean annual retreat of the ice-border for the same time was fixed by means of lines, drawn through the above points of division, parallel to the ice-border of the region, as indicated by terminal moraines or the normals of the glacial striæ.

In this way we get not only a reliable chronological record, to which different kinds of events can be connected, but, at the same time, a no doubt somewhat composite, but very interesting registering of the climatical conditions of the same epoch.

For it is evident that, other things being equal, a slow recession of the ice-border means colder, and a faster recession means more genial conditions; though in making comparisons between different regions it is, of course, necessary to make due allowance for differences in the thickness of the land-ice, in the supply of ice, and in the depth of water influencing the formation of ice-bergs. Still these complications are of less importance at a comparison between adjoining parts of the long investigated line when the question is limited to the finding out of the successive climatic variations. By and by it will, no doubt, also be possible to get out corrections even for such more extended comparisons as those above mentioned.

As to the single line, hitherto investigated, it shows along its southern part, south of the great terminal moraines, a relatively slow ice-recession: in Skåne and Blekinge only some fifty metres pro year and farther northward about hundred metres or somewhat more, thus indicating that the corresponding goti-glacial period was still relatively cold. The great Fennoskandian moraines indicate a marked deterioration of the climate, sufficient to cause the ice-border for a few centuries to stop in its recession or even slightly to advance. But after this epoch the great retreat recommenced and was soon going on with an astonishing velocity and regularity, the annual rate of recession as a rule varying between one and three hundred metres, and being utterly seldom, only for single years, changed into an insignificant, incidental advance. This seems to have been true of nearly the whole last part of the late-glacial ice-recession from the Fennoskandian moraines to the ice-shed, or what I have called the fini-glacial sub-epoch, though a short time before its end there has been a last advance of the ice-border, the exact duration of which is not yet known, though the whole epoch until the last ice-remnant along the ice-shed at first became bipartite, thus marking the end of the ice-age, can be fairly well determined.

* * * * * * *

(1) Th. C. Chamberlin and R. D. Salisbury, Geology, Vol. III, p. 413. New York 1906.

(2) The Swedish word *varv*, subst. (old spelling: hvarf), means as well a circle as a periodical iteration of layers. An international term for the last sense being wanted it seems suitable to use the transcription *varve*, pl. *-s.*, in Engl. and Fr., while in German it might be written *Warw*, pl. *-e.*

(3) Om en postglacial landsänkning i södra och mellersta Sverige, Geol. Fören. Sthlm Förhandl. Bd. 6 (1882), p. 159.

(4) Ibidem, Bd. 7, 1884, p. 3, here only poorly and somewhat erronously reported; autoreport: ibid. 1885, p. 512, where also the first (in april 1884) performed correlation is mentioned.

(5) Geol. Fören. Sthlm Förhandl. Bd. 11 (1889), p. 395. A map showing a group of those moraines was published by the author in »Stockholmstraktens geologi» in the work: Stockholm, Sveriges Hufvudstad, Part. I, p. 13. Sthlm, E. Beckman, 1897.

(6) In the last quoted work, Part I, p. 14—17, with a map p. 4, and Part III, map Pl. 5; and: Om rullstensåsarnas bildningssätt, Geol. Fören. Förhandl. Bd. 19 (1897), p. 366.

(7) Geol. Fören. Förh. Bd. 27 (1905), p. 221.

(8) As to the first of the late-glacial sub-epochs, which properly may be called the Dani-glacial one, or that part of the last ice-recession when the ice-border retired from the extreme limit of the last glaciation past Denmark and to Central Skåne, its duration is not yet known, though from some measurements in 1905 of annual varves in the extinct Lake Steenstrup of the island of Fyen I think there may be a possibility by and by to get some time-estimates by means of the ice-dammed lakes.

Pl. 1.

A Geochronological Standard-line through Sweden

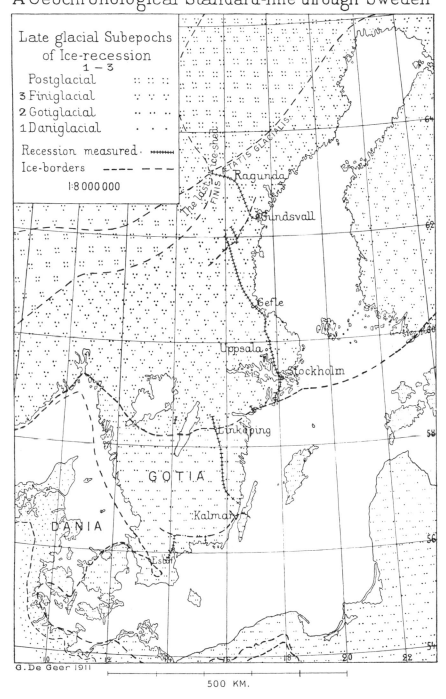

G. De Geer 1911

500 KM.

439

Pl. 2.

A. Map of the annual Ice-recession in the Stockholm-region

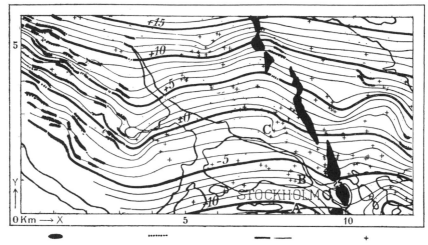

Summer-deltas,   Winter-moraines.   Æquicesses or N.   Measured
glaci-fluvial oses.   limits of varves.   Varve-sections.
±0 .. Ice-border (Æquicess) at Stockholms Högskola and Observatory.

B. Diagrams showing Varve-correlation and Ice-recession between the points A, B, C on the map

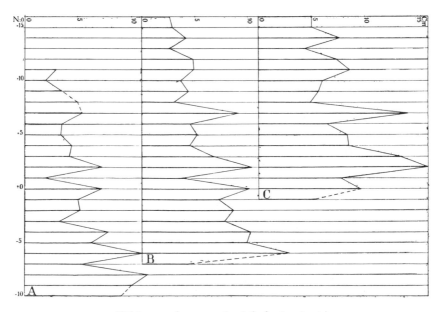

Thickness of varves about 1 : 3 of nat. size.
N:o ±0 ... Varve with N. limit at the æquicess of ±0 on the map.

Reprinted from *Trans. Roy. Soc. Canada*, **23**(4), 49–50, 51–52, 53–55, 57–60 (1929)

An Experimental Study of Varve Deposition

By H. J. Fraser

Presented by R. C. Wallace, F.R.S.C.

1. Introduction

The work of Swedish investigators on the seasonal lamination of late-Glacial and Glacial clays and the value of this stratification in establishing the duration of late Quaternary time has focused considerable attention on this branch of geology. The name varve, from the Swedish word "varv," meaning layer, has been applied by Baron De Geer[1] to the distinctly marked annual deposits of silt, clay and other sediments.

Several types of deposits with laminations suggesting seasonal deposition have been described. Considerable information is now available concerning the chemical and mechanical composition of the most important kind, the varved glacial clay. Yet, there seems to be little exact information regarding the factors which determine the origin of this clay. Antevs[2] has written a paper in which he gives a full theoretical discussion of its origin, but in concluding he states that "the role of each condition and factor for the formation of the varved glacial clay is not well known." In this paper an attempt has been made to determine the relative values of some of the factors affecting the formation of the varved glacial clay.

2. Experiments on the Settling of Various Sized Sand Particles under Different Temperature Conditions

When particles settle in an undisturbed liquid at constant temperature, they rapidly attain a constant maximum velocity, because the acceleration of normal falling is balanced by the resistance of the liquid. Their velocity of fall does not follow any known mathematical formula but varies with their density, size, shape and with the temperature of the water. Holmes[3] gives the following approximate formulae for some of the sizes:

---

[1]De Geer, Gerard. A Geochronology of the last 12,000 years. Compte Rendu Congress Geol. Inter. 11, 1910, Stockholm 1912, pp. 241-253.

[2]Anteve, E. Retreat of the Last Ice-Sheet in Eastern Canada. Geological Survey of Canada, Memoir 146, 1925.

[3]Holmes, A. Petrographic Methods and Calculations, p. 205. D. Van Nostrand Co., 1921.

D—4

Editor's Note: A row of asterisks indicates material deleted owing to lack of space.

For grains larger than .5 mm., $V^2 \propto d(S.G. -1)$

For grains smaller than .2 mm., $V \propto d^2(S.G. -1)$.

Where V is the velocity of fall, d the diameter, and S.G. the specific gravity of the particle.

Particles between .5 and .2 mm. settle with a velocity with agrees with neither of the theoretical formulae given.

*   *   *   *   *   *   *

B. *Method and Results*

Representative samples of each grade were taken for the experiments.   For the larger sizes an individual grain was dropped into the inner tube, and the time taken by it to settle the four feet in the tube was measured.   Twenty-five determinations were made for each grade.   With the finer sizes a small number of grains were dropped into the water and one moving at about the average rate was timed. These measurements were first carried out at a temperature of 2.7°C., then at 10.6° and 20.0°   The results obtained are tabulated.

Size of particle	Average time, in seconds, to fall four feet		
	Temp. 2.7°	10.6°	20.0°
.436 mm..............	17.8	17.2	17.6
.266 "	29.2	27.8	26.4
.190 "	52.6	47.4	41.5
.145 "	97.6	84.2	73.3
.107 "	131.0	109.4	93.0
.082 "	197.5	169.0	146.0

These values were plotted, using the time of fall as ordinate and the temperature as abscissa.   It was found that the values for each size fell on a straight line, none of the values varying from this line by more than .5 seconds, except for size .107 mm. at 10.6° with an error of two seconds.   From the graph it appeared that the time of fall for any size would continuously decrease as the temperature rose to 20°C.

To prove the absence of a maximum retardation at 4°C., some further measurements for the finer sizes were made at 4° and 6.2°. The values obtained were as follows:

Size of particle	Average time, in seconds, to fall four feet	
	Temperature 4.0°C.	6.2°C.
.082....................	. . . .	183.7
.107....................	126.5	122.3
.145....................	91.5	90.5

These values, when plotted, fall on the stright lines found above, and prove conclusively that there is not a maximum retardation at 4.0° but that it occurs at 0°C.   This proved that density of the water is not the controlling factor in the retardation of the fall of these particles, since the density is at a maximum at 4°.   While density is important, the viscosity of the water is the dominant factor in determining the velocity of the falling particles.

3. Experiments on the Settling of Silt and Clay Particles
Under Different Temperature Conditions.

Apparatus and Material

Despite the magnifying effects of the sedimentation tubes used in
the experiments above, it was found impractical to measure the time
of fall of particles much smaller than .082 mm., since following the
course of a single particle was too difficult to permit of accurate
results. To obtain comparative results for the smaller grades of
material a special apparatus was constructed.

*   *   *   *   *   *   *

Results

From the results of successive determinations made on the same
sample it was found that there was some flocculation of the material
with time. Consequently the values obtained from the first determination are the best representation of the rate of falling of unflocculated material.

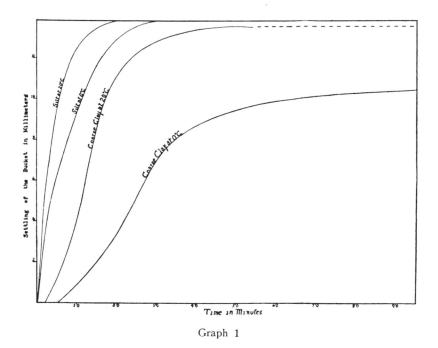

Graph 1

The observed values are expressed in graph 1. For the coarser grades of silt the retardation by temperature of the rate of falling is not very great for 20° to 0°. However, in the finer fractions of the silt the retardation attains considerable proportions, as is shown by the divergence of the curves. The curves for the coarse clay indicate that with low temperatures the velocity of fall of this material is much less than that of the silt. As the fine sizes are approached, the curves of the coarse clay diverge greatly, so that here the difference in the time of fall of the particles would be very considerable. The temperature effect, as shown by the figures for the larger particles and the curves for the smaller particles, proves to be an important factor affecting the velocity of fall of various sized particles.

4. Experiments on the Settling of Varved Clay Under Different Temperature Conditions.

A sample of typical Herb lake varved clay was obtained from the Trapper claim on Herb lake in northeastern Manitoba. The material was thoroughly deflocculated and samples made up of approximately the same weight per cubic centimeter as those used in the previous experiments. This sample consisted mostly of clay, with some silt but very little sand.

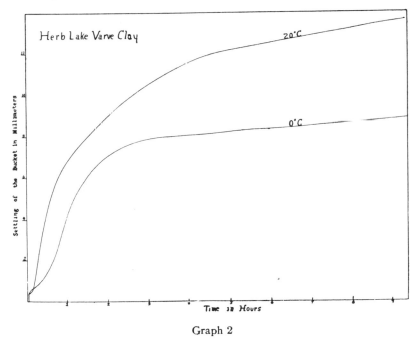

Graph 2

The results of experiments on this clay are given in graph 2.   The curves soon begin to diverge appreciably.   The difference in the time of fall of the smaller sizes is very great as can be seen by a comparison of the slopes of the two curves towards the right hand side of the graph.   Because of experimental difficulties, the measurements were not carried to completion, but were made over a period of ten hours only.

* * * * * * *

6. Experiments on Flocculation at Low Temperatures

Material

As actual sea water was not readily available, a solution was made up according to analysis given by Grabau[4].   Using this solution as a standard, cylinders were filled with solutions having a salinity of 1, 1/2, 1/4, 1/8, 1/16, 1/25, 1/32, 1/50, 1/64, 1/90 and 0/1 that of normal sea water.   The clay used in the experiments was completely deflocculated.   Its composition was approximately 50% silt, 40% coarse clay and 10% fine clay.   The cylinders were placed in a tank having a temperature of 4°C.

Results

In the cylinder of zero salinity a laminated deposit similar to those previously described was formed.   With a salinity of 1/90, layers were formed in the silt component but the finer material was flocculated.

With a salinity of 1/50, the deposit had a laminated appearance, with a number of sets corresponding to each addition of material. Each set consisted of a lower layer of light-coloured, relatively fine-grained material, and an upper component of darker-coloured material consisting of flocculated masses which gave the layer a coarse-grained appearance.   The boundary between light and dark layers of one set was gradational, between dark and light layers of succeeding sets was sharp.   Under the microscope the material in the lower light-coloured layer was coarser in grain than the upper flocculated layer which consisted of flocculated masses of fine-grained material.   This flocculation lamination became very distinct at a salinity of 1/25 and disappeared at about a salinity of 1/8.

With a salinity of 1/4 and 1/2 the deposit appeared fairly uniform with little or no stratification apparent.

Similar experiments carried on with the temperature in the sedimentation cylinders at 20°C. gave the same type of results as above.

7. Application of Results

Glacial lakes differ from temperate lakes in several important features.   Near the ice front at least, they have summer as well as winter inverse temperature stratification with the lower part of the

[4]Grabau, A.   Principles of Stratigraphy, p. 147, 1924.

lake at a temperature of 4°C.   They are fed by glacial streams having a temperature little above 0°C.   The practical absence of erosional phenomena in deposits of glacial lakes indicates quiet water at and near the lake bottom.   The material brought into the lake consists of fine crushed rock, which is unassorted, as mechanical analyses show. Moreover, such fresh ground material or rock-flour is probably to a great extent unflocculated at the time that it reaches the lake.   The glacial lakes may contain a low concentration of electrolytes.   Under these conditions the sedimentary processes beneath the upper water strata would be undisturbed and the rate of settling of the material would not be influenced to any degree by flocculation.

It has been shown in the experimental work that temperature has a very important influence on the rate of settling of particles.   It seems reasonable to assume that in glacial lakes where varves of the diatactic[5] type are formed, the rate of settling of the particles is a direct function of their size and the temperature of the water.   The experimental work has shown that the time of fall of the smaller sized particles is very much increased by low temperatures.   Again it has been shown that, as the particles decrease in size, retardation increases very rapidly.   In a glacial lake this factor would become very important.   During the melting season there is a continuous supply of material.   The coarser sizes would settle relatively rapidly.   The finer sizes would settle very slowly, since the water in the lake would be in the neighbourhood of 4°.   This would allow a very definite grading of the material.   At the close of the melting season the supply of material would diminish to a very small amount, or be entirely stopped.   The result would be that all or nearly all of the coarse material would be desposited during the summer.   The division between the lower and upper parts of the varve would correspond to this change of conditions.

During the winter season the finer material would settle, forming the upper part of the varve.   The manner of sedimentation of the very fine material during the remainder of the winter season is not easily understood.   In water at 4° the fine clay settles at an extremely slow rate, so slow in fact that it is difficult to understand its complete

---

[5]Sauramo, M.   Studies in the Quaternary Varve Sediments in Southern Finland.   Bull. Comm. Geol. de Finlande, No. 60, 1923, pp. 1-164.

Sauramo defines diatactic structure as the effect of sorting of the materials of the varves according to sizes and specific gravity of the particles, the coarest at the bottom and the finest at the top.   Symminct structure is the name used to refer to clay deposited under the control of an electrolyte.   In this case the particles, large and small, go down together and form an unsorted mass due to flocculation of the grains.

sedimentation before spring, unless flocculation takes place. The experimental data indicate that flocculation and subsequent deposition take place much more slowly, especially for the finer sizes, in cold than in temperate water. The stillness of the water during the winter season, coupled with a possible increase in electrolytes[6] would have the effect of increasing the rate of flocculation. It is known that the rate of flocculation increases for increasing numbers of particles per unit volume. As there is an increasing amount of material in suspension in the lake as the season progresses, it is probable that the rate of flocculation would increase towards the close of the melting season.

The maximum salinity permitting the formation of diatactic varves consisting of coarse or fine clay seems to be about 1/50 that of normal sea water. According to Wells[7], coarse suspensions, with particles of from 0.03 to 0.1 mm. in diameter, are in general indifferent to flocculation and the other influences affecting fine sediments, and settle with a velocity dependent on their size, degree of roundness and the temperature of the water. Hence coarse diatactic varves might be formed in saline water.

8. SOME CONDITIONS ESSENTIAL FOR THE FORMATION OF DIATACTIC VARVES.

Some of the essential conditions, as determined by the experimental work are as follows:

1. Unflocculated material of assorted sizes.
2. Sedimentation in water of low temperature.
3. Low concentration of electrolytes.
4. Periodic supply of material. Varves have been reported by Sayles[8] in which the lower part of the varve had an alternating series of coarse and fine layers of silt, above which was the finer part of the varve. Laminations are not necessarily seasonal when the materials are coarse and consequently settle rapidly. A daily variation of supply would suffice to give lamination of deposits in shallow water, if the material was coarse. With finer materials and deeper water a longer period would be required.

---

[6]Antevs considers that there is an insignificant flocculation of the very fine particles in summer and a greater flocculation in winter. Ragnar Eriksson, a Swedish chemical engineer, suggested that the winter flocculation of the colloidal particles takes place because the water becomes slightly acid on account of partial dissolution of the silicic acid of the fine-grained acid rocks.

[7]Wells, R. C. Flocculation of Colloids. Report of the Committee on Sedimentation, April 18, 1923, pp. 50-52.

[8]Sayles, R. W. Report of Progress on Studies on Seasonal Deposition in Glacial Sediments. Report of the Committee on Sedimentation, April 26, 1924, pp. 33-39.

Summary

1. The water temperature and associated properties have an important effect on the velocity of fall of particles smaller than .5 mm. in diameter. As the temperature decreases the velocity of fall of a small particle becomes less, until a minimum is reached at 0°C.

2. As the size of particle decreases, the retardation by temperature of its velocity of fall becomes greater. This gives rise to the definite grading seen in varved deposits.

3. The maximum salinity permitting the formation of varves of coarse clay seems to be about 1/50 that of normal sea-water.

4. Varve-like deposits can be reproduced in the laboratory, if unflocculated material is sedimented in fresh water near a temperature of 0°C.

5. Under certain conditions layer-pairs resembling varves do not necessarily represent a year's deposit, or yearly deposition.

25

Reprinted from Univ. Mass. *Geol. Dept. Contrib. 10,* 1, 2, 6–7, 131–138 (1971)

RHYTHMIC SEDIMENTATION IN GLACIAL
LAKE HITCHCOCK, MASSACHUSETTS-CONNECTICUT

Gail Mowry Ashley

University of British Columbia, Canada

ABSTRACT: Most of the fine-grained bottom sediments of proglacial Lake Hitchcock are rhythmites composed of silt and clay couplets that occur in three textural groups:

Group I   - clay thickness greater than silt thickness
Group II   - clay thickness approximately equal to silt thickness
Group III - clay thickness less than silt thickness

Thin sections from impregnated sediments show flat bedding with a maximum of 40 graded laminae in one 2-inch layer. Erosional contacts and ripple crossbedding are common in Group III, but rare in Groups I and II. The contact between silt and the overlying clay layer in any one varve is gradational in less than 50 percent of the samples. Other sedimentary structures include two distinct types of lebenspuren. Mean grain size of the silt layers (5.5ϕ to 8.5ϕ) depends upon the environment of deposition of the silt within the lake. Mean grain size of the clay layers is much the same everywhere (averaging 10.5ϕ).

Data from 34 localities suggest that the rhythmites are annual (i.e., varves), and the following depositional mechanism is proposed. Sediment was transported by streams from the glacier and from nearby deglaciated uplands. Gravel and sand was deposited on deltas, whereas the finer fraction was carried out into the lake mainly by density underflow. Incoming sediment contained a significant amount of clay that settled out continually but became dominant only during the winter when coarse material was less available.

Varves belonging to Groups I and II generally were formed in water away from inflowing rivers, where little sediment was received directly from density currents. Conversely, Group III varves were formed in water relatively near delta fronts, in a position to receive abundant sediment as it entered the lake.

*   *   *   *   *   *   *

Editor's Note: A row of asterisks indicates material deleted owing to lack of space. Figures appear directly following the article.

CONCLUSIONS

The following is a summary of the important physical properties of the Lake Hitchcock sediments and the processes involved in their deposition.

(1) A rhythmic pattern prevails even though the varves vary between localities in relative thickness of individual layers, total couplet thickness, grain size, color (mineralogy), and sedimentary structures.

(2) Varved clays fill topographic irregularities. Deposits are thickest in the depressions and thinnest over high areas.

(3) Generally, deltas show active growth followed by decreasing growth. Decrease in growth is interpreted to reflect diminishing sediment supply as ice disappeared from each respective drainage basin.

(4) Rhythmic sedimentation occurred in the deltas as well as in the lake deposits.

(5) Near inflowing rivers, varved clays grade shoreward into varved deltaic deposits by gradual thickening of individual silt layers.

(6) Thickness of the silt layers varies considerably and appears to be directly related to proximity of inflowing rivers, while clay layer thickness is relatively constant throughout the lake. This difference implies that the depositional mechanism is different for each layer.

(7) A varve, as a unit, is not a graded bed but consists of two texturally and genetically distinct layers. A couplet is not the result of only one sedimentation pulse.

(8) The silt layer is composed of thin laminations that are commonly graded. Forty graded beds were observed in one 2-inch layer.

(9) Silt layers do not always fine upward, while clay layers do. This gradation within the clay layer suggests that flocculation was not significant in clay sedimentation.

(10) The silt-clay contact varies according to the environment of deposition. Less than 50 percent of the varves have gradational contacts. Group I and III varves rarely have gradational contacts; Group II varves commonly do.

(11) Small scale crossbedding is common in Group III but rare in Groups I and II.

Gail Mowry Ashley

(12) Sedimentary structures within the silt layer appear to be related to grain size: multiple graded beds occur in fine-grained laminations, while crossbedding is found in the coarse-grained beds.

(13) Thin silt laminae sometimes occur in the clay layers; thin clay laminae occasionally occur in the silt layers.

(14) In general, grain size of the silt layer ($M\phi$ varies from 5.5ϕ to 8.5ϕ) depends upon location in the lake. Grain-size distribution of clay is much the same everywhere (average $M = 10.5\phi$).

(15) The range of grain sizes for the silt and clay layers is approximately the same, but each has a decidedly different mode. The positively skewed silt layer is coarser and better sorted than the negatively skewed clay layer.

(16) The clay-silt contact is sharp, though commonly uneven due to burrowing organisms.

(17) Many contacts appear macroscopically regular, but microscopically are erosional.

(18) Lebenspuren were created by two different species of insect larvae, both of which appeared to have lived only part-time in the lake. No evidence of extensive burrowing of permanent infauna has been observed.

(19) The only identified plant remains found in the lake were arctic-alpine species washed in from adjacent land.

(20) The two dominant colors of both silts and clay are olive gray (averaging 5Y 4/1), generally thought to be due to mineralogy of the crystalline and metamorphic uplands, and dark yellowish brown (averaging 10YR 4/2), suggesting the influence of Triassic rocks.

Lacustrine sedimentation

As the glacier retreated up the Connecticut Valley, the large ice mass became a decreasing influence on the southern portions of the elongate lake. Lake Hitchcock probably was not homogeneous in physical characteristics related to seasonal variations in lake water temperature. Most lacustrine circulation is directly related to thermal conditions within a lake. A well-developed thermocline enhances surface circulation. Fall and spring overturns are dependent upon significant annual fluctuations in lake temperature. Overturns and the thermocline are important factors in temperate lakes but are of less importance in subpolar and polar lakes. Although some lake currents probably existed in all parts of Lake Hitchcock, the best circulation occurred in areas farthest from the glacial front.

RHYTHMIC SEDIMENTATION

Concentration of suspended sediment is the most important factor affecting water density, differences in temperature being negligible by comparison. In glacial lake sedimentation, the absolute density of the lake water is not as important as the density contrast between the lake and the inflowing streams. By analogy with modern glacial streams, streams coming directly from the glacier would have a much higher sediment concentration (i.e., were more dense) than streams draining ice-free valleys around the southern end of the lake. Depending upon their relative densities, the major means of sediment distribution would be grouped into underflow, interflow, and overflow.

Using the above suggested limnological conditions as a framework the following mode of deposition is proposed. Sediment was carried to the lake from the glacier or from stagnant ice masses, first directly from the glacier and later by overland streams. Sand and gravel was deposited on the deltas while the finer fraction continued into the lake and flowed at a level determined by its density and that of the lake.

Sediment entered Lake Hitchcock at a number of discrete points. This incoming sediment contained clay that eventually was distributed throughout the lake by currents. The clay settled continuously, unless interrupted by currents, but accumulated in significant amounts only during the winter when coarser material was made less available. The extremely fine-grained winter layer permits the inference that the lake, which was over 200 feet deep in some places, was not cleared of suspended sediment during the winter. Thus the clay composing a winter layer does not necessarily represent the same volume of clay brought in the previous summer. Thickness of the clay layer would be more likely related to concentration of suspended sediment near the lake bottom and length of settling time. Because clay layers tend to be relatively constant in thickness, both of the above factors must have been fairly consistent from year to year.

Most of the sedimentary structures found in the silt layer, such as erosional contacts, crossbedding, and multiple graded beds, are best explained by a bottom current (density underflow). As a stream heavily laden with suspended sediment entered the lake, it flowed down the prodelta slope and out onto the lake floor, depositing sediment as it went. Since streamflow is usually continuous, one can expect that underflow would also be continuous and not like a singlepulse marine turbidity flow. Although flow is continuous, sediment content would certainly vary; multiple graded beds might be explained best by fluctuations in sediment content of the entering stream. Two reasons for these fluctuations could be the diurnal melt cycle or varying runoff due to storms.

Figure 44 shows the postulated density underflow pattern for a portion of one summer. During the rest of the summer and in succeeding years these various flows overlapped and interfingered

Gail Mowry Ashley

as deposition occurred on different areas of the deltas, causing bottom currents to flow in a new direction. A flow pattern such as this would tend to fill in low areas and perhaps flow around highs. Groups I and II varves are found in areas seldom reached by the density underflow, while Group III varves are found in areas reached regularly by underflow.

The summer layer varies greatly in physical characteristics between localities. The clay layer deposited each winter blanketed this complex silt deposit and imprinted a rhythmic nature on the otherwise very diverse sediments.

Contribution No. 10
Department of Geology
University of Massachusetts
Amherst, Massachusetts

Figure 2.   Map of pertinent places mentioned in this study (after Flint, 1959).

454

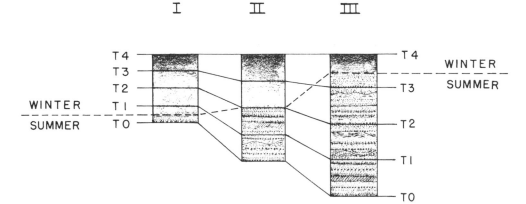

Figure 25.   Proposed chronological relationship between varves
of Groups I, II, and III.   Time-lines T0 through
T4 enclose one varve.

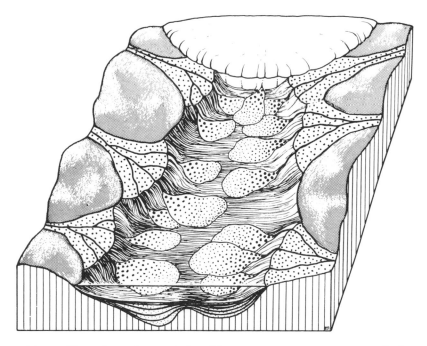

Figure 44.   The density underflow pattern suggested here is
for a portion of one summer.   During the rest of
the summer and in succeeding years, fans would
overlap and interfinger with each other.

Author Citation Index

Subject Index